GEOCHEMICAL PROCESSES, WEATHERING AND
GROUNDWATER RECHARGE IN CATCHMENTS

Geochemical Processes, Weathering and Groundwater Recharge in Catchments

Edited by

OLA M. SAETHER & PATRICE DE CARITAT

Geological Survey of Norway

CRC Press
Taylor & Francis Group
Boca Raton London New York

CRC Press is an imprint of the
Taylor & Francis Group, an **informa** business

A BALKEMA BOOK

Published by:
CRC Press/Balkema
Schipholweg 107C, 2316 XC Leiden, The Netherlands

© 1997 by Taylor & Francis Group, LLC
CRC Press/Balkema is an imprint of the Taylor & Francis Group, an informa business

No claim to original U.S. Government works

ISBN 13: 978-90-5410-641-8 (hbk)
ISBN 13: 978-90-5410-646-3 (ebk)

Visit the Taylor & Francis Web site at
http://www.taylorandfrancis.com

and the CRC Press Web site at
http://www.crcpress.com

Although all care is taken to ensure integrity and the quality of this publication and the
information herein, no responsibility is assumed by the publishers nor the author for any
damage to the property or persons as a result of operation or use of this publication and/or
the information contained herein.

Cover photograph: Automatic water gauging station at the outlet of a monitored catchment in Burte,
west Telemark, southern Norway (Ånund Killingtveit).

Contents

3 CATCHMENT HYDROLOGY
Allan Rodhe & Ånund Killingtveit

4 GROUNDWATER RECHARGE
David N. Lerner

PART 2: TECHNIQUES FOR CATCHMENT STUDIES

5 CHEMICAL ANALYSIS OF ROCKS AND SOILS
Magne Ødegård

13 CHEMICAL CHANGES ATTENDING WATER CYCLING THROUGH
   A CATCHMENT – AN OVERVIEW
Patrice de Caritat & Ola M. Saether

Preface

'Geochemical Processes, Weathering and Groundwater Recharge in Catchments' focuses on the natural processes that take place when dilute water interacts with minerals and organic matter at the surface of the Earth, in soils or in aquifers. This applied topic draws from the fields of geochemistry and hydrology, and is, by necessity, multi-disciplinary, bridging the gap between chemical, physical and engineering sciences. Catchment studies enjoyed a high political profile in Europe and the USA during the 1970's and 1980's as a consequence of concerns about the effects of acid rain on ecosystems. Although political commitment in this particular application regrettably has decreased during recent years, catchment studies still play an essential role in environmental management and in further understanding the impact of human activities on surface and groundwater resources.

The objective of this book is to present in a single volume an overview of the current understanding in catchment sciences. Specialists from various sub-disciplines have contributed their knowledge of their field of research and their vision of its future challenges, with particular emphasis on integrated geochemical-hydrological aspects.

The volume, which stems from a short course offered to students of the Nordic countries, is aimed particularly at graduate students working on catchment processes within Earth, Hydrological, Ecological, or Environmental Science departments around the world. However, it is our hope that professionals and research scientists, also from other disciplines, who are getting into the field of catchment studies, together with those who are already working within one of its sub-disciplines and who wish to be acquainted or stay in touch with progress in others, will find this book both relevant and useful as a reference.

The book is divided in three parts: 1) Catchment processes, 2) Techniques for catchment studies and 3) Integrated catchment studies. The first part introduces concepts relevant to weathering and soils and defines the catchment and its hydrological processes. The second part includes more technical aspects of chemical analysis and sampling of water, rocks and soils, in addition to isotopic tracing of water flow paths and monitoring of discharge from catchments in general. The third part addresses the budgeting of chemical species at the catchment scale, the properties and fate of natural organic matter, the influence of the chemistry of rocks and minerals on water composition, modelling efforts to simulate stream water compositional response to catchment processes, and, finally, the chemical changes attending water circulation within a catchment.

Conceived as a unified and broad-based course from the outset, this book is more than the sum of its chapters. The contributors have made every effort to present their views in an inter-disciplinary perspective, which, we think, they have achieved with scientific rigour and pedagogical flair. We, as editors, have had great pleasure working with them on this book and have learned a great deal from their erudition. We offer our sincere thanks to the Research Academy of the Nordic Countries (NorFA) for financing the short course and to the Geological Survey of Norway for its support. We hope you will enjoy reading this book!

Ola Magne Saether & Patrice de Caritat
Trondheim, August 1996

1. Catchment processes

CHAPTER 1

Weathering processes

JAMES I. DREVER
Department of Geology and Geophysics, University of Wyoming, Laramie, USA

1.1 DEFINITIONS OF WEATHERING

The concept of weathering is relatively simple. Minerals, particularly minerals formed at high temperatures and pressures in the earth's interior, are unstable in contact with dilute waters at the earth's surface and tend to dissolve or transform into minerals that are more stable there. Weathering is a general term for this transformation, including the physical and chemical breakdown of the primary rock, and the accumulation of secondary products as soil. In geochemical and hydrologic studies, however, our interest is not so much in what happens to the minerals themselves, it is rather the effect of weathering reactions on water chemistry. The focus of this chapter will therefore be on the production of solutes rather than on mineralogy. Solutes of particular interest are the *base cations*, Ca^{2+}, Mg^{2+}, Na^+ and K^+. These are important in part because of their role as plant nutrients, but primarily because of their role in neutralizing acidity or generating alkalinity. Weathering of primary (i.e. bedrock-forming) minerals can be described by the general equations:

$$\text{Primary mineral} + n\,H^+ = \text{Uncharged secondary products} + \text{cation}^{n+} \quad (1.1)$$

or:

$$\text{Primary mineral} + n\,CO_2 + m\,H_2O = \text{Uncharged secondary products} +$$
$$\text{cation}^{n+} + n\,HCO_3^- \quad (1.2)$$

Uncharged secondary products may include secondary minerals such as clay minerals or $Al(OH)_3$, and may include uncharged solutes, particularly H_4SiO_4. Alkalinity, and hence chemical weathering, is a major control on the pH (and hence Al concentration and suitability as fish habitat) of surface waters (Fig. 1.1). I shall define *rate of weathering* for the purpose of this chapter as rate of production of dissolved base cations, in units of equivalents per unit area of catchment per unit time, commonly keq ha^{-1} y^{-1}. The base cations need not be transported in runoff: They may also be taken up by plants or adsorbed on a pre-existing exchange site. I would stress, however, that this is a limited definition of weathering, and other definitions will be appropriate when weathering is viewed from other perspectives.

3

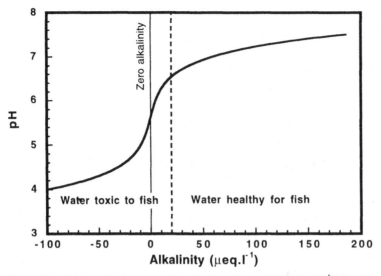

Figure 1.1. Relationship between pH and alkalinity (=[HCO$_3^-$] – [H$^+$]) for dilute waters in equilibrium with atmospheric P$_{CO_2}$ at 10°C. Note the abrupt decrease in pH as alkalinity changes from positive to negative values. The dashed line defining toxicity to fish is approximate. Its exact position depends on the species of fish and the concentrations of other solutes.

Generally speaking, rates of weathering are of interest in environmental studies when they are low – that is to say when acidification or depletion of nutrient cations is a potential problem. This occurs where the bedrock is composed of silicates rather than carbonates, so the following discussion will focus on silicate rather than carbonate rocks.

1.2 TYPES OF WEATHERING REACTION

1.2.1 *Congruent dissolution*

Congruent dissolution means that the entire solid dissolves, leaving no secondary solid phase. An obvious example would be calcite:

$$CaCO_3 + 2H^+ = Ca^{2+} + H_2O + CO_2 \tag{1.3}$$

or forsterite olivine:

$$Mg_2SiO_4 + 4H^+ = 2\,Mg^{2+} + H_4SiO_4 \tag{1.4}$$

Aluminosilicates such as feldspars may dissolve congruently under strongly acidic or extremely dilute conditions:

$$NaAlSi_3O_8 + 4H^+ + 4H_2O = Na^+ + Al^{3+} + 3H_4SiO_4 \text{ (acidic conditions) (1.5)}$$

$$NaAlSi_3O_8 + 8H_2O = Na^+ + Al(OH)_4^- + 3H_4SiO_4 \text{ (neutral conditions) (1.6)}$$

More commonly, because of the low solubility of aluminum hydroxide (Fig. 1.2) and hydrous aluminum silicates such as kaolinite (Fig. 1.3), feldspars dissolve incongruently with the formation of a secondary phase such as gibbsite or kaolinite.

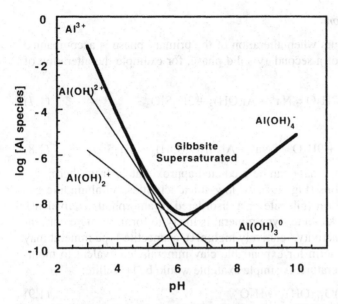

Figure 1.2. Solubility of gibbsite, Al(OH)₃. Note the low solubility at near-neutral pH and the rapid increase in solubility with decreasing pH. The overall shape of the solubility curve reflects the stabilities of the different hydroxy-Al species (data from Wesolowski & Palmer 1994).

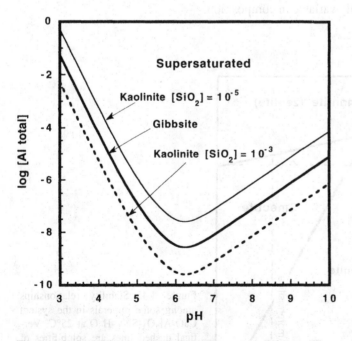

Figure 1.3. Solubility curves of kaolinite at two different silica activities compared to that of gibbsite. The overall shape is the same, but the solubility of kaolinite varies as a function of silica activity ($[SiO_2]$) in solution.

1.2.2 *Incongruent dissolution*

Incongruent dissolution occurs when alteration of the primary phase is accompanied by simultaneous formation of a secondary solid phase, for example the alteration of albite to gibbsite:

$$NaAlSi_3O_8 + H^+ + 7H_2O = Na^+ + Al(OH)_3 + 3H_4SiO_4 \tag{1.7}$$

or kaolinite:

$$2NaAlSi_3O_8 + 2H^+ + 9H_2O = 2Na^+ + Al_2Si_2O_5(OH)_4 + 4H_4SiO_4 \tag{1.8}$$

The identity of the secondary phase can be predicted approximately from thermodynamic equilibrium calculations (Fig. 1.4). At low silica activities, an aluminum hydroxide phase is likely to form (gibbsite or a disordered or amorphous equivalent). At higher silica activities, a kaolinite-type mineral is likely to form. In more concentrated solutions (high silica activity, relatively high pH), a smectite-type mineral may form. Smectite is a general term for expandable clay minerals, equivalent to montmorillonite in some older literature. A simple example would be beidellite:

$$(X)_{0.3}Al_2(Si_{3.7}Al_{0.3})O_{10}(OH)_2 \cdot nH_2O \tag{1.9}$$

where X represents exchangeable cations. Another common variety is montmorillonite (sensu stricto), which has an idealized composition:

$$(X)_{0.3}(Al_{1.7}Mg_{0.3})Si_4O_{10}(OH)_2 \cdot nH_2O \tag{1.10}$$

Natural smectites are highly variable in composition.

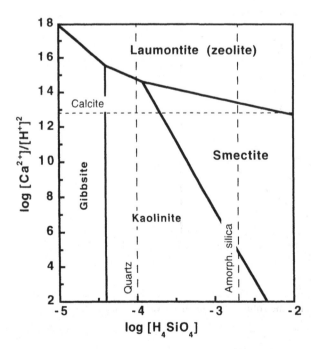

Figure 1.4. Stability relationships among some minerals in the system CaO-Al_2O_3-SiO_2-H_2O at 25°C. Vertical dashed lines are solubilities of silica phases; horizontal dashed line is solubility of calcite at a P_{CO_2} of 10^{-3} (after Drever 1988).

Although natural systems generally conform qualitatively to these predictions, exceptions do occur, and the precise positions of the lines on Figure 1.4 and similar activity diagrams are subject to considerable uncertainty.

We can distinguish two types of incongruent dissolution: in the first type, exemplified by the alteration of albite given above, the primary phase dissolves completely and the secondary products are completely different phases that are precipitated from solution. The second type occurs when ions are leached from a solid phase, but the structure of the solid phase is retained, with a slightly different chemistry. An example would be the alteration of biotite to vermiculite. Biotite can be represented by the formula:

$$K(Mg_{1.5}Fe_{1.5})(AlSi_3)O_{10}(OH)_2 \tag{1.11}$$

When it weathers, some ferrous iron is commonly oxidized to Fe(III). To maintain charge balance, K^+ is lost to solution. The process opens up the mica layers and additional K^+ is lost, being replaced by other ions (typically Ca^{2+} or Mg^{2+}) from solution, resulting in a vermiculite. An idealized formula for such a vermiculite would be:

$$(X)_{0.6}(Mg_{1.5}Fe^{II}_{1.1}Fe^{III}_{0.4})(AlSi_3)O_{10}(OH)_2 \tag{1.12}$$

The fundamental layer structure of the mica is retained, but its chemical composition is changed. Like smectites, vermiculites are highly variable in composition.

As a general statement, incongruent reactions of the first type are relatively easy to model quantitatively, reactions of the second type are much more difficult because thermodynamic equilibrium is rarely attained and stoichiometry is usually not well defined.

1.3 CATION EXCHANGE

Clay minerals, oxides, and solid organic material undergo cation exchange reactions. Ions loosely held by electrostatic forces at the surface of these solids are readily exchanged:

$$2Na^+ + Ca\text{--solid} = 2Na\text{--solid} + Ca^{2+} \tag{1.13}$$

When a soil is 'acidified', base cations on the exchange sites are replaced by hydrogen ions or aluminum ions. For hydrogen ions, the reaction can be written:

$$2H^+ + Ca\text{--solid} = 2H\text{--solid} + Ca^{2+} \tag{1.14}$$

Aluminum uptake is usually a two-stage process: First, hydrogen ions are consumed and Al ions released as acid water interacts with an aluminum hydroxide phase:

$$Al(OH)_3 + 3H^+ = Al^{3+} + 3H_2O \tag{1.15}$$

and then the Al^{3+} displaces base cations from exchange sites:

$$2Al^{3+} + 3Ca\text{--solid} = 2Al\text{--solid} + 3Ca^{2+} \tag{1.16}$$

Soils in high-latitude forests are very commonly acidified in the sense of having exchange sites in the soil dominated by H^+ and Al^{3+}. Acidification can occur as a result of natural processes; it does not have to be related to anthropogenic emissions.

In this discussion, I shall maintain a distinction between ion exchange reactions and chemical weathering, and specifically exclude ion exchange reactions from my definition of weathering. Ion exchange reactions are typically fast, but limited by the number of exchange sites available in a soil. Weathering reactions are typically slow, but the amount of material available for reaction (minerals in the bedrock) is typically very large. From a modeling perspective, ion exchange can be treated as an equilibrium system, whereas weathering is modeled in terms of kinetics.

1.4 MINERAL DISSOLUTION KINETICS

1.4.1 *Relative rates*

There has been a large amount of work done in the laboratory on the rates at which minerals dissolve (for reviews, see papers in Stumm 1990; Sverdrup 1990). A simple way (Lasaga et al. 1994) of comparing rates is to calculate the time it would take for a 1 mm sphere of each mineral to dissolve completely in a dilute solution at pH 5 (Table 1.1). The rates vary enormously. Minerals such as quartz and kaolinite dissolve extremely slowly; feldspars and mafic minerals at intermediate rates, and carbonates very rapidly.

One implication of these differences is that weathering should be highly selective. If a rock contains several minerals, they will not all dissolve at the same rate. Any carbonate minerals will dissolve first, followed by calcic feldspars and ferromagnesian minerals, followed in turn by alkali feldspars. We will see examples of this in the discussion of catchment mass balance (Chapter 9). Also, weathering reactions in

Table 1.1. Approximate time for hypothetical 1 mm diameter spheres of various minerals to dissolve in dilute solution at pH 5 (modified from Lasaga et al. 1994).

Mineral	Lifetime (y)
Quartz	34,000,000
Kaolinite	6,000,000
Muscovite	2,600,000
Epidote	923,000
Microcline	921,000
Albite	575,000
Sanidine	291,000
Gibbsite	276,000
Enstatite	10,100
Diopside	6,800
Forsterite	2,300
Nepheline	211
Anorthite	112
Wollastonite	79
Dolomite	1.6
Calcite	0.1

a catchment will evolve through time. When weathering starts, at the end of glaciation, for example, rapidly weathering minerals such as calcite will be the dominant source of solutes, and the overall weathering rate will be relatively rapid. Over time, these reactive minerals will be depleted in the near-surface environment, and weathering will be dominated by less reactive, but more abundant phases such as feldspars.

1.4.2 *Effect of solution composition on dissolution rates of silicate minerals*

1.4.2.1 *pH*
Far from equilibrium, the dissolution rates of most silicate minerals have the form shown in Figure 1.5. In dissolution experiments, the units are moles per unit surface area of the mineral per unit time. In the acid region, the rate increases exponentially with increasing hydrogen ion concentration:

$$(\text{Rate})_H = k_H (H^+)^n \qquad \text{(acid region) (1.17)}$$

or:

$$\log (\text{Rate})_H = \log k_H - n\,\text{pH} \qquad (1.18)$$

where k_H is a rate constant and n an exponent that is different for different minerals. The numerical value of n is typically about 0.5 (Table 1.2), and the slope of the log (dissolution rate) vs. pH in the acid region is equivalent to $-n$ (Fig. 1.5). An exponent of 0.5 means that for a decrease in pH of one unit, the rate should increase by a factor of about 3 (= $10^{0.5}$).

Above some transition pH, typically 4-5 (Table 1.2), there is generally a pH range in which dissolution rates are independent of pH,

$$(\text{Rate})_{\text{neutral}} = k_N \qquad \text{(neutral region) (1.19)}$$

and then at pH values above 8 or so, they increase with increasing pH:

$$(\text{Rate})_{OH} = k_{OH}(OH^-)^m \qquad \text{(basic region) (1.20)}$$

or:

$$\log (\text{Rate})_{OH} = \log k_{OH} + m(\text{pH} - pK_w) \qquad (1.21)$$

where K_w is the dissociation constant of water, 10^{-14} at 25°C. The exponent m is the

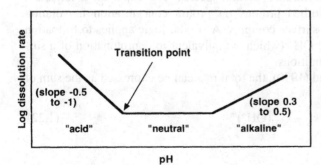

Figure 1.5. Schematic representation of silicate dissolution rates far from equilibrium as a function of pH (after Drever 1994).

Table 1.2. pH of the transition points (see Fig. 1.5) and slope of log (Rate) vs. pH graph in the acid region for silicate minerals (after Drever 1994).

Mineral	Transition pH (acid)	Slope in acid region (= $-n$)	Transition pH (basic)	Source
Albite	4.5	−0.5	7.5	Sverdrup (1990)*
Oligoclase	4.5	−0.5		Oxburgh et al. (1994)
Andesine	4.5	−0.5		Oxburgh et al. (1994)
Bytownite	5	−0.75		Oxburgh et al. (1994)
Anorthite	4.5	−3		Amrhein & Suarez (1988)
K-feldspar	5	−0.5		Schweda (1990)
Forsterite	7**	−0.9	7**	Sverdrup(1990)
Forsterite	4.5 (?)	−0.6	8 (?)	Blum & Lasaga (1988)
Garnet	5.5	−0.9	8	Sverdrup (1990)
Amphiboles	5.5	−0.8		Sverdrup (1990)
Amphibole	<3			Mast & Drever (1987)
Diopside, augite	ca. 6	−0.7 to −0.9	ca. 8	Sverdrup (1990)
Phlogopite, biotite		−0.4		Sverdrup (1990)
Kaolinite	4	−0.4		Wieland & Stumm (1992)

*Sverdrup (1990) represents both original work and a compilation of data from the literature. **No pH-independent region.

slope of the log (dissolution rate) vs. pH in the alkaline region (Fig. 1.5), and typically has a value of about 0.3 to 0.5 (Brady & Walther 1989).

Rates of dissolution are commonly explained by the *transition state theory*. This approach was originally developed to explain the kinetics of gas-phase reactions involving simple molecules. Its extension to silicate dissolution kinetics involves a certain amount of faith. In brief, the rate of dissolution is assumed to be proportional to the concentration of an activated complex at the surface of the solid. The precursor to the activated complex is a *surface complex,* which is formed when an ion (or neutral molecule) is adsorbed from solution onto the mineral surface, forming a complex involving a metal in the solid and the adsorbed species. The principle is illustrated in Figure 1.6.

Adsorption of the solute weakens the bonds between the metal and the bulk of the solid, accelerating its detachment. The rate of dissolution is proportional to the concentration of the surface complex. In the discussion of pH dependence above, adsorption of protons results in a protonated surface complex, and the rate of dissolution is proportional to the concentration of this species. The exponent n reflects a non-linear dependence of adsorbed protons on proton concentration in solution and/or the stoichiometry of the surface complex. A similar logic applies to formation of a surface complex involving OH^- (which is equivalent to deprotonation of a surface-OH group) under basic conditions.

Following Stumm & Wieland (1990), the total rate can be expressed as the sum of the individual rates:

$$(Rate)_{total} = k_H(H^+)^n + k_N + k_{OH}(OH^-)^m \tag{1.22}$$

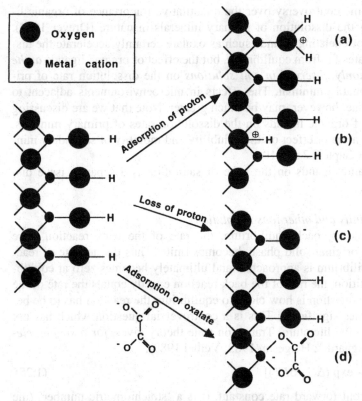

Figure 1.6. Schematic view of possible surface complexes at the surface of an oxide or hydroxide: a-b) Adsorption of protons, c) Deprotonation, d) Complexation of a metal cation by oxalate. Each of these 'complexes' may weaken the bonds between the metal cation and the underlying oxygen atoms, hence accelerating dissolution.

1.4.2.2 *Organic ligands*

Organic ligands may accelerate the dissolution rates of silicates by forming complexes with metals, particularly aluminum, at the mineral surface (Fig. 1.6), and thereby weakening the bonds between the metal and the solid. The most important ligands are those which form chelates with the metals in question. A chelate is a complex in which the ligand binds to the metal through more than one site in the ligand. The effect can be described by the same equations as were used for proton adsorption:

$$(\text{Rate})_{\text{ligand}} = k_{\text{L}}(\text{ligand})^p \tag{1.23}$$

and Equation (1.22) expands to:

$$(\text{Rate})_{\text{total}} = k_{\text{H}}(\text{H}^+)^n + k_{\text{N}} + k_{\text{OH}}(\text{OH}^-)^m + k_{\text{L}}(\text{ligand})^p \tag{1.24}$$

The exponent p again reflects the relationship between adsorbed concentration and concentration in solution and/or the stoichiometry of the complex. The overall effect of organic ligands on dissolution rate will depend on how the magnitude of the term $k_{\text{L}}(\text{ligand})^p$ compares to those of the other terms in the equation.

There has been some controversy over the quantitative importance of organic ligands in accelerating the dissolution of primary minerals in nature (Drever 1994). High concentrations of chelating ligands such as oxalate certainly accelerate the dissolution rates of silicates far from equilibrium, but the effect of organic ligands *at the concentrations commonly observed in soil solutions* on the dissolution rate of primary minerals is probably minimal. The effects in microenvironments adjacent to roots and fungal hyphae, however, may be much greater. Note that we are discussing here only the effect of organic ligands on the dissolution rates of primary minerals. Organic ligands have a major effect on the solubility and mobility of secondary minerals, as discussed in Chapters 2 and 10.

The effect of organic ligands on the state of saturation is a separate issue discussed in Section 1.4.2.3.

1.4.2.3 *Chemical affinity and other ions in solution*

As chemical reactions approach equilibrium, the rate of the back reaction, here reprecipitation of the original solid phase, becomes finite. Thus the net rate of reaction decreases as equilibrium is approached and ultimately becomes zero at equilibrium, where, by definition, the rate of the back reaction exactly equals the rate of the forward reaction. The question is how close to equilibrium the reaction has to be before this effect becomes significant. This is a controversial question which has not been fully resolved in the literature. Transition state theory gives, *for a simple, elementary reaction*, (Aagaard & Helgeson 1982; Velbel 1989):

$$\text{Rate} = k_a \left[1 - \exp \left(\Delta G_r / \sigma RT \right) \right] \qquad (1.25)$$

where k_a is the apparent forward rate constant, σ is a 'stoichiometric number' (the ratio of the rate of destruction of the activated complex to the rate of dissolution of the mineral; if the activated complex has the same formula as the mineral, σ is equal to 1), R is the gas constant and T temperature (K). ΔG_r is the departure from equilibrium, or *chemical affinity*, in units of kJ or kcal per mole. This relationship is shown in Figure 1.7.

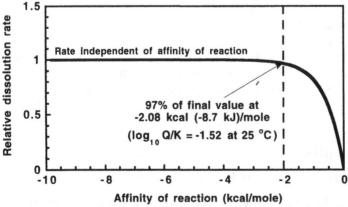

Figure 1.7. Predicted relative dissolution rate of a mineral as a function of chemical affinity (degree of undersaturation) at 25°C. The rate is essentially independent of chemical affinity when the departure from equilibrium is greater than about 2 kcal/mole. σ is assumed to be 1.

Chemical affinity is related to the activity product for the reaction by:

$$\Delta G_r = RT \ln (Q/K_{eq}) \tag{1.26}$$

where Q is the activity product and K_{eq} the equilibrium constant. Thus, for congruent dissolution of albite in acid solution:

$$NaAlSi_3O_8 + 4H^+ + 4H_2O = Na^+ + Al^{3+} + 3H_4SiO_4 \tag{1.27}$$

the activity product, Q, is given by:

$$Q = \frac{[Na^+][Al^{3+}][H_4SiO_4]^3}{[H^+]^4} \tag{1.28}$$

where [ ] represent activities in solution. For undersaturated solutions, the value of Q is less than the equilibrium constant, K_{eq}, and the affinity of reaction is negative.

The term exp $(\Delta G_r/\sigma RT)$ in Equation (1.25) represents the rate of the 'back reaction' – in this case reprecipitation of the primary mineral. It will be effectively negligible compared to the rate of the forward reaction if the rate of the backward reaction is less than 3% (the number is an arbitrary choice) of the forward rate. It will be 3% of the forward reaction if exp $(\Delta G_r/\sigma RT) = 0.03$, i.e. assuming $\sigma = 1$ $(\Delta G_r/\sigma RT)$ $= -3.51 = \ln (Q/K_{eq})$ (see eq. 1.26), or $\log_{10} (Q/K_{eq}) = -1.52$. This yields at 25°C $\Delta G_r = -2.08$ kcal. mol^{-1} (-8.7 kJ. mol^{-1}).

If the departure from equilibrium is greater than this, the back reaction should be negligible and rate should be independent of affinity of reaction. Most solutions in the weathering environment should be much farther from saturation than $\log_{10}$ $(Q/K_{eq}) = -1.52$ with respect to most important primary minerals other than quartz and potassium feldspars (Velbel 1989). According to this model, chemical affinity should have no effect on the dissolution rates of most primary minerals in the weathering environment.

The problem with this model is that it assumes that mineral dissolution behaves as a simple elementary reaction. In experiments conducted by Burch et al. (1993) at 80°C, the dissolution rate of albite seemed to be reduced by chemical affinity effects at much higher degrees of undersaturation than $\log_{10} (Q/K_{eq}) = -1.5$. As a further complication, the dissolution rate of feldspars decreases as the concentration of dissolved Al in solution increases (Chou & Wollast 1985; Sverdrup 1990; Amrhein & Suarez 1992; Gautier et al. 1994; Oelkers et al. 1994). This could be interpreted in terms of a chemical affinity effect or as a specific inhibition caused by Al. It is difficult to separate the two effects. Dissolved Al concentration may affect reaction rate either through its effect on chemical affinity or through some other mechanism that is independent of chemical affinity (e.g. competition between Al and H for adsorption sites: Murphy & Helgeson 1987; Brantley & Stillings 1994). The most detailed results on the effect of Al were obtained from high-temperature experiments – Oelkers et al. (1994) and Gautier et al. (1994) at 150°C; Burch et al. (1993) at 80°C. There is some evidence that these results may not be generally applicable to dissolution reactions at or below 25°C (Drever et al. 1994).

In summary, the effect of solution composition, particularly chemical affinity and Al concentration, on silicate dissolution rates is not well understood. Models of weathering rates in the field have often assumed that pH was the only important variable. This may be adequate in very dilute solutions, but is certainly not adequate as solutions become more concentrated and equilibrium with primary minerals is approached.

1.5 COMPARISONS BETWEEN FIELD AND LABORATORY DISSOLUTION RATES

The rate at which a mineral dissolves is proportional to the area of the mineral in contact with solution. In laboratory experiments, surface areas are normally measured by the BET method (Brunauer et al. 1938), which involves measuring the amount of a gas (commonly nitrogen, less commonly krypton) necessary to form a monolayer on the surface of the mineral. Surface areas can also be computed from the size and shape of grains used in experiments (*geometric surface area*). BET surface areas are always greater than geometric surface areas. The ratio of BET area to geometric area is referred to as the *surface roughness*. Surface roughness values typically increase with increasing grain-size, and change with time as a mineral weathers (Anbeek 1992a,b, 1993; Murphy 1993; Anbeek et al. 1994). As a first approximation, dissolution rate is proportional to *initial* (i.e. prior to reaction) BET area and does not change as surface roughness changes during an experiment (Murphy 1993).

In order to apply laboratory-based dissolution rates to weathering in the field, we need to know the surface areas of all minerals exposed to weathering in the catchment (or soil profile) under study. These are not easy numbers to obtain. One can make some sort of estimate of the geometric surface areas of primary minerals in a soil profile, but one has to make some assumption concerning whether or not they are in contact with percolating water. In catchment-scale studies, Paces (1983) estimated surface area by assuming a plausible fracture pattern in the bedrock. Velbel (1985) assumed a uniform grain-size (1 mm) for minerals in the soil. Both studies estimated that rates in the field, based on the flux of cations leaving a catchment in runoff, were an order of magnitude or more slower than the rate that would be predicted by laboratory experiments. Although their estimates of surface area were crude, they were conservative and probably underestimated the actual wetted surface area of minerals in the catchment and hence overestimated the weathering rate in the field. Swoboda-Colberg & Drever (1993) attempted to measure weathering rates in plot-scale (2 m^2) experiments at a forested site in Maine (USA). They artificially irrigated the plots with acid and collected soil solutions at 25 and 50 cm depths. They also measured the mineralogy and grain-size distribution of the soil to get an accurate estimate of geometrical surface area, and they measured the dissolution rates of minerals from the soil in laboratory reactors. From these experiments they concluded that:

1. Rates in the field were slower by a factor of about 200 compared to rates measured in the laboratory with the same minerals;

2. Untreated minerals from the soil dissolved at rates similar to those of the same size-fraction of crushed fresh 'pure' minerals. Aging of mineral surfaces in the soil was thus not the cause of the discrepancy between field and lab.

Clow (1992) conducted somewhat similar experiments in a small high-elevation catchment on granite in Colorado (USA), using natural precipitation and distilled water rather than acid. He came to the same conclusions: field rates were much slower than laboratory rates, and differences between fresh and weathered minerals was not the cause of the discrepancy. He also showed that the discrepancy was not related to conversions between BET and geometric surface area.

Schnoor (1990) reviewed published studies on field and laboratory dissolution rates and made the interesting observation that silica concentrations, a measure of silicate weathering rate, were generally very similar (within an order of magnitude), about 60 μM, in both natural stream waters and solutions in laboratory dissolution experiments at 20-25°C. Water fluxes and exposed mineral surface areas varied over a far wider range. This suggests that dissolution rates in the field are not controlled by surface reactions at high degrees of undersaturation, as is the case in laboratory experiments, but by either some sort of saturation effect or by a transport process. Drever et al. (1994) visualized the control in terms of two end-members, based on the idea that the porosity in soils and shallow aquifers was very heterogeneous (Fig. 1.8). Under rapid flow conditions, solutions would move largely through high-permeability channels ('macropores'). Solutions in low-permeability channels ('micropores') and intragranular pores would have a relatively long residence time; pH values would be higher and affinities of reaction lower than in the large channels. Weathering rate would be determined by reactions in the large channels (representing a small fraction of the total surface area) and by diffusional transport between the large channels and the small pores of the soil matrix. The other end-member would be represented by slow flow conditions. Concentrations would be more spatially uniform but, because of the implied long contact time, solutions would be more concentrated and closer to saturation with respect to primary phases. Weathering rate would be limited by the affinity of reaction or a related chemical effect. The situation is more complicated in the unsaturated zone because the population of pores that transmit fluid changes as the water saturation changes (Chapter 3).

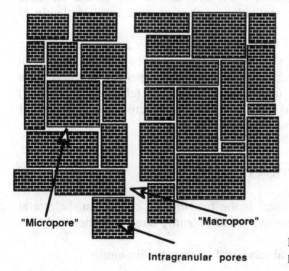

"Micropore"

"Macropore"

Intragranular pores

Figure 1.8. Schematic representation of pores in a soil or shallow aquifer.

1.6 MODELING APPROACHES

The main efforts to model chemical weathering quantitatively have been in conjunction with integrated models for predicting the effect of acid deposition on surface water chemistry (see Chapter 12). Some sort of integrated model is essential in order to calculate the pH of soil solutions, which is generally not the same as the pH of incoming precipitation. Chemical weathering is very important in this context because it is the dominant *long term* (decades or more) control on whether or not a water body becomes acidified under a given acid loading. The short-term variations in the chemistry of a stream or lake, however, are related to changes in input, adsorption processes and biological processes rather than to chemical weathering. What this means is that short-term data (which tend to be the only detailed data available) provide very little information that can be used to calibrate a weathering model. The most common approach (e.g. the MAGIC model, Cosby et al. 1985, 1990) has been to assume that 'weathering' is a single, lumped process that can be described by an equation such as:

$$\text{Rate} = k \, [\text{H}^+]^n \tag{1.29}$$

where n is selected by the operator. This expression is plausible in light of the experimental results discussed above. If, however, rates in the field are effectively determined by transport processes and not surface reaction, the above equation (and/or the choice of n) may not be appropriate. One model that does treat mineral dissolution kinetics explicitly is the PROFILE model (Sverdrup & Warfvinge 1988; Warfvinge & Sverdrup 1992; Altveg et al. 1993). In this model, the mineralogy of each soil layer is input, as is an equation describing the dependence of dissolution rate for each mineral on pH and dissolved Al concentration. The only major 'free parameter' is the specific surface area of the minerals in each soil layer, which must be supplied as an input. The model has been quite successful in describing the chemistry of surface waters in Scandinavia (Warfvinge & Sverdrup 1992).

The models of Henriksen (Henriksen 1984; Brakke et al. 1990) and Kirchner (1992) do not consider weathering explicitly; weathering is included implicitly as part of alkalinity balance.

The problem with essentially all these models is that they are calibrated using short-term data. They should thus, in principle at least, describe well the short-term behavior of the system and the underlying controlling processes. However, because the processes controlling the long-term behavior of the system (chemical weathering) is not really constrained by the short-term calibration procedure, there is no way of knowing whether or not the long-term predictions are accurate.

1.7 FUTURE RESEARCH DIRECTIONS

The most important question at present is the fundamental control on the rate at which silicate minerals weather in nature. We can identify three 'end-member' controlling mechanisms:

1. The dissolution rate of individual minerals far from equilibrium;

2. The inhibition of dissolution by either an approach to saturation or a specific effect of dissolved aluminum;

3. Transport of solutes through pores in the soil and shallow bedrock.

There is very little consensus as to the relative importance of each in specific natural settings. Dissolution rates far from equilibrium (mechanism 1) have been studied quite thoroughly. Mechanism 2 is of fundamental importance, but relatively little research has been done. The few studies to date have generally been at elevated temperatures and high pH, and the applicability of the results to the weathering environment is unclear. We can expect major advances in this area in the next few years. Improved understanding of mechanism 3 will follow from improved understanding of mechanism 2, because this will allow more realistic modeling of the processes controlling solute transport in the weathering environment. I do not believe that there can be a major advance in modeling weathering processes in nature until the question of controlling mechanism is resolved.

REFERENCES

Aagaard, P. & Helgeson, H.C. 1982. Thermodynamic and kinetic constraints on reaction rates among minerals and aqueous solutions, I. Theoretical considerations. *Am. J. Sci.* 282: 237-285.

Altveg, M., Warfvinge, P. & Sverdrup, H. 1993. Profile 3.2: User's guidance for the Apple Macintosh version. Unpublished manuscript, Dept. of Chemical Engineering II, Chemical Center, P.O. Box 124, S-221 00 Lund, Sweden.

Amrhein, C. & Suarez, D.L. 1988. The use of a surface complexation model to describe the kinetics of ligand-promoted dissolution of anorthite. *Geochim. Cosmochim. Acta* 52: 2785-2793.

Amrhein, C. & Suarez, D.L. 1992. Some factors affecting the dissolution kinetics of anorthite at 25°C. *Geochim. Cosmochim. Acta* 56: 1815-1826.

Anbeek, C. 1992a. Surface roughness of minerals and implications for dissolution studies. *Geochim. Cosmochim. Acta* 56: 1461-1469.

Anbeek, C. 1992b. The dependence of dissolution rates on grain size for some fresh and weathered feldspars. *Geochim. Cosmochim. Acta* 56: 3957-3970.

Anbeek, C. 1993. The effect of natural weathering on dissolution rates. *Geochim. Cosmochim. Acta* 57: 4963-4975.

Anbeek, C., Van Breemen, N., Meijer, E.J. & Van Der Plas, L. 1994. The dissolution of naturally weathered feldspar and quartz. *Geochim. Cosmochim. Acta* 58: 4601-4613.

Blum, A.E. & Lasaga, A.C. 1988. The role of surface speciation in the low-temperature dissolution of minerals. *Nature* 331: 431-433.

Brady, P.V. & Walther, J.V. 1989. Controls on silicate dissolution rates in neutral and basic pH solutions at 25°C. *Geochim. Cosmochim. Acta* 53: 2823-2830.

Brakke, D.F., Henriksen, A. & Norton, S.A. 1990. A variable F-factor to explain changes in base cation concentrations as a function of strong acid deposition. *Verh. Internat Verein. Limnol.* 24: 146-149.

Brantley, S.L. & Stillings, L. 1994. An integrated model for feldspar dissolution under acid conditions. *Mineralogical Magazine* 58A: 117-118.

Brunauer, S., Emmett, P.H. & Teller, E. 1938. Adsorption of gases in multimolecular layers. *J. Amer. Chem. Soc.* 60: 309-319.

Burch, T.E., Nagy, K.L. & Lasaga, A.C. 1993. Free energy dependence of albite dissolution kinetics at 80°C and pH 8.8. *Chemical Geology* 105: 137-162.

Chou, L. & Wollast, R. 1985. Steady-state kinetics and dissolution mechanisms of albite. *Amer. J. Sci.* 285: 963-993.

Clow, D.W. 1992. Weathering rates from field and laboratory experiments on naturally weathered soils. Unpublished PhD dissertation, Univ. of Wyoming, Laramie, WY, USA.

Cosby, B.J., Hornberger, G.M., Galloway, J.N. & Wright, R.F. 1985. Modeling the effects of acid deposition: assessment of a lumped parameter model of soil water and stream chemistry. *Water Resources Res.* 21: 51-63.

Cosby, B.J., Jenkins, A., Ferrier, R.C., Miller, J.D. & Walker, T.A.B. 1990. Modelling stream acidification in afforested catchments: long-term monitoring at two sites in central Scotland. *J. Hydrol.* 120: 143-162.

Drever, J.I. 1988. *The Geochemistry of Natural Waters*, 2nd Edition. Englewood Cliffs, Prentice-Hall.

Drever, J.I. 1994. The effect of land plants on the weathering rates of silicate minerals. *Geochim. Cosmochim. Acta* 58: 2325-2332.

Drever, J.I., Murphy, K.M. & Clow, D.W. 1994. Field weathering rates versus laboratory dissolution rates: an update. *Mineralogical Magazine* 58A: 239-240.

Gautier, J.-M., Oelkers, E.H. & Schott, J. 1994. Experimental study of K-feldspar dissolution rates as a function of chemical affinity at 150°C and pH 9. *Geochim. Cosmochim. Acta* 58: 4549-4560.

Henriksen, A. 1984. Changes in base cation concentrations due to freshwater acidification. *Verh. Internat. Verein Limnol.* 22: 692-698.

Kirchner, J.W. 1992. Heterogeneous geochemistry of catchment acidification. *Geochim. Cosmochim. Acta* 56: 2311-2327.

Lasaga, A.C., Soler, J.M., Ganor, J., Burch, T.E. & Nagy, K.L. 1994. Chemical weathering rate laws and global geochemical cycles. *Geochim. Cosmochim. Acta* 58: 2361-2386.

Mast, M.A. & Drever, J.I. 1987. The effect of oxalate on the dissolution rates of oligoclase and tremolite. *Geochim. Cosmochim. Acta* 51: 2559-2568.

Murphy, K.M. 1993. Kinetics of albite dissolution: the effect of grain size. Unpublished M.S. thesis, Univ. of Wyoming, Laramie, WY, USA.

Murphy, W. M. & Helgeson, H. C. 1987. Thermodynamic and kinetic constraints on reaction rates among minerals and aqueous solutions. III. Activated complexes and the pH-dependence of the rates of feldspar, pyroxene, wollastonite, and olivine hydrolysis. *Geochim. Cosmochim. Acta* 51: 3137-3153.

Oelkers, E.H., Schott, J. & Devidal, J.-L. 1994. The effect of aluminum, pH, and chemical affinity on the rates of aluminosilicate dissolution reactions. *Geochim. Cosmochim. Acta* 58: 2011-2024.

Oxburgh, R., Drever, J.I. & Sun, Y.-T. 1994. Mechanism of plagioclase dissolution in acid solution at 25°C. *Geochim. Cosmochim. Acta* 58: 661-669.

Paces, T. 1983. Rate constants of dissolution derived from measurements of mass balance in hydrologic catchments. *Geochim. Cosmochim. Acta* 47: 1855-1863.

Schnoor, J.L. 1990. Kinetics of chemical weathering: A comparison of laboratory and field weathering rates. In W. Stumm (ed.), *Aquatic Chemical Kinetics*: 475-504. New York: Wiley.

Schweda, P. 1990. Kinetics and mechanisms of alkali feldspar dissolution at low temperatures. PhD dissertation, Dept. of Geology and Geochemistry, Stockholm University, Sweden.

Stumm, W. 1990. *Aquatic Chemical Kinetics*. New York: Wiley.

Stumm, W. & Wieland, E. 1990. Dissolution of oxide and silicate minerals: Rates depend on surface speciation. In W. Stumm (ed.), *Aquatic Chemical Kinetics*: 367-400. New York, Wiley.

Sverdrup, H.U. 1990. *The Kinetics of Base Cation Release due to Chemical Weathering*. Lund: Lund University Press.

Sverdrup, H. & Warfvinge, P. 1988. Weathering of primary silicate minerals in the natural soil environment in relation to a chemical weathering model. *Water, Air, and Soil Pollution* 38: 387-408.

Swoboda-Colberg, N.G. & Drever, J.I. 1993. Mineral dissolution rates in plot-scale field and laboratory experiments. *Chemical Geology* 105: 51-69.

Velbel, M.A. 1985. Geochemical mass balances and weathering rates in forested watersheds of the southern Blue Ridge. *Am. J. Sci.* 285: 904-930.

Velbel, M.A. 1989. Effect of chemical affinity on feldspar hydrolysis rates in two natural weathering systems. *Chemical Geology* 78: 245-253.

Warfvinge, P. & Sverdrup, H. 1992. Calculating critical loads of acid deposition with PROFILE – A steady-state soil chemistry model. *Water, Air, and Soil Pollution* 63: 119-143.

Wieland, E. & Stumm, W. 1992. Dissolution kinetics of kaolinite in acid aqueous solutions at 25°C. *Geochim. Cosmochim. Acta* 56: 3339-3363.

Wesolowski, D.J. & Palmer, D.A. 1994. Aluminum speciation and equilibria in aqueous solution: V. Gibbsite solubility at 50°C and pH 3-9 in 0.1 molal NaCl solutions. *Geochim. Cosmochim. Acta* 58: 2947-2970.

CHAPTER 2

Composition, properties and development of Nordic soils

OLE K. BORGGAARD

Chemistry Department, Royal Veterinary and Agricultural University, Frederiksberg, Denmark

2.1 INTRODUCTION

Soils are dynamic, open biogeochemical systems with influxes and effluxes of matter and energy. Many soil-forming (and destroying) processes proceed simultaneously, and the resulting soil reflects the balance of these processes. The definition of soil given in Appendix A points out the five soil forming factors, which according to Jenny (1980) determine the state of the soil:

$$\text{Soil} = f(cl, o, r, p, t) \tag{2.1}$$

This relation is called the 'clorpt' concept, where cl is climate, o is $o$rganisms, r is $r$elief, p is $p$arent material and t is $t$ime. These factors drive the soil forming processes, which may be divided in fundamental processes and soil development processes. The fundamental processes including humus formation, weathering and transport lead to formation of soil from not-soil and to changes of soil development. The fundamental processes include several subprocesses, but are themselves parts of the so-called soil development processes, which are discussed in Section 2.4.

Exact information about the factors affecting the genesis of the soils at a certain site should be based on detailed climatic, topographic and geological maps both of the bedrock and Quaternary deposits, and on careful observations of land-use and vegetation at the site. In relation to Nordic soils the «clorpt»-factors can shortly be summarized in the following way:

Climate (cl): The so-called temperature regimes are *mesic* (mean annual soil temperature 8-15°C) and frigid or cryic (mean annual soil temperature 0-8°C, where the summer temperature is higher under *frigid* than *cryic* conditions). The annual precipitation range from more than 4000 mm in western Norway to 500-700 mm in eastern Denmark, southwestern Sweden, Finland and northern Scandinavia. Despite these variations the so-called moisture regime is *udic* in most well-aerated soils (the soil is not dry in 90 cumulative days per year and is dry less than 45 consecutive days in the 4 months following the summer solstice), whereas it will be *aquic* in water-logged soils (a reducing regime that is virtually free of oxygen as shown by redoximorphic features). The complete definitions of temperature and moisture regimes are given by the Soil Survey Staff (1994).

Organisms (o): Can roughly be divided into two groups according to land-use as woodland and as farmland. Soils in the latter group are seriously affected by liming, fertilization and tillage. Different conifers and deciduous plants undoubtedly affect soils differently. The effect of vegetation on mineral weathering and soil formation has been reviewed by Drever (1994).

Relief (r): The relief exhibits great variability ranging from level or almost level areas through undulated land to very steep mountain slopes. The slope aspect is also important inasmuch as soils on south-facing slopes recieve more sunlight (heat) than soils on north-facing slopes.

Parent material (p): Most soils are developed on glacial deposits originating from the last glaciation (Weichsel Glaciation 117,000-10,000 years ago). The mineral composition of the deposits is determined by the rocks and sea sediments over which the glaciers have passed. The main parent rocks are granites and gneisses, but more nutrient-rich rocks (including limestone) occur at many places. The glacial deposits include glacial till and water-sorted materials of different texture ranging from gravel and sand to clay. Many shallow lakes have turned into peatland, particularly in Finland with nearly 100,000 km^2 of peat deposits. In some areas, especially in the northern regions, where isostatic rebound has elevated previously submerged material above the present sea level, the parent material has been reworked by the sea. More information about the geology of the Quaternary period may be found elsewhere (e.g. Rankama 1965; Ehlers 1983; Catt 1988; Andersen & Borns 1994; Andersen et al. 1995).

Time (t): Soils in the Nordic countries may be considered to have developed after termination of the Weichselian Glaciation, i.e. within the last 15,000-10,000 years. Even soils in southwestern Denmark which was not covered by ice during the last ice age, are considered to have formed since its termination, but on previously leached materials affected by freeze-thaw actions. Soils in northern Scandinavia and Finland, particularly those on uplifted land, are younger than 10,000 years.

Distinct regimes or combinations of processes produce distinctive soils with distinct compositions and properties. True equilibrium is nearly never attained, but soil development may reach a steady-state. The soil processes form soil from not-soil but the processes are themselves functions of the soil change, because they go on in a continuously changing environment. Some processes may initiate other processes. Processes involving very reactive materials are fast, while more inert materials react slowly (Chapter 1).

The composition and properties of a soil can be measured or estimated, but will only give a snapshot of the soil due to its dynamic nature. Since the present soils carry the imprint of processes that were active in the past, such information is fundamental for the interpretation and understanding of soil genesis and of the processes involved. Knowledge about soil development and processes is, in turn, very important in forecasting future soil development, and hence composition and properties, under specified, but most likely changing environmental conditions. Such information is basic to sustainable soil use and management.

Due to the various factors and processes that affect soil formation and development, soils exhibit great variability. Despite the variability it is, however, possible to identify soils that behave more or less the same way. In relation to potential land-use this is very important, because similar soils will have the same needs for management, and experience and knowledge about soils at one place can easily be transferred to similar soils at another place. This is, in short, the underlying reason for classifying soils. The aims of the soil classification is to organize (and to map) different soils in different groups or classes according to their composition and properties. Soils belonging to one class have specific, but similar compositions and properties different from those of soils in other classes. Since soil compositions and properties, and hence classification, are functions of soil genesis, an understanding of and knowledge about soil genesis is very useful for a person who is going to interprete and apply the results of soil classification. On the other hand, soil classification systems should not be based entirely on genesis (as in older soil classification systems), because soil formation processes can seldom be observed and because the processes operating within a soil have been subject to changes over time. In the more elaborated systems such as the FAO/UNESCO system and the American Soil Taxonomy system (FAO/UNESCO 1990; Soil Survey Staff 1994), classification is based on soil compositions and properties that can be measured in the laboratory, but also on properties and features that are assessed in the field.

The purpose of this chapter is to give a short overview of soil compositions, properties, development and classification with emphasis on aspects relevant to soils in Denmark, Finland, Norway and Sweden. The soil classification will be according to the American Soil Taxonomy and FAO/UNESCO systems (FAO/UNESCO 1990; Soil Survey Staff 1994; Appendices A and B). More information on soils in general can be found in numerous textbooks (e.g. Singer & Munns 1987; Catt 1988; Schachtschabel et al. 1989; Brady 1990; Tan 1994), whereas the books of Bohn et al. (1985) and Sposito (1989) cover soil chemistry and those of Birkeland (1974), Duchaufour (1982), Wilding et al. (1983), Buol et al. (1989) and Fanning & Fanning (1989) deal with pedology (soil genesis and classification). Those who are interested in soil analysis and sampling may find appropriate information in Hodgson (1978), Page et al. (1982), Klute (1986), Wilson (1987), FAO (1990), Westerman (1990) and Rowell (1994).

2.2 SOIL COMPOSITION

Soil is a mixture of inorganic and organic solids, water, and air. Solids of different size and shape are surrounded by pores or voids that are full of air, water or both. The spatial arrangement or layering of the solids determines the volume (space) of voids (pores) occupied by air and water; the closer the solids are arranged the smaller the porespace will be. A mineral soil showing good physical conditions for plant growth will have a ratio solid:void ~1 as illustrated in Figure 2.1. Decreased porespace due to increased soil compaction affects plant growth negatively.

The bulk density of a soil is a measure of the mass of a unit volume (voids plus solids) of oven-dry soil (105°C) in natural setting, i.e. in situ. Since soil particle (solid) density exhibits very limited variation around 2.6 Mg/m^3 (e.g. the density of quartz is 2.67 g/cm^3) in most mineral soils, bulk density is a measure of porespace

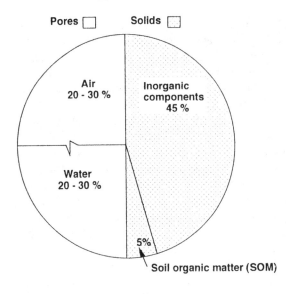

Figure 2.1. Volume composition of a surface soil with good physical plant growth conditions.

or compaction of soil. Bulk densities of various clayey to sandy soils are in the range 1.1-1.8 Mg/m^3 but may exceed 2 Mg/m^3 in very compacted (sub)soils, where root development is seriously impaired. Organic soils have much lower particle and bulk densities. Soil porosity is the fraction of the soil volume occupied by voids and equals 1 – (bulk density/particle density).

Most reactions in soils occur at the surfaces of the solid particles emphasizing the importance of the surface area of the soil solids. The smallest particles will have the largest specific surface area, and hence reactivity. Assuming that the soil particles are spheres, the relationship between surface area and particle size is shown in Figure 2.2 for mineral particles. Obviously, the clay fraction is by far the most reactive inorganic soil fraction.

Soil analyses are normally carried out on the so-called fine-earth fraction, which includes the less than 2 mm particles, since the coarser fractions mainly affect soil properties by acting as 'diluents', because of their very small specific surface area. It must, however, be recognized that minerals bound in the coarser fraction can be released by weathering and that abundant stones and boulders will restrict land exploitation. The fine-earth fraction is divided into three main size separates: sand, silt and clay. Soil texture is the relative proportions of sand, silt and clay in the soil as determined by particle (or grain) size analysis. This analysis is based on a combination of sieving (sand fraction) and sedimentation (silt and clay fractions) performed on the soil sample after removal of soil organic matter and dispersion in water (or diluted sodium pyrophosphate solution). During the last fifteen years modern coultercounter techniques have been developed to measure the particle size distribution of very small samples very accurately. By means of the textural triangle the textural class can then be determined (Fig. 2.3). A textural class is a defined range of particle size distributions with similar behaviour and management needs.

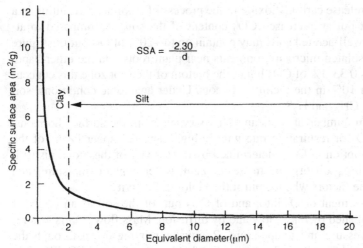

Figure 2.2. The specific surface area of mineral soil materials as a function of particle size.

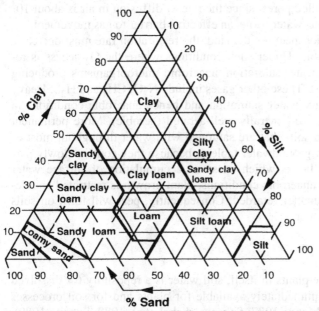

Figure 2.3. Soil textural classes according to the American soil classification system (Soil Survey Staff 1994): Clay < 2 µm, silt 2-50 µm, very fine sand 50-100 µm, fine sand 100-250 µm, coarse sand 250-2000 µm.

2.2.1 *Soil air*

The components of soil air are typically the same as in atmospheric air, which contains 79% N_2, 21% O_2, 0.03% CO_2 etc., but the concentrations are different because of biological activity (Singer & Munns 1987; Schachtschabel et al. 1989; Brady 1990). Microorganisms such as bacteria and fungi and plant roots take up oxygen from the soil air and release carbon dioxide in the process of respiration resulting in a decreased O_2 content but an increased CO_2 content of the soil air compared to atmospheric air. Thus, well-aerated soils may contain up to ~20% of O_2, but this figure can drop to ~2% in isolated microenvironments near plant roots. On the other hand, soil air often contains 0.33-3% of CO_2 but at the bottom of the root zone this concentration may approach 10% in the vicinity of roots. Under anaerobic conditions, soil air may contain N_2O, CH_4 and H_2S.

Due to a maximum number of roots and microorganisms in the surface horizons, the consumption of O_2 or respiration rate will be highest in the upper layers of the soil. Since the consumption of O_2 is determined by the activity of these organisms it depends on temperature, soil organic matter content, water content (moisture) and nutrient supply, i.e. the factors which control the biological activity.

The necessary movement of O_2 into, and of CO_2 out, of the soil as an effect of respiration is accomplished by mass flow and by diffusion. Mass flow occurs as a result of changes in pressure and temperature. Rainwater moving down through the soil can force gas ahead of it and also carry dissolved gases. In general, diffusion, where the gas molecules move by thermal motion in response to the concentration gradients formed in the soil, is, however, the most important process. Diffusion occurs mainly through the air-filled pores, since the rate of diffusion in air is about 10^4 times faster than in water. Thus water forms an effective barrier to gas movement.

If the supply of O_2 does not meet the demand, the respiration rate must decrease, and the soil becomes anaerobic. Under such conditions where the O_2 access is restricted or prevented due to water saturation, anaerobic microorganisms producing other gases than CO_2 can exist. These other gases include N_2O, CH_4 and H_2S. Heavy rainfalls can lead to temporary water saturation and hence anaerobic conditions in most soils even though they are generally well-aerated (aerobic). More permanent anaerobic conditions occur in soils that are saturated with water throughout most of the year because of a shallow groundwater table or because they contain slowly permeable or impermeable soil layers which cause an accumulation of surface water (Section 2.4.2). In soils with anaerobic conditions, plant growth is restricted and decomposition of soil organic matter retarded. Consequently, peat will form on soils permanently saturated with water.

2.2.2 Soil water

Apart from being vital for the plants in itself, soil water is a repository for dissolved solids and gases, which are immediately available for plants and for soil processes (Bohn et al. 1985; Singer & Munns 1987; Schachtschabel et al. 1989; Sposito 1989). The mixture of water and dissolved solids and gases is called the soil solution. In well-aerated soils the main gaseous soil solution components include O_2 and CO_2. The dissolved solids include a great number of different cations (Al^{3+}, Ca^{2+}, Mg^{2+}, K^+, Na^+, NH_4^+, H^+ etc.), anions (HCO_3^-, Cl^-, NO_3^-, SO_4^{2-} etc.) and molecular species such as various organic compounds (dissolved organic matter, DOM).

The dissolution of gases in water is determined by Henry's law:

$$a_{gas} = K_H P_{gas} \qquad (2.2)$$

where a_{gas} is the activity, P_{gas} the pressure of the gas and K_H is a constant. The constants for O_2 and CO_2 are 1.26 and 34.06 mol/m^3 · atm, respectively at 25°C. A partial pressure of 0.2 atm (atmospheric air) will result in a O_2 concentration of 0.25 mM, whereas the considerably higher CO_2 constant suggests soil solution concentrations up to ~1 mM ($P_{CO_2} = 0.03$ atm) or even higher.

2.2.2.1 *Water potentials*
The soil pores are occupied by air or water as shown in Figure 2.1. Fluctuations in water content occur with climate which controls the inputs of water and drainage and evaporation from soil and leaf surfaces which controle losses or outputs. The ability of soil to allow drainage, to hold water for plant use and to hold water so firmly that plants are unable to use it depends on the size, shape and continuity of the soil pores, which, in turn, depend on soil texture, organic matter content and structure.

Soil water is affected by several forces, which determine its availability to the plants and its movement in the soil. These forces include gravitational forces, capillary and adsorptive (suction) forces and the osmotic force (Singer & Munns 1987; Schachtschabel et al. 1989; Brady 1990). While the gravitational forces tend to force water downward, the suction forces act oppositely, that is, the suction forces will retain the water. Due to the various forces, energy is consumed or released when water is moved (flows) from one zone to another. The corresponding potentials can shortly be described in the following way:

Gravitational potential (ψ_z): This potential is a function of the height (z) of water above a reference point (e.g. the groundwater table).

Matrix potential (ψ_m): This potential, which is also called the suction potential, is a function of the attraction of surfaces to water molecules and is most important to unsaturated flow. The matrix potential is inversely proportional to the diameter of the pores. Under saturated conditions ψ_m is replaced by ψ_h, the pressure potential.

Osmotic potential (ψ_o): This potential is a function of the soil solution composition. The osmotic potential only affects water movement through semipermeable membranes and will therefore have little influence on water movement in soils, since such membranes are not likely to occur in soils. Water uptake by the plants through semipermeable membranes in the cells is, however, strongly affected by the osmotic potential.

The sum of these potential (called water potential, ψ) is a quantitative expression for the tendency of water to move, which will be from zones of high water potential to zones where the water potential is lower. For pure water, where the osmotic potential is zero, the so-called hydraulic potential is defined as follows:

$$\psi_H = \psi_z + \psi_m \text{ (or } \psi_h \text{ under saturated conditions)} \tag{2.3}$$

The variation of the potentials at equilibrium and under non-equilibrium conditions at raising (upward) water due to uptake or evaporation and descending (downward) water after rain is shown in Figure 2.4.

The gravitational potential is proportional to the height (z) above the reference point (Fig. 2.4). The matrix potential, i.e. the negative pressure retaining water in the pores under unsaturated conditions, can be measured by a tensiometer or by a pressure membrane apparatus. The tensiometer is limited to suctions > 80 kPa and consists of a porous ceramic cup, which contains water and is to be installed in the soil, connected to a mercury pressure gauge. Under saturated conditions the pressure po-

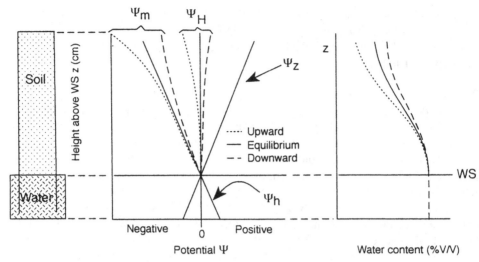

Figure 2.4. Various potentials for pure water under equilibrium ($\psi_z = \psi_m$) and non-equilibrium conditions (with permission from Schachtschabel et al. 1989).

tential (ψ_h) can be determined by a test well (piezometer). Details of these methods may be found elsewhere (Klute 1986; Rowell 1994). The osmotic potential (π) can be determined from the decrease in water vapor pressure or freezing point, or be calculated by the expression: $\pi = cRT$, where c is the concentration of dissolved compounds, R the gas constant and T the temperature. If the concentration exceeds 0.6 M, the osmotic potential corresponds to a negative pressure of less than –15 bar, which corresponds to the wilting point as defined below.

2.2.2.2 *Water content and availability*

The water content in soils is expressed either as gravimetric water, i.e. the mass of water per unit mass of oven-dry soil, or as volumetric water, i.e. the volume of water per unit volume of soil and is often expressed as a fraction or percentage (Fig. 2.4) or as depth of water per unit depth of soil, typically mm water, which allows direct comparison with amounts of rain or irrigation water.

The pores in soils are of different size ranging from the largest macropores to the smallest micropores. The macropores are those which allow rapid drainage of water after heavy rainfall and when these pores are emptied drainage becomes very slow. Further removal of water by evaporation from the soil surface or transpiration from plant leaves will remove water from pores of smaller and smaller size (diameter). The removal of water for instance by the plants will therefore be more and more difficult, since the matrix potential, and suction, is inversely related to the pore diameter. At a certain water content specific for each soil, the plants can no longer extract water from the soil, and they will wilt.

Accordingly, field capacity and wilting point represent the approximate upper and lower limits of water that can, at the same time, be held in soils and be available for plants. At field capacity, the medium and small pores are filled with water, but most of the larger pores are nearly empty. At the wilting point only the smallest pores contain water. The approximate matrix potentials corresponding to field capacity corresponds to –0.1 bar (1 bar ≈ 100 kPa) or –100 cm water column, whereas the approximate matrix potential corresponding to wilting point corresponds to –15 bars or –15000 cm water column. Normally the so-called *pF* value is used, and this is defined in the following way:

$$pF = -\log \text{(cm water column)} \qquad (2.4)$$

Therefore, *pF* = 2 will correspond to the field capacity and *pF* = 4.2 corresponds to the wilting point. In water retention curves, *pF* is plotted against the corresponding water content (Fig. 2.5). Plant-available water is taken as the difference between the water contents at *pF* 2 and *pF* 4.2.

Obviously clayey soil materials retain more water than silty and sandy soil materials, but the available water contents show smaller differences, because micropores dominate in the clayey soils. According to Breuning-Madsen et al. (1992), plant-available water (PAW) in Danish soil materials can be calculated by the expression:

$$\text{PAW (\%V/V)} = 1.79 \times \% \text{ SOM} + 0.07 \times \% \text{ clay} + 0.29 \times \% \text{ silt} + \qquad (2.5)$$
$$0.18 \times \% \text{ fine sand} + 2.56$$

where SOM stands for soil organic matter (see Section 2.2.3). The silt is the 2-20 μm fraction and the fine sand is the 20-200 μm fraction. This expression emphasizes the importance of the silt and fine sand fractions for the capacity of soils to retain water in plant-available form (storage capacity of plant-available water).

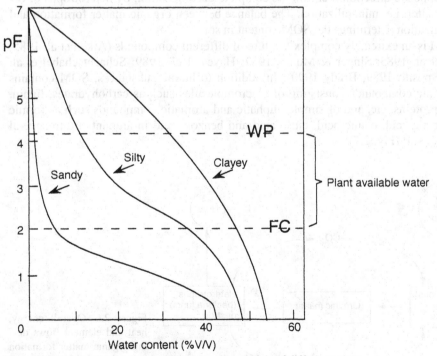

Figure 2.5. Water retention curves for three soil materials of different texture.

The retention curves can also give information about the size of pores and pore size distribution. Thus, the pore diameter (d in μm) is related to pF in the following way:

$$\log (d) = 3.5 - pF \qquad (2.6)$$

The pores are divided into fine pores or micropores with $d < 0.2$ μm corresponding to $pF > 4.2$, medium pores with $d = 0.2\text{-}30$ μm corresponding to pF 4.2-2 and coarse pores or macropores with $d > 30$ μm corresponding to $pF < 2$. Accordingly, plant-available water is retained in the medium pores.

2.2.3 Soil organic matter (SOM)

The soil organic matter (SOM) pool consists of living organisms and dead plant and animal residues and it is controlled by formation and decay processes involving autotrophic and heterotrophic organisms (Fig. 2.6). SOM is per unit mass the most active portion of soils. It is the reservoir for a large pool of carbon and for various essential elements, promotes good soil structure, is a source of cation exchange capacity (CEC) and soil pH buffering and is important for complexation processes and mineral weathering.

The overall formation and decay processes can be simplified as follows:

$$CO_2 + H_2O \Leftrightarrow CH_2O + O_2 \qquad (2.7)$$

From left to right, the process shows the net result of photosynthesis, i.e. production by autotrophs of carbohydrates and oxygen from carbon dioxide and water, whereas in the opposite direction it shows the complete decomposition by heterotrophs of organic matter, i.e. mineralization. The balance between organic matter formation and mineralization determines the SOM content in soils.

SOM is an extremely complex mixture of different compounds (Aiken et al. 1985; Bohn et al. 1985; Singer & Munns 1987; Hayes et al. 1989; Schachtschabel et al. 1989; Sposito 1989; Brady 1990). In addition to humic substances, SOM contains non-humic compounds consisting of macromolecules such as carbohydrates, lignin, lipids, proteins, etc. and of simple aliphatic and aromatic compounds such as formic acid, acetic acid, oxalic acid, citric acid and benzoic acid in amounts up to several mmol/kg soil (Fig. 2.7).

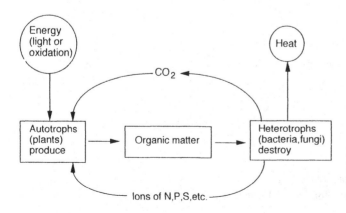

Figure 2.6. Overall energy, heat and element flows during organic matter formation and decomposition.

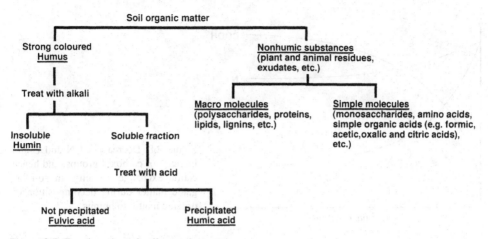

Figure 2.7. Fractionation of soil organic matter.

2.2.3.1 *Humification*

Organic matter decomposition is accomplished through a huge number of chemical and microbiological sub-processes. Partial decomposition leads to formation of humus, which is a dark, heterogeneous mass of colloidal organic compounds. The complex of degradative and synthetic processes leading to humus formation is called humification.

The main elements in SOM include C, O, N, S, P and H. In well-aerated, temperate, fertile soils the mass ratios for the C/N/S/P concentrations are about 100/10/1/1. Since the C/N in fresh plant material such as straw is much higher, humification leads to a relative enrichment of N (Fig. 2.8). The elements are linked together in numerous complex aliphatic and aromatic compounds. Oxygen is present in carboxyl, phenol, quinone and other groups. During humification the oxygen content increases as shown by the increased carboxyl content in Figure 2.8. Since SOM appears to contain ~58% C, the soil SOM content is taken as (100/58) times the percentage of organic C determined by wet or dry combustion (Page et al. 1982).

Humus is often divided into three fractions (Fig. 2.7) according to solubility as follows:

Humin: This fraction is insoluble in both alkali and acid and therefore probably consists of high molecular weight compounds with few functional groups.

Humic acid: This fraction is soluble in alkali but insoluble in acid. It is relatively rich in oxygen-containing groups such as carboxyl and phenol groups.

Fulvic acid: This fraction is soluble in alkali and acid. It is the low molecular weight fraction and it can be considered the most reactive portion of SOM because of its high content of functional groups.

Typically, humic acid and fulvic acid contain 1.5-6.0 and 5.2-11.2 mol/kg of carboxyl groups and 2.1-5.7 and 0.3-5.7 mol/kg of phenol groups (Schnitzer & Khan 1978). In addition to carboxyl and phenol groups, humic and fulvic acids contain several other functional groups such as ether, ester, alcohol and ketone groups. Fulvic acid appears to be less aromatic than humic acid. Typically, fulvic acid have molecular weights in the range 700-2300 daltons, while molecular weights of 2.3 ×

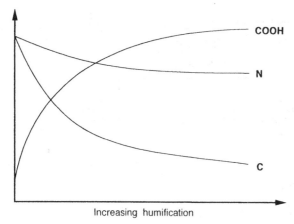

Figure 2.8. Decreases of N and C and increase of carboxyl groups, and hence cation exchange capacity, in soil organic matter during decomposition of organic matter to humus.

10^3 to 1.4×10^6 daltons are commonly reported for humic acid.

Almost all organic matter, whether from litterfall or root turnover, is delivered to the upper layers of the soil where it is subject to decomposition by microorganisms. The SOM content is determined by the production (input) and decomposition of the organic matter. If a steady-state is established so that the annual production equals the annual decomposition, then the SOM content can be calculated by the following equation:

$$(OM\ input) = (OM\ output) = k \times (SOM\ content) \qquad (2.8)$$

In this equation, k is the fraction of SOM, which is decomposed per year. The reciprocal of k ($1/k$) is the residence time. The decomposition rates, and hence the residence times, of the various SOM components are highly different. Compared to fresh organic matter residues with residence times of only a few months or less, humus, or at least part of it, is very resistant to microbial decomposition. The age of the resistant fraction can be > 1000 years. In association with clay silicates and aluminium and iron oxides, SOM seems to be stabilized possibly because it is physically and/or chemically protected against degradation (Christensen 1992).

Since k is the ratio between organic matter input and SOM content, it will change when these quantities change. In addition to the plant species and the biomass export, the input of organic matter depends on several environmental factors such as the water content (precipitation), the temperature and the nutrient content of the soil. Thus, while the annual input of organic matter is about 1 kg/m^2 of organic C in temperate regions and twice this amount can be delivered by humid tropical forests, desert soils and tundra soils receive < 0.1 kg/m^2 of organic C per year. However, the decomposition rate of SOM also depends on these factors because they determine the activity of the microorganisms. The rate of organic matter decomposition increases at increasing soil temperature, but decreases at very low and very high moisture contents exhibiting maximum decomposition at a certain moisture content, which at the same time ensures optimal microbial activity and good oxygen access. In poorly drained and swampy soils, dead plant material can be covered with water leading to peat formation. Peat occurs at many places in the Nordic countries, particularly in Finland.

2.2.3.2 *Soil microorganisms*

In addition to the macroorganisms such as plant roots, earthworms, moles, ants etc. important for mixing of the various soil components, soils contain microorganisms. Soil microorganisms, which include aerobic and anaerobic bacteria, fungi, actinomycetes and algae, are essential for organic matter decomposition and many other soil processes (Schachtschabel et al. 1989; Brady 1990; Tan 1994). While the plow layer (Ap horizon) may contain several Mg per ha of microbial biomass, the B and C horizons are very low in microorganisms.

Soil bacteria are either autotrophic or heterotrophic. The autotrophs obtain their energy from the oxidation of inorganic substances such as ammonium, sulfur and iron compounds and most of their carbon from carbon dioxide. Most bacteria are, however, heterotrophic, that is both energy and carbon come from organic matter. Thus, heterotrophic bacteria, along with fungi and actinomycetes, account for the general breakdown of organic matter in soils. This degradation involves many different bacteria (and other organisms), where specific species are active in different steps of the overall degradation process.

While fungi are rather unaffected by soil pH, SOM decomposing bacteria exhibit maximum activity at pH 6-8 and high calcium concentration. SOM degradation therefore depends on soil pH. Under neutral to slightly alkaline pH conditions with good calcium availability, bacteria are the main SOM decomposers leading to formation of mull, which is well humified SOM with a low C/N ratio (~10). Mull is intimately mixed with the inorganic soil fraction because of earthworm activity. In nutrient-poor soils of low pH, SOM is mainly decomposed by fungi leading to the formation of so-called mor, i.e. poorly humified SOM with a high C/N ratio. Mor accumulates at the soil surface due to absence of soil mixing organisms like earthworms. Mor formation is most likely to occur on sandy soils due to their low buffering capacity against acidification and under certain kinds of vegetation such as spruce and calluna.

Under anaerobic conditions bacteria can decompose organic matter resulting in the formation of methane. If nitrate is present, it may serve as the electron acceptor leading to its reduction to N_2 (and N_2O) in the so-called denitrification process. Nitrogen fixation, either symbiotic or non-symbiotic, where atmospheric nitrogen is converted to plant-available nitrogen compounds, is another very important microbial process. Symbiotic nitrogen fixation by the formation of ammonium is carried out by heterotrophic bacteria living in root nodules of legumes. Non-symbiotic nitrogen fixation can be carried out by certain heterotrophic bacteria and by autotrophic blue-green algae.

Other bacteria are involved in the oxidation of ammonium to nitrate (nitrification), which occurs in two main steps, i.e. oxidation of ammonium to nitrite followed by oxidation of nitrite to nitrate. Normally the first step is rate limiting preventing accumulation of high concentrations of poisonous nitrite in soil. The energy released by the processes is used by the bacteria together with CO_2 in their life processes. Nitrification seriously affects nitrogen mobility in soils because the positively charged ammonium ions are retained in the soil by cation exchange, whereas the negatively charged nitrate ions are prone to leaching. Furthermore, since nitrification is followed by a proton release, application of fertilizers containing ammonium and ammonia will lead to soil acidification.

Bacteria are also involved in formation of pyrite (FeS_2) and other sulphides from reduction of sulphate and iron(III) in the process called sulphidization and in the oxidation of pyrite to iron oxides and sulphuric acid in the process called sulphuricisation (Fanning & Fanning 1989). These process are typically seen in tidal marsh soils.

2.2.4 *Soil minerals*

The inorganic soil fraction consists of primary and secondary minerals. The silt and sand fractions consist largely of primary minerals, i.e. minerals formed under conditions of low oxygen concentration and elevated temperatures and pressures. Primary minerals are inherited from igneous, metamorphic and sedimentary rocks. Some resistant, primary minerals also occur as minor constituents of the clay fraction. The main components of the clay fraction are, however, secondary minerals. These minerals are formed by low temperature reactions under humid and often oxygen-rich conditions and they are either inherited from the parent material or formed by weathering. Weathering processes are covered in details by Drever (Chapter 1). In soils, various silicate minerals constitute the most abundant primary minerals, whereas the secondary minerals include layer silicates and accessory minerals such as carbonates and oxides.

Although all stable elements of the periodic table undoubtedly occur in soils, ten elements account for nearly 99% of the global soil mass. Listed according to decreasing abundance these elements include O, Si, Al, Fe, C, Ca, K, Na, Mg and Ti. Except C and a small part of O, these elements occur as the main building stones of the inorganic soil fraction. The element ions are linked together by ionic and covalent bonds under formation of the different minerals. In mineral formation the most important atomic properties are the valence and radius of the ions.

Minerals can be either crystalline or amorphous. In the crystalline minerals the ions are linked together in a definite order that repeats extensively in three dimensions, whereas the amorphous, or noncrystalline, minerals do not have a definite long-range ordering of the ions. The ionic arrangement can be decisive for the properties of the minerals, as shown for instance by diamond and graphite. However, some soil minerals can show different degree of ordering (or crystallinity) depending on the environmental conditions under which they are formed. The ordering can be from definite long-range ordering through short-range ordering to lack or almost lack of ordering. To designate the different minerals, terms such as well crystallized (or well ordered), poorly crystalline (or poorly ordered) and poorly crystalline to amorphous are used.

Crystallinity is largely reflected by the size of the particles. Accordingly, primary minerals, which constitute the main components of the sand and silt fractions, are normally well crystallized, while secondary minerals are often more or less poorly ordered and dominate in the clay fraction. Since the smallest particles have the largest specific surface areas (Fig. 2.2), a poorly ordered form of a certain mineral will be more reactive than the well crystallized form of the same mineral as demonstrated for the iron oxides (Borggaard 1990). The specific surface areas are determined experimentally by gas adsorption. Thus, adsorption of N_2 is generally used for determination of the outer surface area, while polar gases such as H_2O is used for determination of the sum of outer and inner surface areas. More or less crystalline minerals can be identified by X-ray diffraction and their crystallinity can be assessed by the width of the diffraction lines, since broader lines mean poorer crystallinity. More information about specific area determination and X-ray diffraction may be found in the volumes of Klute (1986), Wilson (1987) and Dixon & Weed (1989).

2.2.4.1 *Silicate minerals*

The composition, structure and formation of various silicate minerals have been compiled in numerous textbooks on general geology, soil science and mineralogy (e.g. Deer et al. 1962, 1963; Gieseking 1975; Newman 1987; Dixon & Weed 1989). The fundamental building block in the structure of these minerals is the silica tetrahedron consisting of four oxygen ions forming covalent bonds to Si^{4+}, giving the basic ionic unit SiO_4^{4-}. Silica tetrahedra can occur as isolated, single units, in single or double chains linked together by shared corners, in sheets and in fully three-dimensional frameworks. The various structures are shown schematically in Figure 2.9. It must be emphasized that the various classes of silicates contain the anions and not the acid forms shown in Figure 2.9. In silicate minerals the silicate structures are held together with divalent cations like Mg^{2+} and Fe^{2+} or trivalent cations including Al^{3+} and Fe^{3+} in octahedral coordination. In the layer silicates the octahedra are linked together in sheets. So-called isomorphous substitution, where one cation is substituted by another cation of comparable size but often of different charge, occurs in many silicates. Thus, part of the Si^{4+} ions may be substituted by Al^{3+} because these ions have similar size.

The primary silicates occurring in soils may shortly be described in the following way:

Olivines: These minerals consist of single silica tetrahedra linked together by divalent cations such as Mg^{2+} and Fe^{2+} in octahedral coordination. Example: $FeMgSiO_4$.

Pyroxenes: These minerals contain single chains of silica tetrahedra forming repeating units of SiO_3^{2-} linked together by a variety of mainly divalent cations in octahedral coordination. Example: $FeSiO_3$.

Amphiboles: These minerals contain double chains of silica tetrahedra forming repeating units of $Si_4O_{11}^{6-}$ linked together by a variety of mainly divalent cations in octahedral coordination. Example: $NaCaMg_3FeAlSi_8O_{22}(OH)_2$.

Micas: These minerals are built up from two sheets of silica tetrahedra fused to each planar side of a sheet of metal cation octahedra, typically containing Al, Mg and Fe ions. The formula of layer silicates is often based on 22 negative charges corresponding to $O_{10}(OH)_2^{22-}$. If the metal cation is trivalent, only two of three possible cationic sites in the octahedral sheet are filled leaving one site empty (vacant) and the sheet is termed dioctahedral. If the metal cation is divalent, all three possible sites are filled and the sheet is trioctahedral. Isomorphic substitution of Al for Si, Fe(III) for Al and Fe or Al for Mg occurs typically in micas. Substitution of an ion of higher charge such as Si^{4+} by an ion of lower charge such as Al^{3+} will result in a negative charge, which must be balanced by a positive ion. Examples: $(Al_2)[AlSi_3]O_{10}(OH)_2$ (muscovite); $K(Mg_2Fe)[AlSi_3]O_{10}(OH)_2$ (biotite). Muscovite is dioctahedral, whereas biotite is trioctahedral. In both minerals one Si per formula unit in the tetrahedral sheets has been substituted by Al resulting in a negative charge, which is balanced by K^+ (the interlayer cation). Due to their layered structure, micas occur in flakes as sheets that pack together as 'books'.

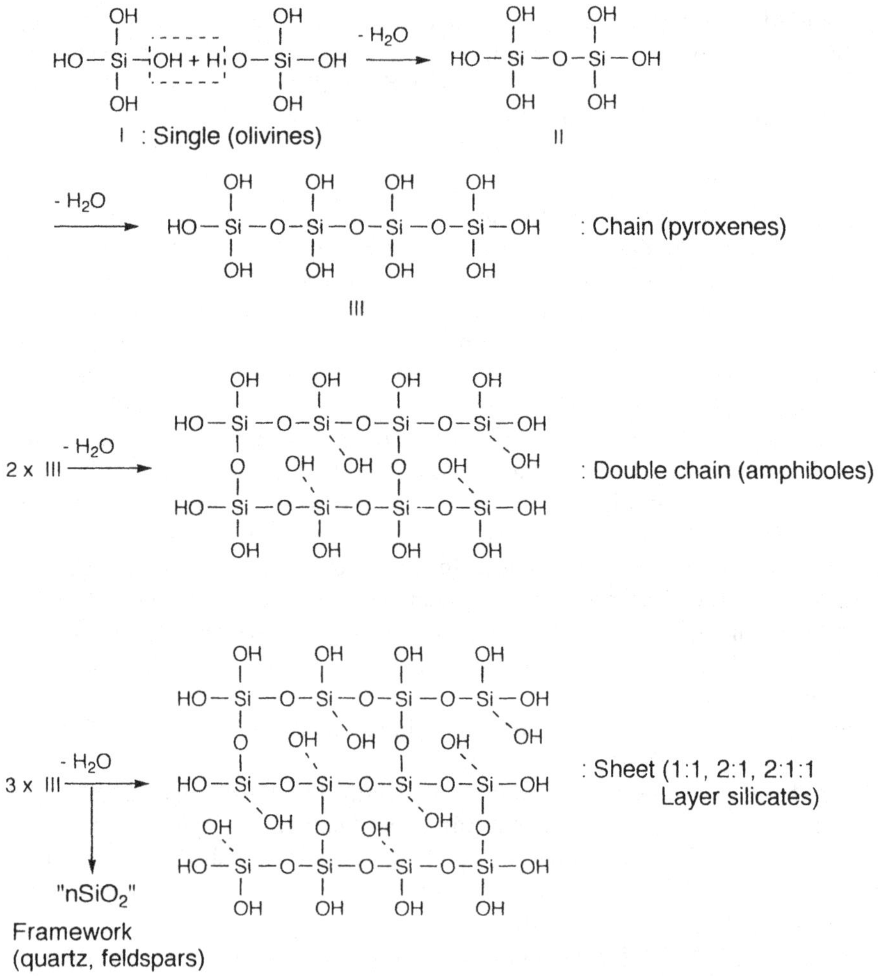

Figure 2.9. Sketch of di- and polymerization of silicic acid forming the basic Si(IV) structures (tetrahedra) as *anions* in various classes of soil silicates.

Framework silicates: These minerals consist of a continuous, three-dimensional framework of silica tetrahedra sharing corners. The general formula of the minerals is: $M_x[Al_xSi_{4-x}O_8]$, where $x = 0,1,2$ for the different framework silicates, which include quartz (Si_4O_8 or SiO_2) and the feldspars such as $KAlSi_3O_8$ (orthoclase, K-feldspar), $NaAlSi_3O_8$ (albite, Na-feldspar), $CaAl_2Si_2O_8$ (anorthite, Ca-feldspar) and mixed Na/Ca-feldspars (plagioclase).

The secondary silicates (often called clay silicates) are layer (or sheet) silicates and consist, like the micas, of sandwiches of tetrahedral and octahedral sheets (Fig. 2.10). The bonding between tetrahedral and octahedral sheets occurs through the apical oxygen ions in the tetrahedral sheet and always produces, because of mismatch, distortion of the anion arrangement in the final layer structure. Three layer types occur. They are distinguished by the number of combined tetrahedral and octahedral sheets. The types are further divided into five mineral groups according to kinds of

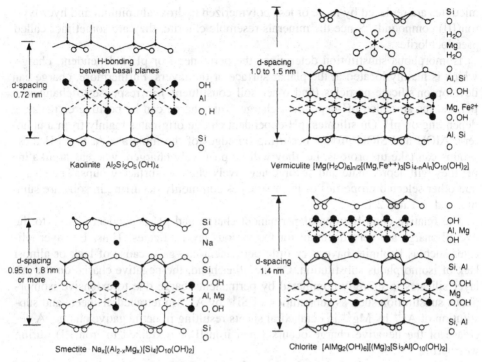

Figure 2.10. Schematic structures of common soil layer silicates.

isomorphic substitution. The kaolin group consists of 1:1 layer silicates with a tetrahedral sheet and an octahedral sheet. The illite, vermiculite and smectite groups are 2:1 layer silicates and are built of an octahedral sheet sandwiched between two tetrahedral sheets. The chlorites are 2:1:1 layer silicates consisting of a 2:1 layer structure with an extra octahedral sheet.

The three 2:1 groups differ from another in two ways: 1) by the layer charge decreasing in the order illite > vermiculite > smectite and 2) by vermiculite exhibiting higher substitution in the tetrahedral sheets than smectite. The smectites, with layer charge of 0.25-0.6 per formula unit due to isomorphous substitution mainly in the octahedral sheet, are freely expanding, that is, the cations balancing the layer charge can readily be exchanged for other cations. In freely expanding minerals such as smectites, the inner planar (interplanar) surfaces are accessible for polar gases such as H_2O resulting in very high specific surface areas. The higher layer charge of the vermiculites, 0.6-0.9 per formula unit, which is caused by substitution in the tetrahedral sheets, results in limiting expansion of these minerals, and hence of cation exchangeability. In illites the layer charge can be up to 1.0 per formula unit as in micas from which they may form. The high charge originating from isomorphous substitution in the tetrahedral sheets results in a strong coulombic attraction for charge-compensating (interlayer) cations. Therefore, the K^+ occurs in a nonexchangeable or fixed form between 2:1 layers in illite. Certain soil horizons, particularly upper Spodosol horizons (Section 2.4.4), often contain so-called hydroxy interlayered minerals (HIM), where the charge-balancing (interlayer) cations in smectites and vermiculites are replaced by more or less polymerized hydroxyaluminium and hydroxyiron(III) compounds. Since the minerals resemble chlorite, they are sometimes called pseudochlorites.

Isomorphous substitution determines the permanent, or pH-independent, charge, which is mainly located at the planar surface of the layers. Furthermore, charge can develop on silicate particles (and other soil constituents) in response to changes in pH, i.e. the so-called pH-dependent charge. This charge can be positive or negative depending on pH. On silicates, pH-dependent charge originates mainly from amphoteric AlOH and SiOH groups occurring on edges of the minerals. At low pH these groups can take up protons, i.e. they will be positively charged, whereas at alkaline pH they will deprotonate and become negatively charged surface groups. The charge and other selected properties of layer silicates commonly occurring in soils are summarized in Table 2.1.

The relative contributions of permanent charge and pH-dependent charge to the overall charge are very different for the various layer silicates. Thus, 1:1 layer silicates such as kaolinite have very little permanent charge because of lack or almost lack of isomorphous substitution. On the other hand, the negative charge on the 2:1 layer silicates is strongly dominated by permanent charge due to extensive isomorphous substitution such as substitution of Si^{4+} by Al^{3+} in tetrahedral sheets and substitution of Al^{3+} by Mg^{2+} in octahedral sheets resulting in net negative charge. A decrease of the negative charge occurs when iron(II) is oxidized to iron(III) during weathering.

It is important to distinguish between the total charge considered above and exchangeable charge. The latter is the charge accounting for exchange reactions, where ions such as Ca^{2+} and Mg^{2+} can be reversibly replaced by other cations (Section 2.3.2.1). The amount of negative charge in cmol(+)/kg that is readily exchangeable is termed the cation exchange capacity (CEC). The origin of this charge is the permanent and pH-dependent charge components but for some layer silicates it is only part of the total charge that is exchangeable. Thus, minerals belonging to the illite group (and the micas) have high layer charge but rather low CEC as shown in Table 2.1 because most of the cations (K^+) on the interplanar surfaces are trapped or fixed in nonhydrated form and are therefore not readily exchangeable. The smectite minerals form the other extreme because these minerals are expansible with ~80% of the total surface area due to interplanar surfaces allowing cations to move in and out between the 2:1 layers. Accordingly, the smectite CEC is high, despite the isomorphous substitution in this mineral group being considerably lower than for the illite group.

Table 2.1. Selected properties of common soil layer silicates.

Property	Kaolinite	Smectite	Vermiculite	Illite	Chlorite
Mineral type	1:1	2:1	2:1	2:1	2:1:1
Layer charge	~0	0.25-0.6	0.6-0.9	0.7-1.0	~1.0
CEC (cmol(+)/kg)	1-10	80-120	120-150	20-50	10-40
Spec. surf. area (m^2/g)	1-20	600-800	400-600	20-150	20-150
d-spacing ($Å = 10^{-10}$m)	7	Variable	10-15	10	14
Expansible	No	Yes	Partly	No	No
pH-dependent charge	Most	Little	Little	Medium	Most

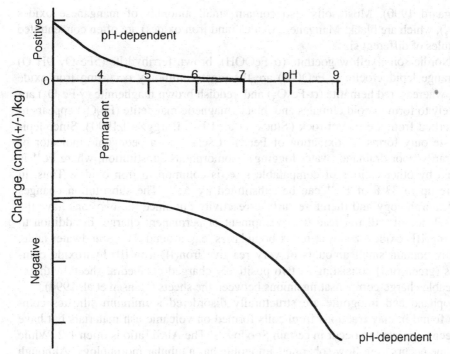

Figure 2.11. Sketch of soil charge versus pH. Note that the positive charge is considerably smaller than the negative charge.

In addition to the surface charge contributed by the layer silicates other soil constituents including SOM and the oxides (Section 2.2.4.2) will also contribute to the surface charge in soils. The charge conveyed by SOM and the oxides is pH-dependent charge. Consequently, soil charge depends on the composition of the soil and is the sum of permanent and pH-dependent charges as indicated in Figure 2.11.

2.2.4.2 *Accessory minerals*
Soils also contain minor amounts of oxides, oxyhydroxides and hydroxides (collectively called oxides) of aluminium, iron, manganese and titanium, which are weathering products retained in the soil due to very low solubility (Deer et al. 1962; Gieseking 1975; Newman 1987; Dixon & Weed 1989). These solids range in degree of ordering from amorphous to crystalline. The most poorly ordered forms will have the largest specific surface area and will therefore be the most reactive in soils. The oxides have only pH-dependent charge properties but they are effective adsorbents for many organic and inorganic compounds (Section 2.3.2.2).

While anatase (TiO_2) is the most commonly occurring soil titanium oxide, gibbsite ($Al(OH)_3$) is the most abundant form of free (nonsilicate) aluminium in soils. Both oxides are white. The aluminium and iron oxides are collectively called sesquioxides. In addition to the better crystallized forms of these oxides, soils also contain poorly crystalline to amorphous oxides. These forms are favoured in environments rich in SOM, particularly humic and fulvic acids, probably because these organic compounds retard crystal development by adsorbing onto oxide surfaces

(Borggaard 1990). Most soils also contain small amounts of manganese oxides (MnO_2), which are black. Manganese oxides (and iron oxides) are often concentrated in nodules of different sizes.

In Nordic soils, yellow goethite (α-FeOOH), brown ferrihydrite ($5Fe_2O_3 \cdot 9H_2O$) and orange lepidocrocite (γ-FeOOH) are the most commonly occurring iron oxide forms, whereas red hematite (α-Fe_2O_3) and reddish brown maghemite (γ-Fe_2O_3) are not likely to form in cold climates and black, magnetic magnetite (Fe_3O_4) appears to be inherited from the parent rock (Stucki et al. 1988; Borggaard 1990). Since lepidocrocite only forms by oxidation of Fe(II), it serves as a pedogenic indicator of temporarily poor drainage (water-logging). Isomorphous substitution, where Fe^{3+} is replaced by other cations of comparable size, is common in iron oxides. Thus, in goethite up to 33% of Fe^{3+} can be substituted by Al^{3+}. The substitution changes crystal morphology and therefore surface reactivity but since the two ions have the same valence, it will not lead to development of permanent charge. In addition to these iron(III) oxides, zones at redox boundaries, e.g. around the groundwater table, probably contain small amounts of very reactive iron(II)-iron(III)-hydroxide compounds (green rust) consisting of two positively charged octahedral sheets with exchangeable charge-compensating anions between the sheets (Hansen et al. 1994).

Allophane and imogolite are structurally disordered aluminium silicates commonly found in clay fractions from soils formed on volcanic ash materials but have also been suggested to occur in certain Spodosols. The Al/Si ratio is often 1-2. While allophane occurs as hollow spherules, imogolite has a tubular morphology. Although minor amounts of these minerals are likely to occur in some Spodosols, they have rarely been identified in Nordic soils (see Gustaffson et al. 1995).

Soils in southern Scandinavia formed on glacial till often contain calcium carbonate because of its occurrence in the parent material. Calcium carbonate is easily weathered and therefore very important in counteracting soil acidification. Pyrite with the ideal formula FeS_2 (and small amounts of FeS) occurs in several subsoils. It will only form and be stable under very reductive conditions as found under anaerobic organic matter decomposition in tidal marsh areas (Section 2.2.3.2). Under aerobic conditions pyrite will oxidize to sulphuric acid and iron oxides (ochre). Although the presence of Fe^{3+} accelerates the process, chemical oxidation of pyrite by O_2 is slow but in the presence of certain autothrophic bacteria the reaction rate increases several orders of magnitude. Oxidation of pyrite at pH < ~3 leads to formation of strongly yellow jarosite ($KFe_3(SO_4)_2(OH)_6$), which therefore can be considered a specific pedoindicator for pyrite.

The mineralogy of Nordic soils is often very complex (Møberg et al. 1988; Olsson & Melkerud 1989; Møberg 1991). Apart from differences in rock composition, the complexity is due to the cold climate and the relative short time of development, which have resulted in limited weathering. The content of weatherable minerals (primary minerals except quartz) in silt and sand fractions can be high and the clay fraction often consists of many minerals. Thus, although illites appear to be the dominant group of clay silicate minerals, smectites, vermiculites, chlorites, HIMs and kaolinite are also commonly occurring, but in highly variable amounts. In addition, the soils contain minor amounts of aluminium, iron, manganese and titanium oxides.

2.3 SOIL PROPERTIES

The physical and chemical properties of soils depend on soil composition and determine the capacity of soils to act as media for plant growth, to support buildings and roads, to retain pollutants etc. The fundamentals of these properties are shortly outlined hereafter, but much more information can be found in numerous textbooks (e.g. Singer & Munns 1987; Schachtschabel et al. 1989; Brady 1990; Rowell 1994).

2.3.1 *Physical properties*

The physical properties concern the soil solids and their arrangement, but also soil air and soil water. Thus, apart from determining the suitability of soils as foundations of building constructions and as production places of plants, the physical properties are fundamental for water movement in the soil and of great importance in controlling wind and water erosion.

Soil texture and soil structure are the basic physical properties, which determine the arrangement of the solid particles and the pore space (volume). Soil texture has been considered in Section 2.2. The subsequent text will therefore concentrate on soil structure but also on the importance of texture and structure for the movement of water in soils. Other aspects of water such as runoff and groundwater recharge are covered by Rodhe & Killingtveit (Chapter 3) and Lerner (Chapter 4). Since micromorphology is the technique that directly shows the arrangement of the solids and pores (voids), it will be shortly introduced.

2.3.1.1 *Micromorphology*
Soils are heterogenous because they consist of different layers (horizons) with more or less different composition. However, even within each horizon voids and different solid components can be recognized. The heterogeneity can be seen by the naked eye, but it is much more comprehensively studied under the microscope using thin sections, which are prepared by cutting and polishing slices of soil impregnated with a resin (FitzPatrick 1984; Douglas 1990). A sketch indicating some of the features that can be studied by thin section analysis is shown in Figure 2.12. The spatial arrangement of solids and voids (pores) is called the soil fabric, the coarser (> 2 µm) organic and inorganic fragments form the soil skeleton and the less than 2 µm organic and inorganic materials constitute the soil plasma.

Features such as selective weathering shown by spots and altered rims and coatings of clay silicates, oxides and humus on coarser grains and on pore and aggregate surfaces are very nicely demonstrated by micromorphology. Furthermore, the size, shape and continuity of pores, which determine water retention (Section 2.2.2.2) and water movement (Section 2.3.1.3) in the soil, is also readily studied by microscopy of thin sections.

2.3.1.2 *Soil structure*
The term soil structure relates to the grouping or arrangement of soil particles (Klute 1986; Brady 1990; Rowell 1994). It describes the overall combination or arrangement of the primary soil particles such as those determined by texture analysis (Section 2.2) into secondary groupings called aggregates or peds. Although some

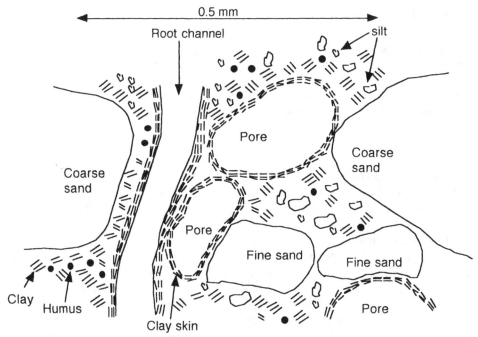

Figure 2.12. Thin section analysis of a small part of a soil showing various micromorphological features. Sand and silt (and coarse SOM) form the skeleton grains, while clay silicates, oxides and humus form the plasma.

soils are dominated by one structure, most soils have several structural types in the different horizons. Soil structure is closely related to soil porosity. Therefore, soil structure is of very big plant-growing and environmental importance. The transfer of gases into and out of the soil, the development of plant roots and the movement of water and dissolved or suspended matter are directly determined by soil structure.

The mechanisms of soil structure formation is poorly understood. However, compression of soil particles into small aggregates occurs because of swelling and shrinking as soils are wetted and dried and because of permeation of plant roots. Organic compounds excreted by plant roots and formed by microorganisms may help to bind the particles into aggregates and to cement the aggregates together enhancing both aggregate formation and stability as seen most clearly in soils under permanent grass. Aluminium and iron oxides are also effective cementing agents in soils as seen by the stable structure of many strongly leached tropical soils. Furthermore, since the behaviour of charged colloids, i.e. whether they are dispersed or flocculated, is strongly affected by the concentration and nature of the cations and pH in the soil solution, soil structure is also dependent of these variables (Newman 1987).

Charged particles are surrounded by a swarm of oppositely charged ions in the diffuse double layer as shown for a clay platelet in Figure 2.13, where the thickness of the diffuse double layer depends on the concentration and valence of the ions or ionic strength in the solution. When two similarly charged particles approach each other so that the diffuse double layers overlap, then the particles will repel each other; the closer the approach the stronger will the repulsion be. However, when the

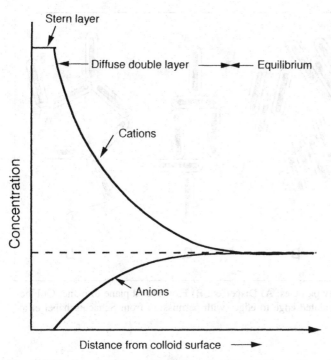

Figure 2.13. Concentrations of cations and anions at the surface of a negatively charged particle. The thickness in mm of the diffuse double layer is $0.303 \times I^{-0.5}$ at 25°C, where I is the ionic strength.

distance between the particles is further reduced they will start to attract each other because of short-range attractive forces. These forces include van der Waals' forces and attraction between positive and negative charges on ions and colloids. When present, polyvalent cations can add to the attraction because they can be adsorbed by both particles. If repulsion is stronger than attraction the particles are dispersed, while stronger attraction than repulsion results in flocculation. Whether repulsion or attraction prevails, i.e. whether the colloids are dispersed or flocculated, mainly depends on the thickness of the diffuse double layer. Since this thickness is a function of ionic strength (Fig. 2.13), high concentrations of di- and trivalent ions such as Ca^{2+} and Al^{3+} will lead to flocculation, whereas lower concentrations of monovalent ions such as Na^+ will favour dispersion.

As a result of flocculation the particles can be oriented in different ways. Thus, the orientation of clay platelets can occur in three ways, that is plane to plane (face to face), plane to edge and edge to edge as shown in Figure 2.14. Since the permanent negative charge on clay minerals such as smectite occurs on the planar surfaces and pH-dependent charge on edges, the plane to edge arrangement is favoured at low pH, where the edges are positively charged. At high pH, flocculation of clay silicates is not favoured because of their negative charge. However, strong drying can force the particles so close together that they are flocculated plane to plane. Other substances such as polyvalent cations, humus and oxides will interact in the flocculation favouring various arrangements. The arrangement of the flocculated particles strongly af-

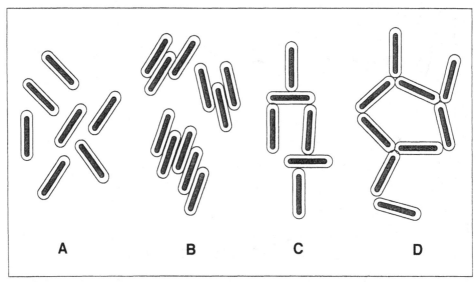

Figure 2.14. Arrangement of clay particles: A) Dispersed; B) Flocculated plane to plane; C) Flocculated plane to edge; D) Flocculated edge to edge (with permission from Schachtschabel et al. 1989).

fects soil porosity. Thus, plane to edge or edge to edge flocculation, which is favoured in the presence of Ca^{2+} (and Al^{3+}), leads to a rather open structure, while plane to plane flocculation results in a more compacted structure. The poor structure found in clay-rich so-called alkali soils in some semiarid regions but also in some marsh soils dominated by Na^+ is due to plane to plane flocculation of the clay particles.

Apart from the single-grain structure found in some poorly developed sandy soils low in SOM, the primary soil particles occur in various aggregate types. Four principal structural types exist including the blocky, the granular, the platy and the prismatic types (FAO 1990). These types are illustrated in Figure 2.15 and can shortly be characterized as follows:

Blocky: This type consists of nearly equidimensional blocks or polyhedrons and includes two subtypes, that is, the angular subtype with faces intersecting at relatively sharp angles and the subangular subtype with faces intersecting at rounded angles. Angular and subangular blocks are commonly found in the more clayey soils.

Granular: This type consists of spheroidal or polyhedral aggregates with curved (rounded) or irregular surfaces. They are more or less porous and often occur in Ap horizons.

Platy: This type consists of aggregates arranged in relatively thin, usually overlapping, horizontal plates, lenses or leaflets. The platy structure can be found in plow pans but is not always a result of soil development. Thus, it can be inherited from the parent materials, particularly those deposited by water and ice.

Prismatic: This type consists of vertically oriented aggregates or pillars with well-defined vertical faces and includes two subtypes, that is, the columnar subtype with rounded caps and the prismatic subtype with horizontal faces intersecting at rela-

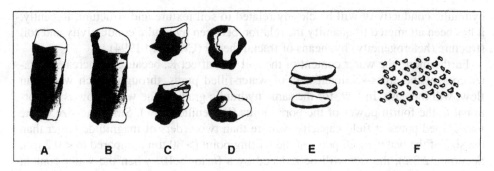

Figure 2.15. Various structural types in mineral soils: A) Prismatic; B) Columnar; C) Angular blocky; D) Subangular blocky; E) Platy; F) Granular.

tively sharp angles. Typically, prismatic structures occur in the B-horizons of clayey alkali soils with low biological activity, and hence they are not common in Nordic soils.

The sizes of the various structural types are different. Typically, granular aggregates have a thickness of 5-50 mm, whereas the angular and subangular blocks are often several times larger and the prismatic aggregates can have diameters of 10 cm and heights of 50 cm or more. In addition to shape and size, stability is also important for characterizing soil structural elements. The structure is denoted as weak, moderate or strong depending on how easily it is distinguished in the soil and on the force needed to break the aggregates. Weak aggregates can succumb to the beating of rain and the rough and tumble of plowing and tilling, while strong aggregates resist disintegration ensuring maintenance of good soil structure. Since large aggregates tend to be unstable in surface soils, particularly in Ap horizons with high biological activity, granular structure is often found in topsoils. At deeper soil depth the biological activity decreases enabling formation of larger aggregates favouring a blocky or prismatic structure in B horizons. In C horizons, which by definition have not been altered by the soil forming processes, the structure is determined by that of the parent materials, that is, in clayey soils the structure often appears to be angular blocky and in sandy soils a single-grain structure is common.

2.3.1.3 *Water movement in soils*

Water moves in the direction of decreasing hydraulic potential (Section 2.2.2.1). Thus, when the hydraulic potentials at two points at distance s are different, water flows through soils in response to the difference in the hydraulic potentials over the distance, the hydraulic gradient ($d\psi_H/ds$). The stationary, one-dimensional rate of water movement (F) can be expressed as a function of the hydraulic gradient (Darcy's law):

$$F = -K (d\psi_H/ds) \tag{2.9}$$

K is the hydraulic conductivity of the soil. The hydraulic conductivity is a measure of the ease with which water can move through the soil and is affected by several factors. It depends on size and shape of the pores, but also on the tortuosity and continuity of the pore system, which may be termed pore geometry. Accordingly, the hydraulic conductivity will be closely related to soil texture and structure. Recently, it has been attempted to quantify the relation between hydraulic conductivity and soil structure (heterogeneity) by means of fractal theory (Crawford 1994).

Furthermore, the water content of the soil will affect K, because at increasing water content the cross-sectional area of water-filled pores, through which water can flow, will increase. In fact, for the same hydraulic gradient the water flow is proportional to the fourth power of the pore radius (Poiseuille's law). Since the size of the water-filled pores at field capacity is more than two orders of magnitude larger than the size of the water-filled pores at the wilting point (> 30 μm compared to < 0.2 μm, Section 2.2.2.2), the rate will be reduced by a factor > 10^8 when the water content decreases from that corresponding to field capacity to the content at the wilting point. When all pores are saturated with water the hydraulic conductivity will be at maximum, termed saturated hydraulic conductivity. Since sandy soils are dominated by larger pores, their hydraulic conductivity will be high at high water contents. However, the hydraulic conductivity of sandy soils decreases sharply when the water content falls below field capacity because most of the pores are emptied resulting in poor connection between the pores that are still water-filled. In soils with higher clay contents the hydraulic conductivity will not decrease so sharply due to larger amounts of retained water. Although clayey soils with few macropores have low hydraulic conductivities, they are able to maintain moderate conductivities over a wide range of water content as a result of the continuous network of small pores which retain water.

Nordic soils are subject to water percolation because precipitation exceeds evapotranspiration. Since the different soil horizons and layers are nearly never uniform in texture and structure, they will normally have different hydraulic conductivities, which will affect the movement of water through the soils. A sand layer underlying a clayey soil will retard the vertical water movement, since the macropores of the sand offer less attraction for the water than does the finer textured materials. Heavy rainfall may then lead to saturation with water of the upper horizons, which can result in oxygen depletion and lowering of the redox potential. Compacted and cemented soil layers can introduce similar effects (Section 2.4.2). On the other hand, some clayey soils have fractures running from upper layers to a depth of several meters, which are filled with more sandy materials (Jørgensen & Fredericia 1992). The water movement through such fractures can be considerably faster than through the clayey matrix. Continuous, vertical pores will have the same effect. Consequently, in many, particularly clayey, soils, water will move through different parts of the soil at different rates as schematically shown in Figure 2.16. This flow pattern is termed dual flow, bypass or preferential flow (Booltink & Bouma 1991; Jørgensen & Fredericia 1992). After drying, inorganic and, in particular, organic soil components are difficult to wet, that is, the components repel the water, forcing water and solutes to flow via preferential paths through the unsaturated soil (Bisdom et al. 1993). Preferential flow may lead to transportation of solutes and colloids by the running water in soils, where the soil matrix has the capacity to retain such compounds. Thus, clayey soils were formerly considered impermeable or slowly permeable for various pollutants, but recent investigations have indicated readily transport of pesticides and clay minerals to deep soil layers through such a macropore system (Jørgensen & Fredericia 1992).

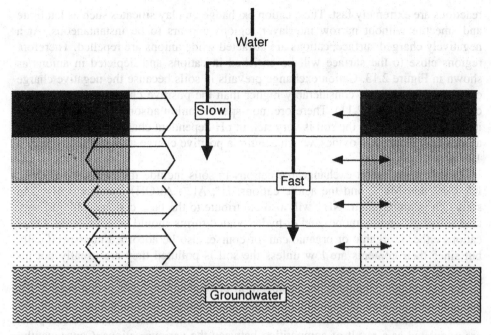

Figure 2.16. Schematic diagram of a soil with a macropore system allowing preferential flow. The thickness of the horizontal arrows indicates extent of reactions.

2.3.2 *Chemical properties*

Among the most important soil properties are the adsorption properties, i.e. the capacity of organic and inorganic solid soil components to act as sinks in non-specific and specific adsorption of anions, cations and uncharged molecules (Bohn et al. 1985; Schachtschabel et al. 1989; Sposito 1989; Brady 1990; McBride 1994). Adsorption is sometimes called sorption or retention. Non-specific adsorption (or outer-sphere complexation) is due to coulombic attraction between adsorbate (the solute) and adsorbent (the solid) with at least one water molecule interposed, whereas in specific adsorption (or inner-sphere complexation) the adsorbate is directly bound to the adsorbent with no water molecules interposed (Sposito 1989). Several mathematical models have been developed in order to describe non-specific and specific adsorption (Stumm 1987; Goldberg 1993).

2.3.2.1 *Cation exchange*
Non-specific adsorption reduces the loss of ions by leaching while keeping them available for plant uptake. Non-specifically adsorbed ions can be replaced or exchanged by other similarly charged ions. More precisely, ion exchange is the capability of soil particles to reversibly take up non-specifically adsorbable anions and cations from solution. Typically, non-specifically adsorbable ions include the anions of strong acids, e.g. Cl^- and NO_3^-, and cations of strong bases, e.g. Na^+ and K^+. The ion exchange properties in soils are attributed to the permanent and pH-dependent charge on soil particles. While layer silicates have both permanent and pH-dependent charge properties, SOM and oxides only have pH-dependent charge. Ion exchange

reactions are extremely fast. Thus, cation exchange on clay silicates such as kaolinite and smectite without narrow interlayer regions appears to be instantaneous. At a negatively charged surface, cations are attracted while anions are repelled. Therefore regions close to the surface will be enriched in cations and depleted in anions as shown in Figure 2.13. Cation exchange prevails in soils because the negative charge on the soil particles is considerably higher than the positive charge at naturally occurring soil pH (Fig. 2.11). Therefore, non-specific anion adsorption will be of limited importance unless the soil is very rich in pH-dependent charge components such as aluminium and iron oxides, which acquire a positive charge at low pH (Borggaard 1990).

The most important exchangeable cations in soils include the base cations, Na^+, K^+, Ca^{2+} and Mg^{2+}, and the acidic cations, H^+, Al^{3+}, $Al(OH)^{2+}$ and $Al(OH)_2^+$. In soils rich in ammonium, NH_4^+ will also contribute to the base cations, while in very acid soils Fe^{3+} and its mono- and di-hydroxylated forms should be considered. Other cations either inorganic or organic can, of course, also be adsorbed non-specifically but since their contents are low unless the soil is polluted they are usually not included in the exchangeable cations.

As a first approximation, the composition of the exchangeable cation pool at the soil particle surfaces is determined by the composition and concentrations of the cations in solution, that is, the exchange sites (sometimes called exchange complex) are populated as a result of competition between the various cations. Consequently,

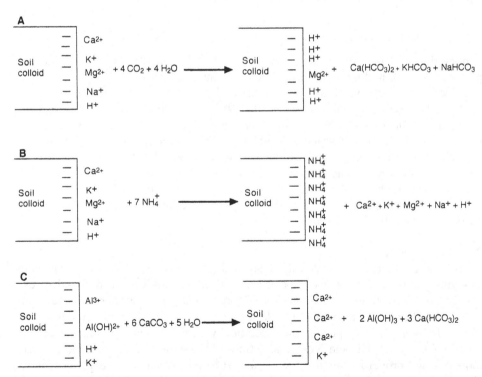

Figure 2.17. Various cation exchange reactions: A) Acidification by CO_2; B) Saturation with ammonium ions; C) Exchange of acidic cations by liming.

the composition of cations at exchange sites on the solid soil components will change if the composition of the solution is changed. The origin of the changes can be natural or anthropogenic. Thus, CO_2 from the atmosphere and from root respiration and SOM mineralization will increase the amount of H^+ at the exchange sites on the expense of base cations in a neutral to alkaline soil. On the other hand, the plant-growing properties of acid soils dominated by acidic cations can be improved by liming. In the laboratory, the exchangeable cation pool can be determined after replacements of the adsorbed cations by another cation such as NH_4^+ added in high concentration, often 1 M. Reaction schemes for these exchanges are shown in Figure 2.17.

The cations are held at the negative particle surface by coulombic forces suggesting the composition of adsorbed cations to be a function of concentration only for ions of the same valence. However, the attraction is also affected by the size of the hydrated cation, the larger the hydrated radius of the ion the less tightly will it be held. This is shown by the relative replaceability in the so-called lyotropic series: $Na^+ > K^+ \sim NH_4^+ > Mg^{2+} > Ca^{2+} > Al^{3+}$. Compared to the other exchangeable cations, H^+ behaves somewhat differently. Adsorption sites on pH-dependent charge components must be deprotonated ('titrated') before they are available for exchange reactions. Thus, the carboxylic and phenolic hydrogen ions on SOM can not simply be replaced by the cations of a neutral salt. Furthermore, clay silicates saturated with H^+ are unstable and rearrange within rather short time to Al-saturated clay. Therefore H^+ will occur mainly as titratable acidity at normal soil pH.

The cation exchange capacity (CEC) of a soil is the amount in cmol(+)/kg of base and acidic cations that can be adsorbed by the soil under given conditions:

$$CEC = \Sigma \text{ (base cations)} + \Sigma \text{ (acidic cations)} \qquad (2.10)$$

Since CEC is a function of soil particle charge, which is partly pH-dependent, CEC will also be dependent on pH. Consequently, CEC will increase at increasing pH as shown in Figure 2.18, where a soil sample has been titrated by $Ca(OH)_2$. The pH de-

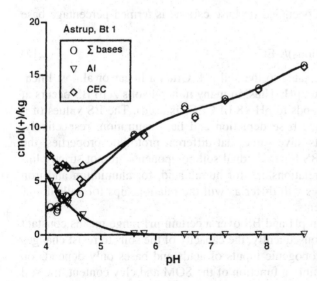

Figure 2.18. The amounts of exchangeable base and aluminium cations and CEC plotted against pH for an originally acid Bt horizon from a Danish forest soil. The soil sample was titrated by $Ca(OH)_2$.

pendency of CEC is determined by the composition of the soil. In soils dominated by permanent charge components such as smectite the increase is limited, whereas in soils rich in SOM and other pH-dependent charge components CEC will exhibit strong pH dependency.

Since CEC depends on the pH of the solution it is necessary to specify the pH to which a certain CEC corresponds or at which it has been determined. In the laboratory, CEC is determined using various salt solutions either unbuffered or buffered at a certain pH (Page et al. 1982; Rowell 1994). The so-called effective or actual CEC (ECEC) is determined by use of a salt solution, e.g. 1 M KCl or NH_4NO_3. After exchange, the concentrations of base cations and aluminium can be determined by atomic absorption and emission spectroscopy (Chapter 5). The sum of the cations is often taken as a measure of ECEC. This method is mainly used with acid soils such as forest soils. Ammonium acetate is commonly used to give the CEC at pH 7 (CEC_7). In this method, the exchange complex is saturated with NH_4^+ from 1 M ammonium acetate followed by replacement of NH_4^+ with another cation and quantification of replaced NH_4^+, which corresponds to CEC_7. Since pH is ~8 in an aqueous suspension of $CaCO_3$, the CEC at pH ~8 (CEC_8) determined using a buffer solution at that pH can be considered a reference value at least for soils that were calcareous after the last deglaciation (many soils in southern Scandinavia) and before soil development caused carbonate dissolution in upper soil layers (Section 2.4.3). For most (non-arable) Nordic soils, these cation exchange capacities will decrease in the order: $CEC_8 > CEC_7 > ECEC$.

Soil CEC is largely determined by the small soil particles including the colloidal (< 2 μm) SOM and clay minerals. It will therefore be dependent on abundance and composition of these fractions. Since the composition of both the SOM and the clay minerals can be highly variable, their CEC will show great variability. Thus, the CEC values of the different secondary silicates are shown in Table 2.1 to be highly different. However, for Danish surface soils the following equation often gives a fair estimate of CEC_8:

$$CEC_8 = 4 \times \% \text{ organic C} + 0.5 \times \% \text{ clay} \tag{2.11}$$

The percentage of CEC that is occupied by base cations is termed percentage base saturation (BS):

$$BS = 100 \times \Sigma \text{ (base cations)/CEC} \tag{2.12}$$

The base saturation is closely related to the soil pH. Often a linear or almost linear relation is found between BS and pH. Thus, in many mineral soils zero BS occurs at pH ~4 while BS = 100 corresponds to pH ~8 (if CEC_8 is used). The BS values of 0 and 100 correspond to complete base depletion and base saturation, respectively. Since different soil components have somewhat different protolytic properties, the relationships between pH and BS for individual soil components are not strictly linear and coinciding. Thus, the relationships for humic acid, for aluminium and iron oxides and for 2:1 layer silicates will differ as will the relationships for soils dominated by each of these components.

A linear relationship between pH and BS over a certain pH range means constant buffer capacity in that range. Consequently, the capacity of the soils to resist changes in pH due to natural and anthropogenic inputs of acids and bases only depends on CEC. Since CEC, in turn, is mainly a function of the SOM and clay content, the soil buffer capacity will increase at increasing contents of SOM and/or clay. Buffering of pH in soils is important because pH affects many properties and processes in soils. In summary, specific and non-specific adsorption, and hence availability and mobility of numerous organic and inorganic compounds of plant-growing and environmental importance can be seriously affected by pH changes (Section 2.3.2.2). Low pH is detrimental to soil bacteria while fungi are less affected (Section 2.2.3.2). Soil structure (Section 2.3.1.2), and hence the transport of water and dissolved and suspended compounds may also respond to pH changes.

2.3.2.2 *Specific adsorption*

Although trace (or heavy) element cations such as Co^{2+}, Cu^{2+}, Cd^{2+}, Pb^{2+} etc. can be adsorbed by cation exchange in soils, their strong adsorption even in the presence of high concentrations of exchangeable base cations neccessitates an alternative mechanism, that is specific adsorption. Rather than the permanently charged sites on planar surfaces of layer silicates, it is the amphoteric AlOH and FeOH groups on aluminium and iron oxides and edges of silicates together with carboxyl and phenol groups on SOM, i.e. the pH-dependent charge sites, which are the main adsorption sites for these metals.

At iow pH the amphoteric groups are positively charged, whereas at alkaline pH they will be negatively charged as shown schematically for an iron oxide in Figure 2.19. The pH at which the charge is zero is called the zero point of charge (ZPC) and can be determined by acid-base titration at different ionic strengths (Sposito 1989; Borggaard 1990). Pure aluminium and iron oxides will have ZPC near 7 (Borggaard 1990) but it is affected by specific adsorption (Sposito 1989). An important characteristic of specific cation (and anion) adsorption is that adsorption can occur at pH

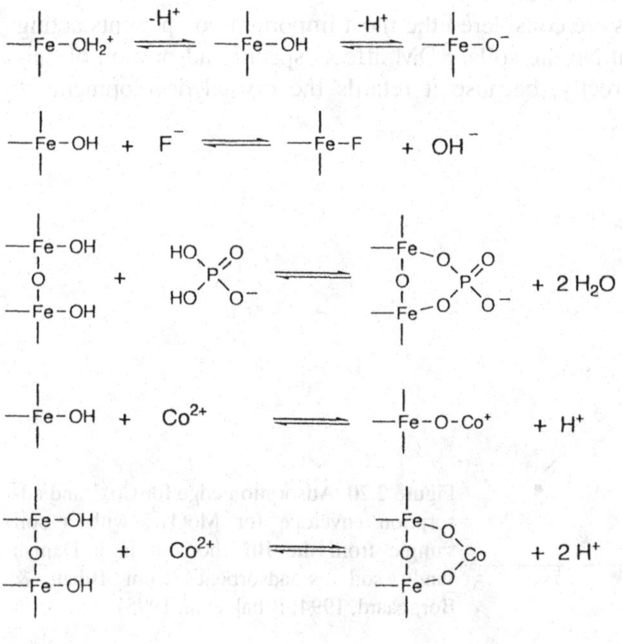

Figure 2.19. Dissociation and specific anion and cation adsorption processes at an iron oxide surface.

values where adsorbent and adsorbate have similar charge. Hence heavy metal cations can also be adsorbed at pH below adsorbent ZPC.

The specific adsorption of metal cations such as Co^{2+}, Cu^{2+}, Cd^{2+} and Pb^{2+} is followed by a release of protons, which for iron oxides corresponds to a release of 1-2 protons per metal ion adsorbed (Borggaard 1990) as shown for Co^{2+} adsorption in Figure 2.19. Accordingly, specific metal cation adsorption is strongly dependent of pH. The amount of metal adsorbed raises within a narrow pH range from negligible adsorption at low pH to high adsorption at higher pH. The curve resulting from plotting amounts adsorbed versus pH is called an adsorption edge. An example of an adsorption edge, that is the adsorption of Co^{2+} by a Danish sandy soil sample, is shown in Figure 2.20. The position of the adsorption edge along the pH axis depends on the metal and on the adsorbent. Thus, while the adsorption edges for Cu^{2+} and Pb^{2+} occur at acid pH, the adsorption edge for Cd^{2+} is at neutral to weakly alkaline pH. Consequently, the hazardous metals Cd^{2+} and Pb^{2+} will behave differently. Lead will be retained by the solid soil components, and hence move very slowly in soils unless the soil pH is very low, whereas a moderate pH drop in a neutral soil can cause release of Cd^{2+}. Furthermore, the position of the adsorption edge is affected by the concentration of the metal ion and the reaction time (Bibak et al. 1995) and other variables.

In contrast to their limited to negligible non-specific anion adsorption properties, most soils including those in Nordic countries have significant specific anion adsorption properties. Thus, anions of weak organic and inorganic acids such as arsenite, citrate, fluoride, molybdate, oxalate, phosphate and selenite can be retained by specific adsorption (Parfitt 1978; Borggaard 1990). Adsorption occurs as a result of ligand exhange as shown for fluoride and phosphate in Figure 2.19. The pH dependency of specific anion adsorption exhibits the so-called adsorption envelope with maximum adsorption around pH corresponding to the pK of the adsorbate and decreasing adsorption at lower and higher pH values as illustrated for adsorption of molybdate by a soil sample in Figure 2.20.

Aluminium and iron oxides are considered the most important components acting as specific anion adsorbents in Nordic soils. SOM affects specific adsorption of anions, such as phosphate, indirectly, because it retards the crystal development of

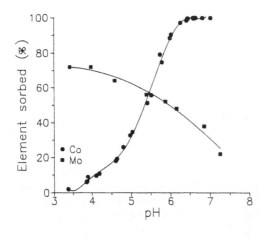

Figure 2.20. Adsorption edge for Co^{2+} and adsorption envelope for MoO_4^{2-} with a soil sample from the Bhs horizon in a Danish sandy soil as adsorbent (from Bibak & Borggaard, 1994; Bibak et al. 1995).

aluminium and iron oxides resulting in poorly crystalline oxides with high specific surface area and hence high reactivity (Borggaard 1990). However, metallic anions such as molybdate have been shown to react directly with SOM, probably by complexation of the metal (e.g. Mo) to organic ligands (Bibak & Borggaard 1994).

2.4 SOIL DEVELOPMENT PROCESSES

Except for the processes leading to peat formation, which were briefly considered in Section 2.2.3, the most important processes in Nordic soils include decalcification, gleization, lessivage and podzolization. While lessivage and podzolization are very complex processes that are yet only partly understood, decalcification and gleization appear less complicated. These processes are shortly described below but more information can be found elsewhere (Birkeland 1974; Petersen 1976; De Coninck 1980; Duchaufour 1982; Wilding et al. 1983; Boul et al. 1989; Fanning & Fanning 1989; Gustafsson et al. 1995). Definition, description and classification of soils and soil horizons involved in or affected by these processes are given in Appendices A and B.

2.4.1 *Decalcification*

The term decalcification is specifically used for the eluviation of carbonates within a soil. The process may lead to the complete removal of carbonate from the entire soil profile, as is commonly seen in humid areas, or be followed by reprecipitation, which leads to accumulation of carbonates as observed in more dry regions. Because of the humid conditions in Nordic countries, reprecipitation (or calcification) is not likely to be important in these soils.

Generally, the decalcification process involves two steps, i.e. the dissolution step and the transport step. The first step is the dissolution of readily soluble carbonates such as calcium carbonate:

$$CaCO_3 + CO_2 + H_2O \Rightarrow Ca^{2+} + 2HCO_3^- \qquad (2.13)$$

Although decalcification in the strict sense is due to dissolution by CO_2, other natural and anthropogenic acids will also contribute. Since CO_2 and other acid compounds come from the atmosphere and from biological activity in the soil, decalcification will start in the uppermost calcareous soil layers. The second step of the decalcification process is the transport of the ions in water. Due to the surplus precipitation, and hence prevailing downward movement of water in Nordic soils, dissolved carbonate will be leached from upper soil layers. Consequently, the carbonate front will occur at deeper and deeper depth at increasing soil development as indicated in Figure 2.21, which will be considered in more details in Section 2.4.3.

Decalcification is a prerequisite for other soil processes such as silicate weathering, lessivage and podzolization. Acid-induced weathering of silicates such as feldspars and dioctahedral micas (muscovite) will be very limited in calcareous soil materials because of the much higher reactivity of carbonates compared to these minerals (see Chapter 1). Clay dispersion, and hence lessivage will be retarded in the presence of the rather high Ca^{2+} concentration that occurs in calcium carbonate-containing soils. The absence of carbonates and other easily weatherable minerals is

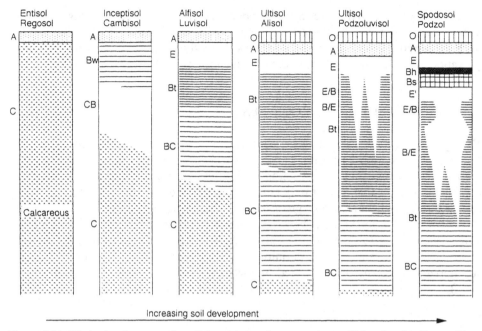

Figure 2.21. Likely development of a soil formed on calcareous, clayey till in a humid climate. The soils occurring at different stages of development are tentatively classified according to the Soil Taxonomy and FAO/UNESCO systems (Appendix B).

a prerequisite for the downward movement of humus and aluminium and iron oxides that characterizes podzolization. Accordingly, the Spodosol or Podzol (Appendix B) stage is the most advanced one in the development sequence in Figure 2.21.

2.4.2 Gleization

Gleization processes are considered in explaining the origin of soil features, especially colour patterns, formed in response to wetness-induced reduction of iron and manganese in some or all parts of a soil. By reduction iron and manganese are solubilized because of formation of Fe(II) and Mn(II), which can move from one zone to another within the soil by diffusion or can be leached out of the soil. Typically, Fe(II) and Mn(II) move to zones of higher redox potential (aerated zones), where they precipitate as iron and manganese oxides. The processes tend to give low chroma (typically gray) colours to those parts of the soil from which iron has been removed and high chroma (typically brown, orange, yellow) to the parts, where the iron oxides have accumulated. Manganese oxides are black. The different colouring is termed mottling. Soil colour is considered in Appendix A3.

Reduction of iron and manganese oxides consumes protons and, reversely, oxidation of Fe(II) and Mn(II) releases protons, as shown for the iron system:

$$FeOOH + e^- + 3H^+ \Leftrightarrow Fe^{2+} + 2H_2O \tag{2.14}$$

The electrons needed for the reduction of Fe(III) and Mn(IV) most likely come from oxidation of SOM to compounds such as CO_2 or HCO_3^- (mineralization) but this oxidation depends on microbial catalysis. On the other hand, the oxidation of Fe(II) and Mn(II) is spontaneous in the presence of oxygen. Possibly, green rust (Section 2.2.4.2) is an intermediate product in the oxidation of Fe(II). According to the redox equation, the redox potentials for the Fe(III)/Fe(II) and Mn(IV)/Mn(II) systems depend on pH as shown in Figure 2.22. It should be noticed that the straight lines for the two systems are not coinciding, suggesting that oxidation and reduction occur at different redox potentials, and hence different environments. At the same pH, lower redox potentials are needed to reduce iron oxides compared to manganese oxides. On the other hand, at redox potentials high enough to cause iron oxide formation, manganese may still exist as Mn(II). Therefore, enrichments of iron oxides and manganese oxides in soils often occur separated. This is schematically shown in Figure 2.23 by separated rims and nodules of the two oxides. On the other hand, iron oxides are, as outlined in Section 2.3.2.2, effective adsorbents for heavy metal ions including Mn^{2+}. Soil iron oxide nodules are therefore often enriched in manganese because, once they are formed, the iron oxides will adsorb Mn(II).

The lowering of the redox potential in soils are due to exclusion of oxygen and microbial activity. Such anaerobic conditions can occur when the soil is water-logged either permanently or temporarily. Soils that are permanently water-logged because of a shallow groundwater table will be gley soils, whereas pseudogley soils are formed due to surface water accumulation (episaturation) over a slowly permeable soil horizon (perched or stagnant water table). Slowly permeable soil horizons or layers can be dense argillic (argic) horizons and placic horizons and various pans such as fragipans (horizons and pans are described in Appendix A). The main differ-

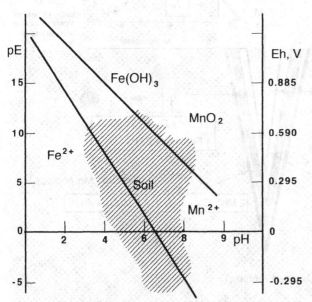

Figure 2.22. Stability fields for the Fe(III)/Fe(II) and Mn(IV)/Mn(II) systems. Shading indicates soil measurements. The straight lines correspond to: $pE = 16.0 - 3pH + pFe(II)$ and $pE = 20.7 - 2pH + 0.5pMn(II)$, where $pFe(II) = pMn(II) = 4$.

Gley

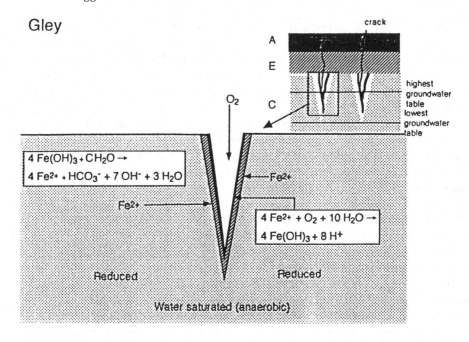

Pseudogley

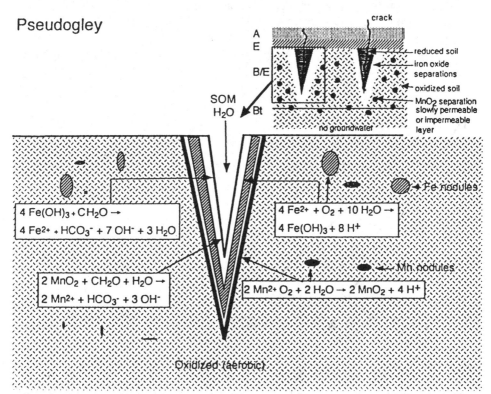

Figure 2.23. Schematic diagram showing iron and manganese oxide depletions and concentrations in gley and pseudogley.

ence between gley soils and pseudogley soils is that while the soil matrix is reduced in gley soils, it is oxidized in pseudogley soils (Fig. 2.23). Gley soils can contain pyrite, particularly those occurring in marsh areas (Section 2.2.4.2).

Gleization processes are important in many Nordic soils. Clayey soils are often mottled due to low hydraulic conductivity, and hence temporary water saturation (pseudogley). Soils at low places in the landscape often suffer from shallow ground-water tables suggesting that they may be gley soils. The extent of gleization is quantified in the FAO/UNESCO and Soil Taxonomy systems (FAO/UNESCO 1990; Soil Survey Staff 1994; Appendix B) by so-called gleyic and stagnic properties and aquic conditions. Soils with gleyic properties will be classified as Gleysols according to the FAO/UNESCO system, whereas the Soil Taxonomy system addresses aquic properties at the suborder level, e.g. Aqualfs and Aquods.

2.4.3 *Lessivage (clay migration)*

Lessivage includes all the processes leading to migration by eluviation and illuviation of clay minerals from upper to deeper soil layers. If the clay accumulation exceeds certain limits, the B (illuvial) horizon will fulfil the requirements of an argillic (argic) horizon (Appendix A2.2). Lessivage consists of two main steps, that is eluviation by leaching water after clay dispersion and deposition by flocculation or other mechanisms of the clay particles. The result of clay migration is reflected by the clay distribution shown in Figure 2.24.

In general, two similarly charged particles such as two negatively charged smectite platelets will affect each other by two opposite forces, mass attraction and charge

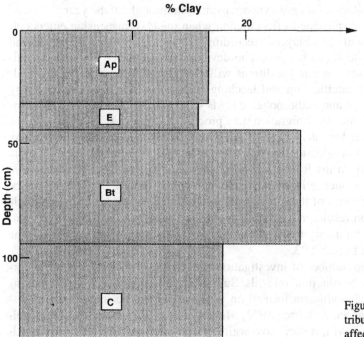

Figure 2.24. The clay distribution in a Danish soil affected by lessivage.

repulsion as discussed in Section 2.3.1.2. The result of these forces is determined by the thickness of the electric double layer. Since mass attraction is due to short-range forces, repulsion will prevail if double layers of the particles are thick. Flocculation results if the particles are attracted to each other, whereas repulsion causes dispersion. Thickness of the electric double layer is a function of ionic strength, i.e. the concentration and charge of the ions. Clay minerals will disperse in solutions with low concentrations of, in particular, monovalent ions, but will flocculate at high concentrations, especially if the ions are di- or trivalent. Saturation with monovalent ions, particularly Na^+, promotes dispersion, whereas Ca^{2+}- and Al^{3+}-saturation will encourage flocculation. Therefore, clay minerals in calcareous soils will only show limited dispersion until calcium carbonate has been removed (by decalcification) and part of the Ca^{2+} on the clay particles has been replaced by monovalent ions as indicated in Figure 2.21. On the other hand, since Al^{3+}-saturation leads to flocculation, lessivage will not be important in very acid soils. SOM can affect clay silicate dispersion, and hence lessivage because it can complex di- and tri-valent ions or because it can be adsorbed onto clay particles and iron oxides.

The composition of the clay fraction and the amount of precipitation (leaching) stongly affects lessivage. Smectites, which have a high charge and are the smallest clay particles, will most easily be dispersed, and hence transported, whereas kaolinite migrates only slowly. Consequently, the argillic (argic) horizon will be particularly enriched in fine clay. Lessivage appears to be most pronounced in a climate with dry and wet seasons. By drying, cracks will form in the soils. When the rain comes, the water with its suspended clay will move into these cracks. Since the soil matrix may still be dry, the water will be removed by capillary withdrawal into the matrix leading to formation of clay cutans (argillans) on pore walls and aggregate surfaces (Appendix A). Therefore, argillic horizons are not likely to form under very wet conditions, where precipitation exceeds evapotranspiration throughout the year.

Clay migration will stop due to flocculation when the clay suspension enters calcium carbonate-containing soil layers. Accordingly, clay accumulation is often found in or just above a calcareous horizon. The development sequence for soils on calcareous till under udic moisture conditions will therefore start with removal of calcium carbonate by decalcification and leaching from upper soil layers followed by lessivage. Since natural and anthropogenic acids will cause removal of $CaCO_3$ at increasing depth over time clay, migration may proceed. However, at a certain stage of development the upper boundary of the argillic horizon tends to degrade, which may lead to formation of a glossic horizon (Appendix A2.2) with tongues of albic (bleached) material from the E (eluvial) horizon extending into the argillic horizon. Such a development sequence is schematically shown in Figure 2.21. The difference of hydraulic conductivities of the more sandy tongues and the more clayey materials of the argillic horizon results in gleization (pseudogley), which can be seen as iron-enriched rims at the boundary between the two materials resembling very much the pseudogley sketch in Figure 2.23.

Despite the limited number of investigations, lessivage is undoubtedly an important process in many Nordic mineral soils. Supporting this suggestion, investigations of soils in southern Scandinavia formed on till have shown the importance of lessivage (Breuning-Madsen & Jensen 1992; Møberg et al. 1988). Soils that are developed on calcareous till will, if they have argillic (argic) horizons, be classified as Alfisols (Luvisols). In contrast, clayey soils with argillic (argic) horizons developed on more acidic deposits will be Ultisols (Alisols). Soil classification is treated in Appendix B and the development sequence shown in Figure 2.21 indicates main features that may be seen during the delopment of such soils.

2.4.4 *Podzolization*

Podzolization is the process leading to the formation of Spodosols or Podzols (Appendix B), which are typically formed on coarse-textured materials in a cold and humid climate. In fact, it is not a single process, but rather a series of processes translocating aluminium with or without iron from upper to deeper soil layers by leaching water, under the influence of protons and organic compounds. This transport can lead to the formation of a spodic B horizon (Appendix A2.2). A sketch of a typical Spodosol profile from the Danish Klosterhede soil described in Appendix A3 is shown in Figure 2.25 and the distribution of SOM and oxalate-extractable aluminium and iron, which is a measure of non-silicate aluminium and iron or so-called free oxides, for the same soil is shown in Figure 2.26. The enrichment of the B horizon with SOM and aluminium and iron oxides is obvious.

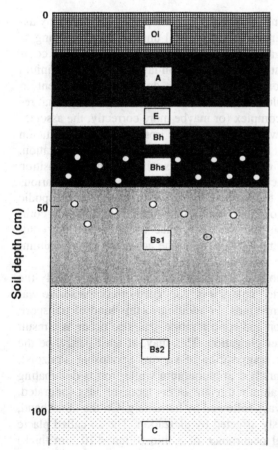

Figure 2.25. Sketch of the Klosterhede soil affected by podzolization and described in Table 2.3 (Appendix A3).

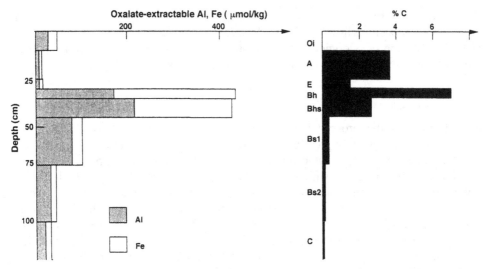

Figure 2.26. The distribution of organic carbon and oxalate-extractable ('free') aluminium and iron in the Klosterhede soil affected by podzolization and described in Table 2.3 (Appendix A3).

Although several theories have been suggested to explain podzolization, many aspects of Spodosols (Podzols) and their formation are still unresolved. According to the organic complexation theory negatively charged, water soluble organic compounds (fulvic acid) formed in base-cation-depleted topsoil layers take up aluminium and iron from inorganic soil components during their downward movement in leaching water. Increasing uptake of aluminium and iron ions results in a gradual reduction of the negative charge of the complex (or maybe more correctly, the association) leading eventually to saturation with aluminium and iron and then precipitation of the association. An increasing pH at increasing depth may encourage precipitation. Continued production and leaching of organic compounds with varying composition (molecular weight and number of functionalities) will result in formation of various spodic subhorizons (Bh, Bhs, Bs) and the gradual downward migration of the spodic horizon. Other theories are focusing on microbial degradation of the organic compounds as the cause for precipitation and the occurrence of imogolite-like compounds (Section 2.2.4.2), found in some Spodosols, as the active carrier of aluminium.

The mechanisms leading to transport of aluminium and iron are only partly the same. Aluminium can not be transported unless it is associated or complexed to water soluble organic (or inorganic) compounds. In addition to this kind of transport, iron movement, and hence its depletion and concentration, can also occur as a result of redox processes, as discussed under gleization. This may, at least partly, be the reason for the great variability in the composition of various Spodosols (Podzols). Some spodic horizons are strongly enriched in aluminium while iron is dominating in others. Spodic horizons high in aluminium are often more or less indurated, whereas a high iron content is not always followed by hardening. On the other hand, hardening is common in soils obviously affected by gleization. The so-called placic horizon (Appendix A2.2) is a special occurrence of a thin (often < 10 mm thick) iron-indurated soil layer, presumably formed by iron reduction followed by downward diffusion and oxidation at a sharp redox boundary between an upper, reduced spodic subhorizon and a lower, oxidized subhorizon. Placic horizons seem to occur only in well developed ('mature') Spodosols (Podzols) with rather strongly cemented spodic horizons of low hydraulic conductivity. Since placic horizons are barriers to water, peat formation can be seen on top of soils with placic horizons.

Podzolization is restricted to strongly leached (acidic) soils poor in plant nutrients. Such conditions occur in soils formed on sandy materials in a cold and humid climate (cf. also final stage in Figure 2.21). Formation of poorly humified SOM (mor, Section 2.2.3.2) appears to be closely linked to podzolization, possibly because the acidic and nutrient-poor soil environment both retards SOM decomposition and favours the formation of the fulvic-acid-like organic compounds involved in the transport. Mor formation, and hence podzolization, is encouraged by certain plants such as spruce and calluna, while other plants, particularly oak, seem to retard podzolization or maybe even revert it, that is depodzolization (Nielsen et al. 1987). Although the effect of oak is uncertain, it is beyond any doubt that amendments such as liming and fertilization and tillage will cause depodzolization of Spodosols (Podzols).

Many Nordic soils occur on coarse textured materials. Since the vegetation is often spruce or other conifers and the climates are cold and humid, podzolization will be the dominating process in many Nordic soils, at least in the more well-drained soils. Accordingly, Spodosols (Podzols) or more or less podzolized soils are commonly occurring in Denmark, Finland, Norway and Sweden.

2.5 SUMMARY WITH CONCLUSIONS

Composition, properties and development of soils are reviewed with emphasis on aspects relevant to soils in Denmark, Finland, Norway and Sweden. The factors affecting soil genesis include climate, parent material, relief, organisms and time, and can exhibit great variability from place to place. However, Nordic soils may in general be considered to have formed on clayey to sandy till and outwash materials from the Weichselian Glaciation within the last 15,000 to 10,000 years in cold humid climates under the influence of different forms of vegetation and land-use.

Soils are dynamic, open biogeochemical systems consisting of mixtures of air, water and inorganic and organic solids. Soil air contains the same gases as atmospheric air, but in different concentrations because of biological activity, which leads to increased CO_2 contents but decreased amounts of O_2 in well-aerated soils. In waterlogged (reduced) soils, other gases such as CH_4 and H_2S can form. Soil water is fundamental for plant growth, but also for the processes occurring in soils, and hence for soil development because the percolating water with its content of gases and inorganic and organic ions and compounds (soil solution) is the driving force of the soil reactions.

Soil organic matter (SOM) consists of large and small molecules from animals, plants and microorganisms and their decomposition products such as humus (humin, humic acid and fulvic acid) formed by chemical and microbial processes (humification). It contains many nutrients and is decisive for most chemical and physical soil properties such as ion exchange, complexation, water retention and structure. Functional groups such as carboxyl and phenol groups, particularly on humus, are involved in the processes. As acids of different strength, these groups will be more and more dissociated at increasing pH resulting in the negative pH-dependent charge of the organic particles. The SOM content is determined by the balance between the organic matter input from autotrophs (plants) and mineralization by heterotrophs (bacteria, fungi), the activity of which are determined by precipitation, temperature and nutrient status of the soil. Various SOM compounds exhibit great difference in decomposition rates ranging from less than one year to > 1000 years.

During soil formation primary minerals such as olivines, pyroxenes, amphiboles, micas and feldspars weather to secondary silicate minerals (also called clay silicates) such as kaolinite, illite, vermiculite, smectite and hydroxy-interlayered minerals (HIM) and accessory minerals such as various slightly soluble aluminium, iron, manganese and titanium oxides. The silicates consist of Si^{4+} (and Al^{3+}) in tetrahedral coordination and metal cations such as Al^{3+}, Mg^{2+}, Fe^{2+}/Fe^{3+} etc. in octahedral coordination, which are linked together in various ways forming the different mineral groups. Isomorphous substitution, where cations of higher valence are replaced by cations of lower valence but with similar size, such as Al^{3+} substituting for Si^{4+} and Mg^{2+} substituting for Al^{3+}, accounts for development of negative charge (so-called permanent charge), which is compensated for by adsorption of cations. Such adsorbed cations can be more or less readily exchanged by other cations. The charge corresponding to the readily exchangeable cations contributes to the cation exchange capacity (CEC) of soils. For the secondary silicates the CEC will decrease in the order: vermiculite > smectite > illite > HIM > kaolinite. These minerals form the main constituents of the clay fraction in Nordic soils, while the silt and sand fractions are composed of quartz and feldspar together with highly variable amounts of more easily weatherable minerals such as olivine, pyroxene, amphibole and mica depending on the parent material and degree of weathering. Gibbsite ($Al(OH)_3$)) has been identified in some Danish soils, while poorly crystalline aluminium oxides are common in most podzolized soils, which also contain more or less crystalline iron oxides. In fact, small amounts of iron oxides occur in most Nordic soils as brownish ferrihydrite ($5Fe_2O_3 \cdot 9H_2O$), yellow goethite (α-FeOOH) and/or orange lepidocrocite (γ-FeOOH). In strongly reduced soils, pyrite (FeS_2) can often be found.

The main physical soil properties can be attributed to soil texture and structure, which determine soil porosity and pore geometry and, therefore, the retention and movement of water in soils. Soil structure depends on soil composition and relates to the combination of primary particles as determined by particle size analysis into aggregates or peds with SOM, aluminium and iron oxides and combinations thereof acting as binding agents. Structure also depends on dispersion and flocculation, which results from repulsion and attraction, respectively of charged colloids such as the negatively charged clay silicates. Accordingly, various aggregate or structural types occur leading to blocky, granular, platy and prismatic structures. Poorly developed sandy soils often have single-grain structure.

The most important chemical soil properties relate to non-specific adsorption of readily exchangeable cations such as Ca^{2+}, Mg^{2+}, K^+ and Na^+ (base cations) and H^+, Al^{3+}, $Al(OH)^{2+}$ and $Al(OH)_2^+$ (acidic cations) and specific adsorption of heavy metal cations such as Cd^{2+}, Co^{2+}, Cu^{2+} and Pb^{2+} and anions of more or less weak acids such as citrate, oxalate, phosphate and selenite. The cation exchange properties can mainly be attributed to the permanent and pH-dependent charge on the colloidal SOM and clay silicate particles. In fact, the CEC, which is the sum of acidic and base cations, is a function of the contents of clay and SOM. Since the charge depends on pH, CEC will increase at increasing pH. The percentage of the sum of base cations to CEC is termed base saturation and is often found to be an almost linear function of pH. Specific adsorption of cations and anions occur at pH-dependent sites on the soil particles. Therefore, SOM and clay silicates are important specific adsorbents, but aluminium and iron oxides appear to be more important and in sandy subsoils they are the dominant or only adsorbents.

The most important soil development processes in Nordic mineral soils are decalcification, gleization, lessivage (clay migration) and podzolization. Decalcification occurs in calcareous soils, where readily soluble carbonates such as $CaCO_3$ react with CO_2 and other acid compounds in aqueous solutions under formation of soluble HCO_3^-, which is transported to greater depth or out of the soil by leaching water. Decalcification initiates other soil processes such as lessivage. Gleization occurs in soils and soil horizons that are permanently or temporarily saturated with water and denotes the redox processes by which iron(III) and Mn(IV) in slightly soluble oxides are reduced to soluble Fe^{2+} and Mn^{2+}, which move to more oxygen-rich zones, where they precipitate after oxidation as yellow, orange or brown iron oxides and black manganese oxides, which are especially enriched in the boundary zones between oxidized and reduced parts. Lessivage includes the processes leading to migration of clay minerals from upper to deeper soil layers. In addition to the transport by leaching water, lessivage requires dispersion and deposition of the clay silicates. Dispersion and deposition (flocculation) depend on the concentration and nature of the cations. Monovalent cations, especially Na^+, in low concentration will encourage dispersion, while higher concentrations of, in particular di- and trivalent cations will reduce dispersion or cause deposition of dispersed clay. Podzolization typically occurs in acidic, base-cation-depleted sandy soils and includes the group of processes causing translocation by leaching water, under the influence of protons and organic compounds, of aluminium with or without iron from upper to deeper soil layers. The processes lead to enrichment in SOM and aluminium and iron oxides of the layer of deposition, which can be more or less cemented. The processes of podzolization are not unanimously agreed upon but it appears to be important that dissolved, negatively charged organic compounds (fulvic acid) during the downward movement take up more and more positively charged aluminium and iron eventually leading to saturation and precipitation of the metal-organic association (complex). In addition to translocation by this mechanism, iron can also move because of changing redox conditions as for gleization.

Soils in Denmark, Finland, Norway and Sweden can be classified as Histosols, Spodosols, Ultisols, (Mollisols), Alfisols, Inceptisols and Entisols according to the American Soil Taxonomy system and as Histosols, Anthrosols, Leptosols, Fluvisols, Gleysols, Arenosols, Regosols, Podzols, Podzoluvisols, Alisols, Luvisols and Cambisols according to the FAO/UNESCO system.

2.6 FUTURE RESEARCH DIRECTIONS

In order to ensure at the same time both a high productivity of food and fiber and a good quality of the aquatic environment, a high level of knowledge about the different soils is mandatory. However, despite the achievements shortly outlined above, many aspects of soils including those in Nordic countries are still unresolved or poorly understood. The intention here is not to point out all gaps in knowledge about Nordic soils, but only to indicate a few directions for future research on the soil development processes gleization, lessivage and podzolization. It must, however, be emphasized that such research must be closely linked to and will depend on improved knowledge about the chemical and mineralogical composition of the soil constituents and the chemical and physical soil properties.

Identification of the very reactive green rust in soils affected by gleization is an obvious scientific challenge because green rust can reduce nitrate and, hence, pollution of the aquatic environment. The enrichment with iron oxides at the boundary between the reduced and oxidized parts in pseudogley may be important for adsorption of phosphate and other chemicals. The importance of these enrichments should be studied further, particularly in soils where the temporarily reduced zones consist of coarse textured materials in vertical, continuous cracks or fractures running from upper soil layers to the groundwater because such zones can act as preferential pathways for chemicals and clay particles dissolved or suspended in leaching water. Improved knowledge should also be provided about the precise mechanisms behind clay translocation by lessivage including the factors determining or influencing it and of the extent and importance of the process in various soil types. Studies on the significance of lessivage in sandy soils, where podzolization is considered the main process, are also needed because lessivage appears to occur in such soils and may or may not be important for podzolization in the strict sense, that is, the translocation of aluminium, iron and humus from upper to deeper soil layers.

Many other aspects of podzolization are unresolved or need to be elucidated. In fact, the whole process is still a matter for dispute including the role of saturation of the complex (association) formed between water-soluble organic compounds and aluminium and iron for the deposition in the B horizon compared to microbial degradation of the organic compounds leading to precipitation. The significance of aluminium compared to iron in the translocation and for the degree of cementation of the spodic horizon needs clarification. Studies have indicated that calluna and spruce encourage podzolization while oak inhibits or even reverts the process but more work is needed to substantiate the indications and to determine the mechanistic reasons for the difference of various kinds of vegetation. Research on podzolization is particularly important because of its significance in very many Nordic soils.

APPENDIX A: DEFINITION, TERMINOLOGY, HORIZONS AND DESCRIPTION OF SOIL

Soil is a term understood by almost everyone, but the meaning of it varies among different people, e.g. the farmer, the engineer, the forester and the gardener think of soil in different ways and for different purposes, a thinking that deviates from that of the soil scientist. Experts in soil genesis and classification (pedology) liken soils to organisms and talk about young soils and old soils, about soil bodies and about inherited versus acquired (through soil development) properties. In this appendix definition, terminology, horizons and description of soil is introduced, while soil classification according to the Soil Taxonomy and FAO/UNESCO systems is shortly outlined in Appendix B.

A1 *Definition and terminology*

Soil is comprehensively defined by the Soil Survey Staff (1994). In short this definition is:

Soil is the collective term for the natural bodies, made up of mineral and organic materials, that cover much of the earth's surface, contain living matter and can support vegetation, and have in places been changed by human activity. The upper limit of soil is air or shallow water. Its horizontal boundaries are where it grades to deep water or to barren areas of rock or ice. Soil consists of the horizons near the earth's surface which, in contrast to the underlying rock material, have been altered by the interactions, over time, between climate, relief, parent materials, and living organisms. Soil grades at its lower boundary to hard rock or to earthy materials virtually devoid of animals, roots, or other marks of biologic activity. Thus the lower limit of soil is normally the lower limit of biologic activity, but if biological activity or current pedogenic processes extend to depths greater than 200 cm, the lower limit of the soil can arbitrarily be set at 200 cm.

The distinction between mineral and organic materials can be seen in Figure 2.27. It may be seen that the limit between the two kinds of materials depends on the clay content of the mineral materials. In the definition not-soils are defined indirectly. Thus, active sand dune, glaciers, bare rock and the like are not-soil. For a land surface to be recognized as a soil it must be capable of supporting vegetation but plants need not actually be present.

A soil is a three-dimensional component of the landscape. The pedon is the smallest volume of the soil, which can be described and sampled to represent the nature and arrangement of its horizons and variability in other properties that are preserved in samples. The surface exposure of a pedon ranges from 1 to 10 m^2, depending on the variability of the soil layers or horizons (Buol et al. 1989; Fanning & Fanning 1989). Contiguous pedons all falling within the defined range of the lowest category of soil classification (Appendix B) is a true individual soil body and is termed a polypedon. The solum consists of a set of related horizons such as O, A, E and B horizons. The C horizon, where the materials reflect the composition and properties of the parent materials, is normally not considered part of the solum. A soil profile is a vertical section through the soil, that is the 2-dimensional vertical surface of a pedon from top to bottom. The meaning of some of these terms is illustrated in Figure 2.28 for a hypothetical soil formed on weathered rock.

A2 *Horizons*

Two kinds of soil horizons (layers) are considered, that is, genetic horizons and diagnostic horizons. They are not equivalent. While genetic horizon designations used in the so-called ABC sys-

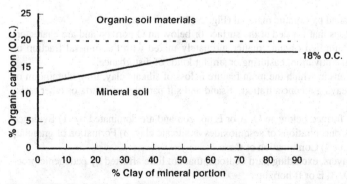

Figure 2.27. Definition of mineral and organic soil materials by the Soil Survey Staff (1994). Solid lines are for materials usually saturated with water (or artificially drained) and the dashed line is for dry materials.

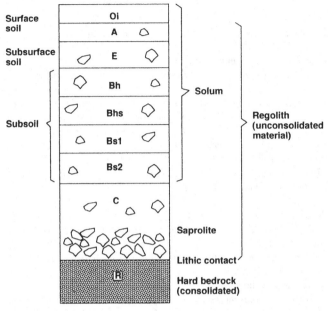

Figure 2.28. Sketch of the profile of a hypothetical, natural soil showing various designations. Horizon boundaries are normally not clear and smooth but more or less diffuse and wavy. Strictly, the term surface soil (or topsoil) is restricted to the plow layer (Ap) in cultivated soils.

tem express qualitative judgements about believed changes (genesis), diagnostic horizons are quantitatively defined features used in soil classification (Appendix B). The definitions of the diagnostic horizons, which will be considered in more details in a later section (Appendix A2.2), rely on combinations of field information (soil description, Appendix A3) and laboratory measurements.

A2.1 *Genetic horizons*

The ABC system is especially useful in making soil (profile) descriptions, e.g. according to the FAO guidelines (FAO 1990). In the ABC system, horizon designations are based on field examination alone. For the genetic master O, A, E, B and C horizons and R layers the designations are shortly described as follows:

O horizons: Layers dominated by organic material (Fig. 2.27).

A horizons: Mineral horizons that formed at the surface or below an O horizon and are characterized by an accumulation of humified organic matter intimately mixed with the mineral fraction or have properties resulting from cultivation, pasturing or similar kinds of disturbance.

E horizons: Mineral horizons in which the main feature is loss of silicate clay, iron, aluminium or some combination of these, leaving a concentration of sand and silt particles of quartz or other resistant materials.

B horizons: Horizons that formed below an O, A or E horizon and are dominated by: 1) Evidence of removal of carbonates; 2) Concentrations of sesquioxides or silicate clay; 3) Formation of granular, blocky or prismatic structure; or 4) Combination of these.

C horizons: Horizons or layers, excluding hard bedrock, that are little affected by pedogenic processes and lack properties of O, A, E or B horizons.

R layers: Hard bedrock including granite, basalt, quartzite and indurated limestone or sandstone that is sufficiently coherent to make hand digging impractical (hardness > 3 on Mohs scale).

Two kinds of transitional horizons occur. In one, the properties of an overlying or underlying horizon are superimposed on properties of the other throughout the transition zone (i.e. AB, BC). In the other, distinct parts that are characteristic of one master horizon are recognizable and enclose parts characteristic of a second recognizable master horizon (i.e. E/B, B/C). Thus, AB is a horizon with characteristics of both an overlying A horizon and an underlying B horizon, but which is more like the A than the B; E/B is a horizon comprised of individual parts of E and B horizon components in which the E component is dominant and surrounds the B materials.

An obvious change in the mineral material, e.g. as seen in a soil on glacial till covered by eolian sand, is considered a lithological discontinuity and the underlying horizons are designated by a number prefix, e.g. A, E, 2B, 2C, if the discontinuity occurs at the E to B boundary.

Lower-case letters are used as suffixes to designate specific kinds of master horizons and layers as indicated in Figure 2.28. The subordinate distinctions most relevant for Nordic soils can shortly be described as follows: a = highly decomposed organic matter; b = buried soil horizon; e = intermediately decomposed organic matter; f = frozen soil; g = strong gleying; h = illuvial accumulation of organic matter; i = slightly decomposed organic matter; k = accumulation of carbonates; m = strong cementation; p = plowing or other disturbance; r = weathered or soft bedrock; s = illuvial accumulation of sesquioxides; t = accumulation of clay; w = colour or structural B; x = fragipan character.

A2.2 *Diagnostic horizons*

The diagnostic surface horizons or epipedons are shortly defined in Table 2.2 based on the characteristics of the mollic epipedon. Some diagnostic subsurface horizons that can be found in Nordic soils are briefly described as follows:

Agric horizon: This horizon is an illuvial horizon, which occurs directly under the plow layer, has formed under cultivation and contains significant amounts of illuvial silt, clay and humus.

Albic horizon: This horizon is an eluvial horizon (typically E) consisting of albic materials, i.e. light coloured materials with high colour value and low chroma (Appendix A3).

Argillic horizon: This horizon is an illuvial B horizon formed by illuviation of clay silicates (for formation see Section 2.4.3). In general, it contains at least 1.2 times as much clay as an overlying eluvial horizon or 3% or more clay if the overlying horizon contains < 15% clay or 8% or more if it contains > 40% clay. It should be at least 0.1 times as thick as all overlying horizons. Clay cutans may occur on peds (see Section 2.3.1.2).

Table 2.2. Simplified key to epipedons according to the definitions of the Soil Survey Staff (1994). In addition to those listed, two more epipedons may occur, i.e. the melanic (organic matter-rich, developed on vulcanic materials) and the plaggen (man-made, >50 cm thick) epipedons. The definitions agree with those of the FAO/UNESCO system (FAO/UNESCO 1990), where the anthropic and plaggen epipedons are put together and called fimic.

Property	Mollic	Umbric	Anthropic	Histic	Ochric
1. SOM	+	+	+	+++	+/−
2. Colour	+	+	+	+	+/−
3. Structure	+	+	+	+	+/−
4. Base sat.	+	−	+/−	+/−	+/−
5. Depth	+	+	+	+/−	+/−
6. Moisture	+	+	+/−	+	+/−
7. P content	+	+	−	+/−	+/−
8. *n* value	+	+	+	+/−	+/−

1. More than 0.6% organic carbon; 2. Dark coloured; 3. Not hard and massive at the same time; 4. More than 50% base saturation; 5. More than 25 cm (18 cm) deep; 6. Not aridic moisture regime (very dry); 7. Less than 250 mg P_2O_5/kg (< 110 mg P/kg); 8. *n* value less than 0.7 (measure of trafficability).

Cambic horizon: This horizon is a structured B horizon of very fine sand, loamy very fine sand or finer texture being too weakly developed to meet the requirements for argillic and spodic horizons.

Fragipan: This horizon is a subsoil layer with hard to very hard consistence when dry, and brittle when moist.

Glossic horizon: This horizon is ≥ 5 cm thick and occurs between an overlying albic horizon and an underlying argillic horizon. It developes as a result of degradation of an argillic horizon, where tongues of albic materials extend into the argillic horizon (Fig. 2.21).

Placic horizon: This horizon is a 2-10 mm thick, strongly cemented (by iron-humus complexes) layer commonly occurring within 50 cm of the soil surface.

Spodic horizon: This horizon is a ≥ 2.5 cm thick, illuvial B horizon consisting of spodic materials, i.e. illuvial active materials composed of humus and aluminium, and sometimes iron oxides. The processes leading to formation of spodic horizons are outlined in Section 2.4.4.

Sulfuric horizon: This horizon is a ≥ 15 cm thick mineral or organic layer that has a pH < 3.5 and is toxic to plant roots. The low pH is caused by oxidation of pyrite, which also may result in formation of yellow jarosite (Section 2.2.4.2).

According to the FAO/UNESCO system (FAO/UNESCO 1990) these horizons are defined similarly, although small differences occur, e.g. the argillic (termed argic) and the cambic horizons contain ≥ 8% clay according to the FAO/UNESCO system.

A3 *Profile description*

Apart from information about the soil forming factors at the site in question (geology, climate etc.), a soil profile description contains a careful description of the different soil horizons including colour, texture, structure, consistence, voids, cutans, roots, cementation, nodules/concretions, rock fragments (stones), faunal activity and horizon boundary (FAO 1990). An example of a profile description is shown in Table 2.3 for a sandy soil.

Soil colours are mainly determined by free or nonsilicate bound iron compounds (iron oxides), which give yellow, orange and red colours under aerobic conditions and green and blue colours in water-logged soils together with the black and brown colours of manganese oxides and organic matter. The colours of the soil matrix and mottles (if present) are measured by comparison with Munsell (1975) colour charts, which quantifies the colour by its *hue* (red, yellow, green, blue), *value* (lightness) and *chroma* (strength) arranged as (hue value/chroma), e.g. the notation of a yellowish brown coloured soil layer will be (10YR 5/4). The moisture content (dry, moist) is given, since it affect the colour. Soil texture classes of the fine earth fraction (< 2 mm) are shown in Figure 2.3. Soil structure has been considered in Section 2.3.1.2. Grade (stability), size and type of the structural units are used in describing soil structure. Soil consistence is recorded in dry, moist and wet states and is as structure determined by the composition of the organic and inorganic fractions. Voids (pores) include all space in the soil and is described in terms of type, size and abundance. Cutans (coatings) consist of translocated clay silicates, humus and aluminium and iron oxides (sesquioxides), or mixtures thereof, deposited on ped and pore surfaces. They are described according to abundance, contrast, nature and location. Abundance, size and orientation of roots are recorded. The occurrence of cementation or compaction, in pans or otherwise, is described according to continuity, structure, nature of cementing agent and degree. Nodules and concretions are described in terms of abundance, kind, size, shape, hardness, nature and colour. Stones and rock fragments are described according to abundance, size, shape, weathering state and nature. Abundance and kind of faunal activity is recorded. Artifacts such as pieces of bricks and tiles are also recorded.

APPENDIX B: CLASSIFICATION

Soil classification according to the FAO/UNESCO and Soil Taxonomy systems is based on the diagnostic surface horizons (epipedons, Table 2.2), diagnostic subsurface horizons and diagnostic soil characteristics or properties, which are quantitatively defined features determined by field observations and laboratory data (FAO/UNESCO 1990; Soil Survey Staff 1994) and shortly considered in Appendix A2.2 for those likely to occur in Nordic soils. The basic aspects of classification according to the two systems are oulined subsequently and illustated by a Danish soil.

For the Danish Klosterhede soil, the profile description is shown in Table 2.3 and some laboratory data are plotted in Figure 2.26. This very sandy forest soil is formed on glacial outwash. It has a thick O horizon of almost undecomposed organic materials, which according to Section 2.2.3.2 can be designated a mor layer. The soil is strongly affected by podzolization and has a spodic B horizon.

Table 2.3. Description of a profile (Klosterhede) from a Danish sandy soil in western Jutland.

Location:	32V MH 630589 (GI: Klosterhede Plantage 1115 IV SØ)
Classification:	USDA: Typic Haplorthod, sandy, mixed/siliceous, mesic, noncemented
	FAO/Unesco: Haplic Podzol
Climate:	Udic, mesic
Elevation:	27 m.a.s.
Parent material:	Glacial fluvial sand
Land-form:	Almost flat
Vegetation-Landuse:	Picea abies (planted 1923)
Drainage:	Well drained
Erosion:	None
Groundwater:	Not observed
Remarks:	Upper layers somewhat mixed when the trees were planted

Description of individual soil horizons

Oi (0-10)*: Very dark brown (10YR 2/2) mainly undecomposed needles and lichens

A (10-25): Very dark gray (10YR 3/1, moist) sand; fine, weak granular structure; non sticky, non plastic, friable, loose; many fine and medium, interstitial voids; many, very fine, fine and medium roots; few, small and medium, slightly weathered stones (flint, granite); clear, wavy boundary

E (25-30): Dark gray (10YR 4/1, moist) sand; porous massive; non sticky, non plastic, friable; few, fine and medium, interstitial voids; few fine and medium roots; few, small and medium, slightly weathered stones (flint, granite); clear, wavy boundary

Bh (30-35): Black (5YR 2/1, moist) sand; porous massive; non sticky, non plastic, friable; broken, thin humus cutans between and on mineral grains; few, fine and medium, interstitial voids; many, very fine, fine and medium roots; few, small and medium, slightly weathered stones (flint, granit); gradual, smooth boundary

Bhs (35-45): Dark reddish brown (5YR 3/2, moist) matrix with few, medium to coarse, distinct, clear brownish yellow (10YR 6/8, moist) mottles; sand; porous massive; non sticky, non plastic, loose; broken, thin humus and sesquioxide cutans between mineral grains; few, fine and medium, interstitial voids; common, very fine and fine roots; few, small and medium, slightly weathered stones (flint, gneiss); diffuse, smooth boundary

Bs1 (45-70): Dark brown (7,5YR 3/3, moist) matrix with few, medium, faint, diffuse, very dark grayish brown (10YR 3/2, moist) mottles; sand; porous massive; non sticky, non plastic, loose; broken, thin sesquioxide cutans on mineral grains; few, fine, interstitial voids; few, very fine roots; few, small and medium, slightly weathered stones (flint, gneiss); diffuse, smooth boundary

Bs2 (70-100): Yellowish brown (10YR 5/4, moist) sand; porous massive; non sticky, non plastic, loose; broken, thin sesquioxide cutans on mineral grains; few, very fine, interstitial voids; few, small and medium, slightly weathered stones (flint, gneiss); diffuse, smooth boundary

C (100–): Pale brown (10YR 6/4, moist) sand; porous massive; non sticky, non plastic, loose; few, very fine, interstitial voids; few, small and medium, slightly weathered stones (flint, gneiss)

*Very often the upper boundary is set at the boundary between the organic layer and the mineral soil, i.e. Oi (-10-0), A (0-15), E (15-20) etc.

B1 *Soil Taxonomy system*

The system contains six categories or levels as examplified for the Klosterhede soil in Figure 2.29. The differentiating characteristics used in the various categories are of the following types:

Order: This division is based largely on soil-forming processes as indicated by presence or absence of diagnostic horizons.

Suborder: The subdivision of orders emphasizes properties that suggest genetic homogeneity including wetness, soil temperature and moisture regimes and major parent material. For Histosols organic fiber decomposition stage is used in the differentiation.

Great group: The subdivision of suborders is according to kind, arrangement and degree of expression of horizons. Presence of placic horizons and fragipans is considered at this level as is the climate, if not accounted for in a higher category.

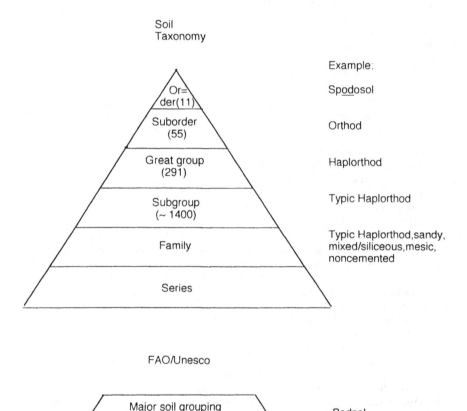

Figure 2.29. The categories of the Soil Taxonomy and the FAO/UNESCO soil classification systems and number of units in each category. Detailed guidelines for classification at the subunit level in the FAO/UNESCO system is still under consideration. Application of the classification systems is examplified by the Klosterhede soil described in Table 2.3.

Table 2.4. Simplified key to soil orders. Italicized orders are those found in Nordic countries.

If the soil has:		Order	Formative element
1	Organic material to a depth of more than 40 (60) cm	*Histosols*	ist
2	Other soils with a spodic horizon within 2 m	*Spodosols*	od
3	Other soils with > 35 cm of andic soil properties	Andisols	and
4	Other soils with an oxic horizon within 1.5 m and no kandic horizon, or contain ≥ 40% clay in the surface 18 cm and have a kandic horizon within 1.5 m	Oxisols	ox
5	Other soils with more than 30% clay in all horizons; some cracks when dry at 50 cm	Vertisols	ert
6	Other soils that are dry more than 50% of the year and have an ochric or anthropic epipedon	Aridisols	id
7	Other soils that have an argillic or kandic horizon but a *BS at pH 8.2 less than 35% at a depth of 1.8 m	*Ultisols*	ult
8	Other soils that have a mollic epipedon	*Mollisols*	oll
9	Other soils that have an argillic or kandic horizon	*Alfisols*	alf
10	Other soils that have an umbric, mollic, or plaggen epipedon, or a cambic horizon	*Inceptisols*	ept
11	Other soils	*Entisols*	ent

*BS (base saturation) of 35% at pH 8.2 corresponds to BS of 50% at pH 7 (ammonium acetate).

Subgroup: This subdivision of great groups arranges soils in relation to the central concept of the great group and to properties indicating intergradation to other great groups, suborders and orders.

Family: This subdivision of subgroups is according to texture, mineralogy, soil temperature, depth and other properties affecting plant growth.

Series: This subdivision of families is mainly based on kind and thickness of horizons and other properties, particularly those affecting soil use, not accounted for at higher levels. Series are not very much used outside the USA.

Practically, the classification is carried out according to the key. A simplified key to soil orders is shown in Table 2.4. It is very important to follow through the key in order from the beginning (the top) in attempting to 'key out' the identification and name of an unknown soil. If not, misclassification will result. Classification at the suborder, great group and subgroup levels is carried out in the same way, i.e. by keying, whereas classification of soil families is indicated by listing texture, temperature etc., as shown in Table 2.3 for the Klosterhede soil. Naming of soils according to the Soil Taxonomy system is carried out by putting together formative elements for the order, suborder and great group, but in the opposite order as exemplified by the Klosterhede soil in Figure 2.29. The formative elements of the orders are included in Table 2.4. Subgroups are identified by adjectives and families by texture, mineralogy, etc. as outlined in Figure 2.29.

B2 *FAO/UNESCO system*

The system contains only two categories, i.e. the major soil grouping and the soil unit, with a third category, the soil subunit, under consideration (Fig. 2.29). With only 153 units at the soil unit level the FAO/UNESCO system is obviously less elaborated than the Soil Taxonomy system with >1000 units already at the subgroup level. Classification according to the FAO/UNESCO system is based on criteria similar to those used in classification according to the Soil Taxonomy system, e.g. on presence and absence of diagnostic horizons and properties.

The major soil groupings most relevant for Nordic soils may briefly be summarized in the following way:

Histosols: Include organic soils, i.e. soils containing organic materials to a depth of at least 40 cm (or 60 cm if fibers).

Anthrosols: Include soils profoundly modified by human activities (man-made soils).

Leptosols: Include shallow soils formed on rocks and lack diagnostic subsurface horizons.

Fluvisols: Include soils with fluvic properties, i.e. exhibit stratification due to water deposition of the materials.

Gleysols: Include soils with gleyic properties, i.e. exhibit evidence of reduction processes (these processes are considered in Section 2.4.2).

Arenosols: Include poorly developed soils containing < 8% clay and an ochric epipedon.

Regosols: Include poorly developed soils containing > 8% clay.

Podzols: Include soils with a spodic horizon.

Podzoluvisols: Include soils with an argic B horizon showing an irregular or broken upper boundary resulting from deep tonguing of the E into the B horizon (Fig. 2.21).

Alisols: Include soils with an argic B horizon and a CEC ≥ 24 cmol(+)/kg clay and a base saturation < 50%.

Luvisols: Include soils with an argic B horizon and a CEC ≥ 24 cmol(+)/kg clay and a base saturation ≥ 50%.

Cambisols: Include soils with a cambic B horizon.

Classification is carried out by means of a key in the same way as outlined above for classification according to the Soil Taxonomy system (Appendix B1). The soil name consists of the major grouping designation and an adjective indicating the subdivision at the soil unit level as shown for the Klosterhede soil in Figure 2.29. An updated description of soils of the various major soil groupings and some modifications of the groupings is under launching (Spaargaren 1994).

It must be emphasized that soil classification should be based on correct use of the entire classification key (FAO/UNESCO 1990; Soil Survey Staff 1994) and never on short excerpts like those outlined above. Furthermore, both systems are (and have been) under current revision as a result of improved knowledge about soils. In doing soil classification, it is therefore important to state which edition of each of the two classification systems that has been used for an actual classification, because the naming of a particular soil according to different editions may be different, which may lead to misinterpretation.

Soils in Denmark, Finland, Norway and Sweden can be classified as Histosols, Spodosols, Ultisols, (Mollisols), Alfisols, Inceptisols and Entisols according to the Soil Taxonomy system and as Histosols, Anthrosols, Leptosols, Fluvisols, Gleysols, Arenosols, Regosols, Podzols, Podzoluvisols, Alisols, Luvisols and Cambisols according to the FAO/UNESCO system (FAO/UNESCO 1981; Møberg et al. 1988; Olsson & Melkerud 1989; Møberg & Breuning-Madsen 1991; Rasmussen et al. 1991; Breuning-Madsen & Jensen 1992; Skyllberg 1993).

ACKNOWLEDGEMENT

The author thanks Per Jensen for making the drawings and Maria Strandberg for writing the tables.

REFERENCES

Aiken, G.R., McKnight, D.M., Wershaw, R.L. & MacCarthy, P. (eds) 1985. *Humic Substances in Soil, Sediment, and Water.* Wiley-Interscience, New York.

Andersen, B.G. & Borns, H.W. 1994. *The Ice Age World,* Scandinavian University Press.

Andersen, B.G., Mangerud, J., Sørensen, R., Reite, A., Sveian, H., Thoresen, M. & Bergstrøm, B. 1995. Younger Dryas ice-marginal deposits in Norway. *Quaternary International,* 28: 147-169.

Bibak, A. & Borggaard, O.K. 1994. Molybdenum adsorption by aluminium and iron oxides and humic acid. *Soil Sci.,* 158: 323-328.

Bibak, A., Gerth, J. & Borggaard, O.K. 1995. Retention of cobalt by pure and foreign-element associated goethites. *Clays Clay Miner.,* 43: 141-149.

Birkeland, P.W. 1974. *Pedology, Weathering, and Geomorphological Research.* Oxford, New York.

Bisdom, E.B.A., Dekker, L.W. & Shoute, J.F.T. 1993. Water repellency of sieve fractions from sandy soils and relationships with organic materials and soil structure. *Geoderma* 56: 105-118.

Bohn, H.L., McNeal, B.L. & O'Connor, G.A. 1985. *Soil Chemistry.* 2nd Edition, Wiley, New York.

Booltink, H.W.G. & Bouma, J. 1991. Physical and morphological characterization of bypass flow in a well-structured clay soil. *Soil Sci. Soc. Am. J.* 55: 1249-1254.

Borggaard, O.K. 1990. *Dissolution and Adsorption Properties of Soil Iron Oxides.* DSR-Forlag, Copenhagen.

Brady, N.C. 1990. *The Nature and Properties of Soils.* 10th Edition, Macmillan, New York.

Breuning-Madsen, H. & Jensen, N.H. 1992. Pedological regional variations in well-drained soils, Denmark. *Geografisk Tidsskrift* 92: 61-69.

Breuning-Madsen, H., Nørr, A.H. & Holst, K.A. 1992. Atlas over Denmark. *The Danish Soil Classification. Series 1,* Vol. 3, The Royal Danish Geographical Society, Reitzel, Copenhagen.

Buol, S.W., Hole, F.D. & McCracken, R.J. 1989. *Soil Genesis and Classification. 3rd Edition,* Iowa State University Press, Ames.

Catt, J.A. 1988. *Quaternary Geology for Scientists and Engineers.* Ellis Horwood, Chichester.

Christensen, B.T. Physical fractionation of soil and organic matter in primary particles and density separates. *Adv. Soil Sci.* 20: 1-87.

Crawford, J.W. 1994. The relationship between structure and the hydraulic conductivity of soil. *Europ. J. Soil Sci.* 45: 493-502.

De Coninck, F. 1980. Major mechanisms in formation of spodic horizons. *Geoderma* 24: 101-128.

Deer, W.A., Howie, R.A. & Zussman, J. 1962 and 1963. *Rock Forming Minerals.* Vol. 1: *Ortho- and Ring Silicates;* Vol. 2: *Chain Silicates;* Vol. 3: *Sheet Silicates;* Vol. 4: *Framework Silicates;* Vol. 5: *Non-Silicates.* Longmans, London.

Dixon, J.B. & Weed, S.B. (eds) 1989. *Minerals in Soil Environments.* 2nd Edition, Soil Science Society of America, Madison.

Douglas, L.A. (ed.) 1990. *Soil Micromorphology: A Basic and Applied Science.* Elsevier, Amsterdam.

Drever, J.I. 1994. The effect of land plants on weathering rates of silicate minerals. *Geochim. Cosmochim. Acta* 58: 2325-2332.

Duchaufour, P. 1982. *Pedology.* (Translated by T.R. Paton), Allen & Unwin, London.

Ehlers, J. (ed.) 1983. *Glacial Deposits in North-west Europe.* Balkema, Rotterdam.

Fanning, D.S. & Fanning, M.C.B. 1989. *Soil. Morphology, Genesis, and Classification.* Wiley, New York.

FAO 1990. *Guidelines for Soil Description.* 3rd Edition, Food and Agriculture Organization of the United Nations, Rome.

FAO/UNESCO 1981. *Soil Map of the World.* Vol. 5, Europe, Food and Agriculture Organization of the United Nations, Paris.

FAO/UNESCO 1990. *Soil Map of the World.* Revised legend, World Soil Resources Report 60, Food and Agriculture Organization of the United Nations, Rome.

FitzPatrick, E.A. 1984. *Micromorphology of Soils.* Chapman and Hall, London.

Gieseking, J.E. (ed.) 1975. *Soil Components.* Vol. 2: *Inorganic Components.* Springer, New York.

Goldberg, S. 1993. Use of surface complexation models in soil chemical systems. *Adv. Agron.* 47, 233-329.

Gustafsson, J.P., Bhattacharya, P., Bain, D.C., Fraser, A.R. & McHardy, W.J. 1995. Podzolisation mechanisms and the synthesis of imogolite in soils of northern Scandinavia. *Geoderma* 66 (3-4): 167-184.

Hansen, H.C.B., Borggaard, O.K. & Sørensen, J. 1994. Evaluation of the free energy of formation of Fe(II)-Fe(III) hydroxide-sulphate (green rust) and its reduction of nitrite. *Geochim. Cosmochim. Acta* 58: 2599-2608.

Hayes, M.H.B., MacCarthy, P., Malcolm, R.L. & Swift, R.S. (eds) 1989. *Humic Substances II. In Search of Structure*. Wiley-Interscience, Chichester.

Hodgson, J.M. 1978. *Soil Sampling and Soil Description*. Clarendon, Oxford.

Jenny, H. 1980. *The Soil Resource: Origin and Behavior*. Springer-Verlag, New York.

Jørgensen, P.R. & Fredericia, J. 1992. Migration of nutrients, pesticides and heavy metals in fractured clayey till. *Géotechnique* 42: 67-77.

McBride, M.B. 1994. *Environmental Chemistry of Soils*. Oxford, New York.

Klute, A. (Ed.) 1986. *Method of Soil Analysis. Part 1: Physical and Mineralogical Methods*. 2nd Edition, Soil Science Society of America, Madison.

Munsell 1975. *Soil Color Charts*. Macbeth, a division of Kollmorgen Corp., 2441 North Calvert Street, Baltimore, Maryland 21218.

Møberg, J.P. 1991. Formation and development of the clay fraction in Danish soils. *Folia Geographica Danica* 19: 49-61.

Møberg, J.P., Petersen, L. & Rasmussen, K. 1988. Constituents of some widely distributed soils in Denmark. *Geoderma* 42: 295-316.

Møberg, J.P. & Breuning-Madsen, H. (eds) 1991. Soil Research in Denmark. *Folia Geographica Danica* 19, Reitzel, Copenhagen.

Newman, A.C.D. (ed.) 1987. *Chemistry of Clays and Clay Minerals*. (Mineralogical Society), Monograph No 6, Longman, London.

Nielsen, K.E., Dalsgaard, K. & Nørnberg, P. 1987. Effects on soils of an oak invasion of a calluna heath, Denmark, I. Morphology and chemistry. *Geoderma* 41: 79-95.

Olsson, M. & Melkerud, P.-A. 1989. Chemical and mineralogical changes during genesis of a Podzol from till in southern Sweden. *Geoderma* 45: 267-287.

Page, A.L., Miller, R.H. & Keeney, D.R. (eds) 1982. *Method of Soil Analysis. Part 2: Chemical and Microbiological Properties*. 2nd Edition, Soil Science Society of America, Madison.

Parfitt, R.L. 1978. Anion adsorption by soils and soil materials. *Adv. Agron.* 30: 1-50.

Petersen, L. 1976. *Podzols and Podzolization*. DSR-Forlag, Copenhagen.

Rankama, K. (ed.) 1965. *The Geological Systems. Vol.1: The Quaternary*. Interscience, New York.

Rasmussen, K., Sippola, J., Urvas, L., Låg, J. Troedsson, T. & Wiberg, M. 1991. *Soil Map of Denmark, Finland, Norway and Sweden, Scale 1:2000000*. Landbruksforlaget, Oslo.

Rowell, D.L. 1994. *Soil Science: Methods and Applications*. Longman, Essex.

Schachtschabel, P., Blume, H.-P., Brümmer, G., Hartge, K.-H. & Schwertmann, U. 1989. *Lehrbuch der Bodenkunde*. 12th Edition, Ferdinand Enke, Stuttgart.

Schnitzer, M. & Khan, S.U. 1978. *Soil Organic Matter*. Elsevier, Amsterdam.

Singer, M.J. & Munns, D.N. 1987. *Soils. An Introduction*. Macmillan, New York.

Skyllberg, U. 1993. *Acid-base Properties of Humus Layers in Northern Coniferous Forests*. Dissertation, Swedish University of Agricultural Sciences, Umeå.

Soil Survey Staff 1994. *Keys to Soil Taxonomy*. 6th Edition, (SMSS technical monograph) No. 19, Pocahontas, Blacksburg.

Spaargaren, O.C. (ed.) 1994. *World Reference Base for Soil Resources*. ISSS, ISRIC, FAO, Wageningen/Rome.

Sposito, G. 1989. *The Chemistry of Soils*. Oxford University Press, New York.

Stucki, J.W., Goodman, B.A. & Schwertmann, U. (eds) 1988. *Iron in Soils and Clay Minerals*. Reidel, Dordrecht.

Stumm, W. W. (ed.) 1987. *Aquatic Surface Chemistry*. Wiley, New York.

Tan, K.H. 1994. *Environmental Soil Science*. Marcel Dekker, New York.

Westerman, R.L. (ed.) 1990. *Soil Testing and Plant Analysis*. 3rd Edition, Soil Science Society of America, Madison.

Wilding, L.P., Smeck, N.E. & Hall, G.F. (eds) 1983. Pedogenesis and Soil Taxonomy. I. Concepts and Interactions. II. The Soil Orders. Elsevier, Amsterdam.

Wilson, M.J. (ed.) 1987. *A Handbook of Determinative Methods in Clay Mineralogy*. Blackie, Glasgow.

Geochemical processes, weathering and groundwater recharge in catchments
O. M. Saether & P. de Caritat (eds) © 1997 Balkema

CHAPTER 3

Catchment hydrology

ALLAN RODHE
Department of Earth Sciences, Uppsala University, Uppsala, Sweden

ÅNUND KILLINGTVEIT
Department of Hydraulic and Environmental Engineering, Norwegian University of Science and Technology, Trondheim, Norway

3.1 INTRODUCTION

In 1674, Pierre Perrault published 'Sur l'origine des fontaines'. From measurements of precipitation and estimates of stream runoff, he showed that the volume of precipitation over the Seine River catchment was more than enough to feed the river with water. He was the first to establish a mass budget for water in a catchment, which was to become the basis of modern hydrology: Input – Output = Change in Storage. But today, three hundred years after Perrault, many basic questions still remain to be answered concerning how water flows through a catchment, which paths it follows and how long it stays in various parts of the catchment. The answers to these questions are poorly understood and are the focus for much of today's hydrological research.

3.2 THE CATCHMENT

For any point along a water course, and for any other point in the landscape, a catchment (or watershed or drainage basin) can be defined. It is the area within which all water flowing through the point as surface runoff or groundwater flow has been collected from precipitation. The catchment is enclosed by a *water divide*. The catchment is a fundamental unit in the study of runoff processes in hydrology and in any water management practice. The boundaries of surface (or topographic) drainage system and subsurface (or groundwater) drainage systems need not be the same, but in Nordic moraine terrain with a shallow groundwater table (due to shallow soil depths and relatively impermeable bedrock) it is usually not necessary to distinguish between the two. Figure 3.1 illustrates the concept of a topographic water divide and catchment area.

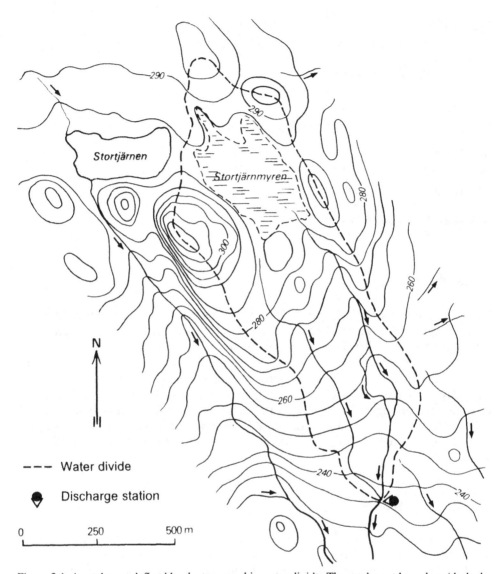

Figure 3.1. A catchment defined by the topographic water divide. The catchment boundary (dashed line) crosses the elevation contours (in m) at right angles and generally coincides with the natural ridgelines; black arrows indicate downstream flow direction (from Grip & Rodhe 1994).

3.3 WATER BALANCE

Precipitation falling within a catchment boundary can be stored in the catchment temporarily, return to the atmosphere as evapotranspiration or be transported out of the catchment as stream or groundwater runoff. No water can disappear. This can be expressed by the water balance equation, which in its most general form reads:

$$P = E + Q + \Delta S \qquad\qquad (3.1)$$

where P = precipitation, E = evapotranspiration, Q = runoff, ΔS = change in storage (which can be either positive or negative).

This is sometimes referred to as the short term or *dynamic* water balance. The water balance is most often expressed in mm/time where 1 mm water depth = 1 liter/m^2. ΔS can conveniently be divided into change in storage in *surface water* (ponds, streams, lakes), *soil water, groundwater, snow* and *glaciers*. Much of hydrology is concerned with the determination of these flows and changes in storage in different time and spatial scales. The effect of the storage term is to damp and delay the time variability of the outflow terms (E and Q). A rainfall event may last for a few hours, whereas the runoff and evapotranspiration may continue for several days, fed by the declining storage in the catchment. Some of the most important processes affecting the input to and processes within a catchment, are illustrated in Figure 3.2. Over long periods (several years), ΔS will be negligible compared to the flow terms, and the equation may be simplified to the long-term or *static* water balance equation:

$$P = E + Q \tag{3.2}$$

A long-term water balance can be computed based on measurements of at least two of the components in Equation (3.2), usually P and Q. Actual evaporation from catchments is very difficult and expensive to measure directly, and is therefore often computed by using the water balance equation. In the last part of this chapter a set of water balance maps for the Nordic countries is presented. These are all based on measured runoff and precipitation.

In the study of groundwater recharge and streamflow runoff, however, much more detailed measurements must be made in order to observe internal processes in the catchment. If the whole catchment is considered, the average areal precipitation and evaporation must be computed, based on observed values at a number of observation stations, sometimes referred to as *point observations*. The many problems involved in the measurement of, and computation of areal values of precipitation and evaporation will not be covered in this text. A detailed treatment of these problems can be found in many hydrological textbooks, e.g. Eagleson (1972), Shaw (1988) and Ward & Robinson (1990). For the rest of this chapter the main focus will be on processes transforming precipitation over the catchment into stream runoff – usually called runoff processes.

3.4 RUNOFF PROCESSES IN THE CATCHMENT

If one wants to study only the relationship between input (precipitation) and output (runoff), rather simple models may be sufficient. It is not necessary to know in detail what happens within the catchment, which can therefore be treated almost like a 'black box' provided that the input(s) and output are observed during a calibration period so that the model parameters may be calibrated. An array of such models or computational tools are available, ranging from simple equations such as the rational formula via unit-hydrograph techniques (Shaw 1988) to lumped conceptual hydrological models such as the HBV-model (Bergström 1976).

If calibration is not possible, or if the catchment properties are changing, for example due to human influence, more complex and physically correct models may be

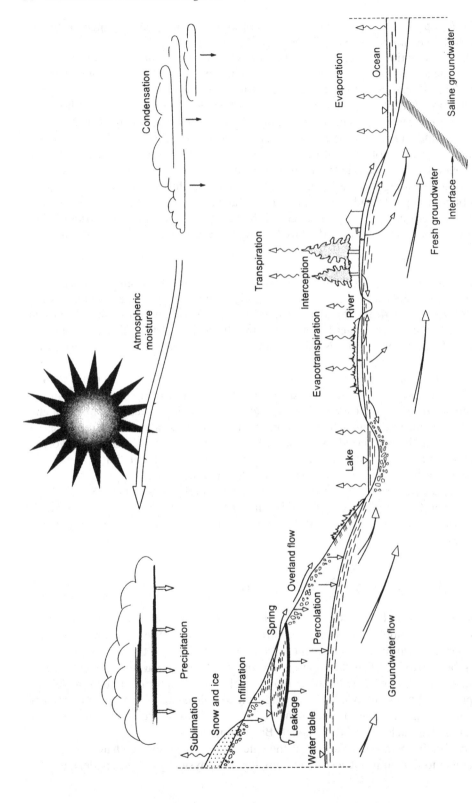

Figure 3.2. Important processes in the *water cycle* (from Bear 1979). Reproduced by permission of McGraw-Hill.

Table 3.1. Information on catchment properties needed for different levels of hydrologic modeling.

Model		Catchment information
$Q = f(P)$	With calibration	Black box
$Q = f(P)$	Without calibration	From catchment physiography, need to know the response of different types of sub-areas, where the flow impulses are given
$c_{out} = f(P, c_{in})$		Need to know bio-geochemical environment, flow, flow-paths and transit times

needed. In order to understand and forecast the *chemical* changes taking place in the water during its flow through a catchment, it is necessary to know the *flow-paths* and *transit times* for water in various biogeochemical (micro)environments. This can be illustrated as in Table 3.1, where $f(x)$ denotes a function or model of x, c is the concentration of a chemical constituent, and Q and P are as defined in Equation (3.1).

The temporal and spatial distribution of storages and flows within a catchment depends on the complex interaction of topography, geology, soil types, land use and the variability of meteorological factors. Water can be stored for a shorter or longer time in the catchment as snow, surface water, in lakes and rivers, in soil water and in groundwater, and then contribute to stream flow at a later time, or return to the atmosphere as evaporation. In order to understand and model the spatial and temporal distribution of water storage and water fluxes within the catchment, a very detailed description of the catchment may be needed.

The HBV-model, a simplified model for precipitation-runoff computations which is widely used in Nordic countries, is described at the end of this text. The following sections, however, are principally used to describe in detail the most important physical processes influencing soil water and groundwater storage and flow, and their relevance for runoff generation and groundwater recharge.

3.5 SOIL WATER STORAGE AND FLOW PROCESSES

A knowledge of the factors controlling storage and movement of water in the soil is important in order to understand a wide range of processes of hydrological importance. The soil system generally consists of three phases: solid particles, air and water as illustrated in Figure 3.3. The groundwater table is defined as the level at which the pressure of the water equals that of the atmosphere. It is the level to which water rises (after a sufficently long time for equilibrium to be established) in a pit or borehole in the ground. Below the groundwater table, in the *groundwater zone* or the *saturated zone*, all pores are filled with water and the pressure of the water is positive relative to the atmosphere. Above the groundwater table, in the *soil water zone* or the *unsaturated zone*, there is both water and air in the pores and the pressure is negative relative to the atmosphere. The upper part of the soil water zone, the root zone, can be seen as the key factor for the partitioning of water reaching the ground surface from rainfall and snowmelt into evapotranspiration, overland flow and groundwater recharge. The movement of water in the soil is governed by meteorological conditions, vegetation, soil properties and the initial water content in the soil.

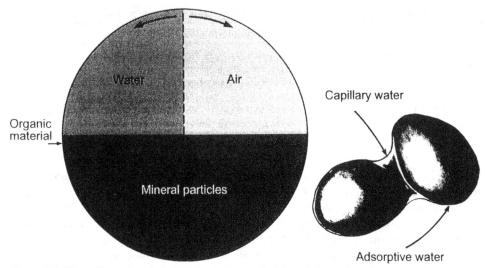

Figure 3.3. The soil matrix consists of three phases: Solid particles (mineral and organic), air and water. For most soils the particles occupy about 50% of the volume while air and water make up the other 50%. Water is held by capillary forces between the soil particles and by adsorption as a thin film around the particles (based on Grip & Rodhe 1994).

The following section deals with the general principles of flow in the unsaturated zone, with the forces acting upon the water, and with the resulting movement in the vertical direction, upwards or downwards. In order to understand the behaviour of soil water, it is first necessary to describe the physical properties of the soil particles and the influence of the soil matrix upon storage and movement of water.

The main forces acting upon water in the soil are *gravity* and *retention* forces. Retention forces can be divided into *capillary*, *adsorption* and *osmotic* forces. Both capillary and adsorption forces exert a tension or *suction* on the soil water, leading to a negative pressure relative to the atmospheric pressure. It is usually not possible to distinguish these forces, and since they both act to hold water in the soil matrix, their combined effect is called the *matrix suction*. The osmotic pressure is generally much less important than the other matrix forces, and is not discussed further here. The total suction holding water in a soil is the sum of these three retention forces, and their combined magnitude is strongly related to the physical structure and the moisture content of the soil.

The relationship between moisture content and soil moisture suction for a given soil is called the *soil moisture retention* or *soil moisture characteristic* curve (see Fig. 3.4). The mechanism of water retention varies with water content. At high water contents (low suctions) it depends primarily on the capillary surface tension, and hence on the pore size distribution and soil structure. At lower water contents (high suction) water retention is increasingly due to adsorption, and is therefore influenced more by soil texture and the specific surface area of the soil. Clay and till soil contain a large number of fine pores, and therefore tend to have higher water contents at a given suction than sandy or silty soils, as seen in Figure 3.4.

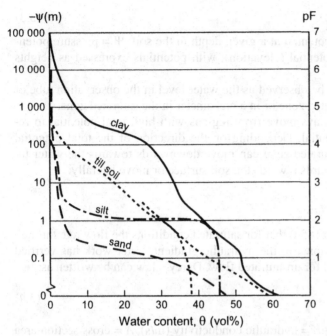

Figure 3.4. Typical soil moisture retention curves for different soil types. Matrix suction, $-\psi$, is given in m on the left y-axis and in pF = $\log_{10} (-\psi)$, where ψ is in cm, on the right axis. The shape of the different curves are closely related to the pore size distribution. For sand most of the pores are of the same (large) size, and they are drained at a fairly low suction – less than 100 cm (pF = 2). Most pores in the silt have the same size and they are therefore drained at the same suction. Till and clay have pores of widely varying sizes, so that water content decreases gradually over a large range of suction values (from Grip & Rodhe 1994).

3.6 MATHEMATICAL MODELLING OF SOIL WATER MOVEMENT

3.6.1 *Soil water potential*

Water in the soil, both above and below the groundwater level, moves as a function of external forces acting upon the water particles. In the mathematical formulation of soil water flow, the concept of *soil water potential* is very useful. It can be defined as '...*the amount of work that must be done per unit quantity of pure water in order to transport reversibly and isothermally an infinitesemal quantity of water from a pool of pure water at a specified elevation at atmospheric pressure to the soil water (at the point under consideration)*' (Ward & Robinson 1990). Since work is needed to extract water from an unsaturated soil, the pressure potential is negative in the unsaturated zone. From the definition it further follows that the pressure potential is zero at the groundwater table and positive in the saturated zone. The total soil water potential at a given depth in the soil comprises the sum of several components, of which only the gravitational (elevation) potential and the pressure potential usually need to be considered:

$$\Phi = \Psi + z \tag{3.3}$$

where Φ = total soil water potential at a given depth in the soil, Ψ = pressure potential and z = gravitational potential (elevation), with potentials expressed as heights (m).

The total potential Φ can be observed as the water level in the observation tube of a tensiometer in the unsaturated zone or in a piezometer in the saturated zone.

Water in the soil will always move from regions with high total potential to regions with lower total potential. Depending on the direction of the total potential gradient, water in the unsaturated zone can move downwards towards the water table, be stationary, move upwards towards the soil surface or move laterally.

3.6.2 *Water flow – Darcy's law*

It was first shown by Darcy (1856) that for saturated conditions the flow rate through a porous medium is proportional to the hydraulic gradient. Later work has verified that this relation is also valid for unsaturated flow. Darcy's law can be written as:

$$Q = -K \cdot A \cdot \frac{d\Phi}{dx} \tag{3.4}$$

where Q = discharge (m^3/s), K = hydraulic conductivity (m/s), A = cross section area (m^2), $d\Phi/dx$ = total potential gradient (hydraulic gradient) (m/m).

The hydraulic conductivity K varies with the water content of the soil. The variation depends upon the pore size distribution of the soil. The hydraulic conductivity of a given soil increases rapidly with increasing water content, mainly due to the increasing radius of the water-filled pores, and reaches its maximum when the soil is saturated.

3.6.3 *Drainage equilibrium*

The direction of the water flow in the unsaturated zone can be analyzed by using the soil water potential and Darcy's law together, as illustrated in Figure 3.5. If there is no vertical flow, then the vertical flow, $Q_z = 0$ and $d\Phi/dz = 0$ giving the total potential Φ = constant. This implies that $\psi + z$ = constant, Equation (3.3), or $-\psi = z$ + constant. If we take the groundwater level, where $\psi = 0$, as the reference level for z, then the constant will become 0, and the pressure potential will be equal to the elevation potential:

$$-\psi = z \tag{3.5}$$

At equilibrium the suction (negative pressure) is equal to the height above the water table. This situation is described by Equation (3.5) and corresponds to alternative b in Figure 3.5. With the potential gradient $d\Phi/dz > 0$ the flow will be directed downwards (alternative c in Figure 3.5) and with the gradient $d\Phi/dz < 0$ the flow will be directed upwards (alternative a in Figure 3.5). The equilibrium relationship ($-\psi = z$) may hold to a certain level above the water table, say a few decimeters in sand and up to a few meters in more fine-grained soils. At greater elevations the hydraulic conductivity has declined so much, due to emptying of the larger pores, that the capillary contact is broken. Then ψ tends to a constant and $d\Phi/dz$ is 1 m/m.

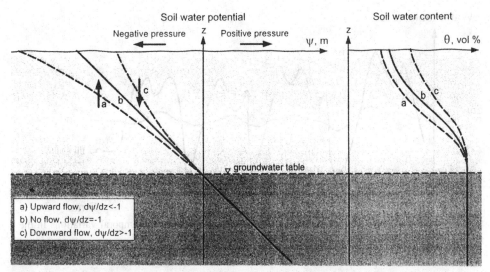

Figure 3.5. Water content and water pressure in the soil water zone and groundwater zone. Three different situations leading to a) Flow upward towards the ground surface; b) Equilibrium – no flow; c) Flow down towards the groundwater zone (from Grip & Rodhe 1994).

The equality between suction and height above the water table during drainage equilibrium implies that the depth of a shallow water table may have a strong influence on the water content in the entire unsaturated zone. One consequence of this influence is that the water content of the surface layers normally increases down a hillslope, as the groundwater table approaches the ground surface.

3.7 GROUNDWATER STORAGE AND FLOW

3.7.1 *Aquifers and aquitards*

Groundwater was previously defined as water in soils and rocks that are fully saturated. Groundwater is fed by the percolation of water from the soil moisture zone, or by infiltration from rivers and other water bodies. The upper boundary of the groundwater zone varies according to whether the groundwater is confined or unconfined.

An *unconfined aquifer* has a free water table and the groundwater table is the upper boundary of the groundwater zone. In the case of a *confined aquifer* the aquifer is overlain by a comparatively impermeable layer, called an *aquitard*. The groundwater level, defined as the water level in an observation well, is above the upper boundary of the aquifer. Only unconfined aquifers are treated here, due to their importance for runoff processes and water balances in catchments.

3.7.2 *Storage coefficient*

The storage coefficient or storativity (*M*) of an aquifer is defined as '...*the volume of*

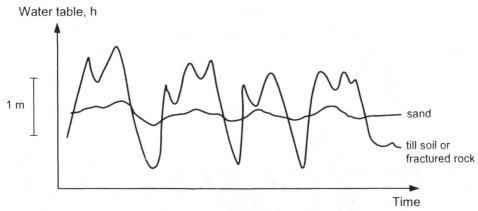

Figure 3.6. $\Delta h = \Delta S/M$, i.e. a small storage coefficient (M) gives large temporal variations in the groundwater level (Δh) for a given change in groundwater storage (ΔS).

water that the aquifer releases from, or takes into, storage per unit surface area and per unit change in head' (Ward & Robinson 1990). Mathematically this can be expressed as:

$$M = \Delta S \, / \, \Delta h \qquad\qquad (3.6)$$

where M = storage coefficient (or storativity), ΔS = change in storage (m^3/m^2 = m) and Δh = change in groundwater level (m).

The storage coefficient in an unconfined aquifer is equal to the *effective porosity* or *specific yield*. The effective porosity equals porosity minus field capacity in the soil. It indicates the percentage of water in the aquifer that can theoretically be utilized for groundwater extraction. The storage coefficient in confined aquifers is not directly related to the water retention characteristics of the soil, but is determined by the capacity of the soil skeleton to resist pressure changes in the water. Some typical examples of storage coefficients are given below:

  – Confined aquifer: 0.001;
  – Coarse till soil: 0.02-0.05;
  – Coarse sand: up to 0.20.

An aquifer with small storage coefficient (M) will give a large temporal variation in the groundwater level for a given groundwater recharge, as seen from Equation (3.6) and in Figure 3.6. For a till soil aquifer the coefficient of storage can typically be assumed to be 0.025. This means that a groundwater recharge of 20 mm of water will give an increase in groundwater level of 20/0.025 = 800 mm or 0.8 m!

3.7.3 *Groundwater flow*

The following simplification, first suggested by Dupuit and later elaborated by Forchheimer, is often used in the computation of groundwater flow in unconfined aquifers where a thin permeable aquifer rests on an impermeable bed: The potential Φ is assumed to be constant with depth. This means that there will be no vertical groundwater flow ($d\Phi/dz = 0$), and all flow lines are horizontal. The gradient in total potential will then be equal to the slope of the groundwater table. This can be written mathematically as:

$$d\Phi/dx = dh/dx \tag{3.7}$$

where h = height of groundwater table.

Applying Equation (3.7) into Darcy's law gives:

$$Q = -K \cdot A \cdot \frac{dh}{dx} \tag{3.8a}$$

or

$$Q = -T \cdot b \cdot \frac{dh}{dx} \tag{3.8b}$$

where T = transmissivity of the aquifer (m^2/s), and b = width of the cross section (m).

The transmissivity is an integrated measure of the water conducting capacity of the aquifer, and can in principle be computed by:

$$T = \int_{bottom}^{top} K(z)\, dz \tag{3.9}$$

The direction of the groundwater flow is easily determined from the groundwater levels, being perpendicular to the elevation contours of the groundwater table. The flow rate, however, is difficult to determine since there are no methods available for direct measurements. Furthermore, calculations based on Darcy's law need a knowledge of K which can vary with several orders of magnitudes in natural soils (from 10^{-3} m/s in coarse sand down to around 10^{-10} m/s in clay). The advantage of using transmissivity is that this parameter, expressing the ability of the whole aquifer thickness to conduct groundwater, can be determined from field experiments (pumping tests).

3.7.4 *Flow velocity*

When discussing the velocity of groundwater and soil water, one must distinguish between three concepts, all having the dimension of velocity. The *Darcian velocity* is the flow rate per cross sectional area of the ground, i.e., $(m^3/s)/m^2$ = m/s. The *particle velocity* is the velocity of an imagined water particle through the ground (as seen at a macroscopic scale). This is the transport velocity of a perfect tracer with the water. The *pressure propagation velocity* is the velocity by which a flow change is propagated, determining, for instance, the time lag between infiltration and groundwater level rise.

1. Darcian velocity: $v_{Darcy} = \dfrac{Q}{A}$

2. Particle velocity: $v_{particle} = \dfrac{Q}{A \cdot \theta}$

where θ is volumetric water content or, for groundwater, the porosity.

Typical particle velocities are:

soil water	1-2 m/year
groundwater	loose deposits 0.01-0.1 m/day
fractured rock	0.1-10 m/day
forest stream	0.2 m/s
river	1 m/s

3. Pressure propagation velocity: $v_{pressure}$

$$v_{pressure} \gg v_{particle} > v_{Darcy}$$

3.7.5 *Preferential flowpaths – macropore flow*

In a structured soil there may be a system of comparatively large interconnected pores which can transmit water much more rapidly than the soil matrix. Such *macropores* may account for a large fraction of the flow through a cross section and may give a much higher particle velocity than the expression given above. Macropores may be derived from biological activity (earthworms, burrowing animals, plant roots etc.) or from mechanical activity (drying/wetting, freezing/melting) creating cracks and fissures in the soil. A prerequisite for a macropore to conduct water is, of course, that the pore contains groundwater. In the groundwater zone all pores are water-filled. Interconnected macropores will then conduct water very efficiently and to a large degree determine the hydraulic conductivity of the soil. In the unsaturated zone, on the other hand, the contribution to the flow and solute transport by macropores is not so certain (Fig. 3.7). At low capillary potentials, i.e. when the soil is comparatively dry, the large pores are empty and do not contribute to the flow. At moderate rate of water input to the ground surface, the pores will rapidly be emptied by flow through the walls in response to capillary potential gradients to the surrounding soil matrix. But at high rates of water inflow and/or small saturated hydraulic con-

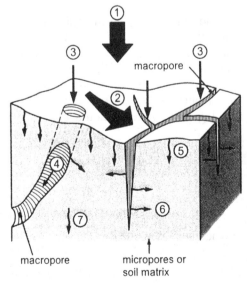

Figure 3.7. Flowpaths during infiltration: 1) Precipitation; 2) Overland flow; 3) Infiltration into macropores; 4) Flow within the macropores; 5) Infiltration into the soil matrix from the ground surface; 6) Infiltration into the soil matrix from macropores; 7) Flow within the soil matrix (from Germann 1986). Copyright 1986, John Wiley & Sons Ltd., reproduced by permission of the publisher.

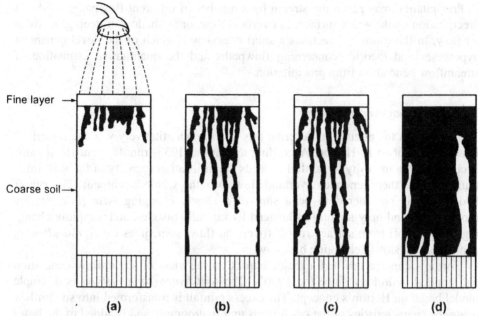

Fine layer

Coarse soil

(a)   (b)   (c)   (d)

Figure 3.8. Unstable unsaturated flow may occur when a coarse layer underlies a fine layer. Small saturated areas builds up temporarily in the fine layer, giving local breakthroughs into the coarse layer. The flow pattern in the coarse layer has a characteristic form, called fingering.

ductivity of the soil matrix, saturated or near saturated zones will build up around the pores, which then can remain water-filled and make a major contribution to the flow. The effect on the particle velocity will be that the velocity is very low at low or moderate rainfall intensities, but will increase very rapidly during heavy rainfall.

Concentrations of the vertical flow in the unsaturated zone may also occur in a layered soil, where a fine layer overlies a coarse layer. Water cannot enter the coarse layer until the capillary potential is suffciently high (a small negative number). A near saturated zone will build up at the bottom of the fine layer and, when the capillary potential exceeds a certain value, local breakthrough will occur into the coarse layer. This unstable flow gives a characteristic flow pattern in the coarse layer called fingering (Fig. 3.8).

3.8 STREAMFLOW GENERATION

Streamflow or *runoff* comprises the water moving in rivers and channels in a catchment. At a general or system level, a catchment can be regarded as a system receiving inputs of precipitation and transforming these into outputs as runoff and evaporation. Some of the internal processes in this transformation have already been described in this text, especially the processes in soil water and groundwater. Almost since Perrault's (1674) time the key question for hydrologists has been:

How is rainfall or snowmelt over a catchment transformed into stream runoff?

Precipitation may reach the stream by a number of different flowpaths: as direct precipitation on the water surface, as overland flow, or as shallow or deep groundwater flow. In the following sections a short overview is given of the development of hypotheses and theories concerning flowpaths and the quantitative estimation of streamflow generation from precipitation.

3.8.1 *Hortonian overland flow*

The pioneer in the attempt to describe this relation quantitatively was the American hydrologist Robert E. Horton. According to Horton (1933) runoff events in streams occur when the intensity of rainfall exceeds the infiltration capacity of the soil. Infiltration excess then generates overland flow over the whole catchment. The runoff hydrograph is considered to be a sum of a slowly changing *base flow*, fed by groundwater and only slightly influenced by rainfall episodes, and a rapidly changing direct runoff from surface runoff (overland flow) sometimes called *quickflow* in contrast to the slowly changing baseflow.

Horton's concept has had a major impact on the view of streamflow generation and applied hydrology since the 1930's. The *unit hydrograph method* is a simple model based on Horton's concept. The excess rainfall is transformed into streamflow using the characteristics of the catchments unit hydrograph, and is added to the baseflow, which is assumed to come from groundwater flow. The unit hydrograph method is well suited to the description of stream runoff in many catchments, but when it comes to water chemistry it fails completely.

3.8.2 *Variable source area*

Alternatives to Horton's method were gradually developed during the 1960's, based on increasing doubts whether Hortonian overland flow was really a correct model for runoff in catchments, especially in humid areas. One important criticism came from the lack of direct observations of overland flow, even during intense precipitation events. It was also verified that the infiltration capacity of natural catchments is normally greater than the intensity of rainfall or snowmelt; maximum streamflow seems more related to the total rainfall volume than to the rainfall intensity. This may be illustrated by some results from investigations in Sweden. Figure 3.9 shows typical measured infiltration capacities in clay and till soil. In most till soils, the infiltration capacity is much higher than rainfall intensities, which rarely exceeds 8-10 mm/h as shown in Figure 3.10. Snowmelt intensities are even lower, and only in extreme cases exceed 5 mm/h.

Hewlett (1961) and Hewlett & Hibbert (1967) formulated an early alternative to Hortons method, based on a different model for the generation of quickflow. Their theory was modified and extended by Betson (1964) and Dunne & Black (1970), to become what is now generally known as the *variable source area* model.

These models are all based on the fact that the infiltration capacity is normally greater than the rainfall intensity over most of the catchment, and that direct surface runoff can therefore only occur where the soil is saturated up to the soil surface. This occurs close to rivers and other water bodies, where the water table rises to the surface. Precipitation falling on these areas can not infiltrate, and a direct surface runoff

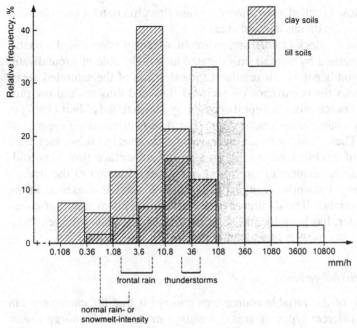

Figure 3.9. Results from infiltration capacity measurements in Swedish clay and till soils. In most of the till soils the infiltration capacity exceeds normal rain and snowmelt intensities, while in clay soils the infiltration capacity may be too low, resulting in Hortonian overland flow (from Grip & Rodhe 1994).

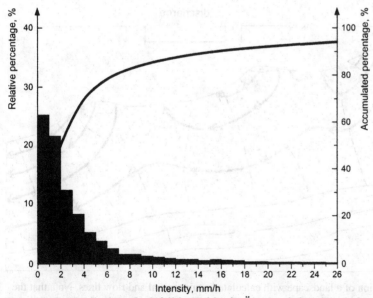

Figure 3.10. Distribution of rainfall intensities in Östersund, Sweden, during a 9-year period. The bar graph (left y-axis) shows the percentage of rain falling within given intensity intervals. The line graph (right y-axis) shows the accumulated percentage of rain falling below certain intensities. Almost 50% of the rain fell with an intensity below 2 mm/h (from Grip & Rodhe 1994).

occurs. This type of flow is called *saturation overland flow,* in contrast to *Hortonian overland flow* which occurs on unsaturated areas.

According to Dunne & Black (1970), stormflow in streams is dominated by saturation overland flow, caused by rainfall on saturated areas. The role of groundwater in streamflow generation is passive. It regulates the extension of the saturated areas, which in turn determines the occurrence of overland flow and thus streamflow production. The variable source area concept discussed by Hewlett & Hibbert (1967) is different. The variable source areas constitute saturated or near-saturated areas connected to the stream. These areas transmit unsaturated and saturated subsurface flow from the upper parts of the hillslopes to the stream. The subsurface flow is considered to be an important component of stormflow and a large fraction of the streamwater discharged during a storm is assumed to consist of water that existed in the catchment before the rainfall. The dominating role of this pre-event water, often interpreted as groundwater, has been identified by several later studies using environmental isotopes, e.g. Rodhe (1987) (see also Chapter 7).

3.8.3 *Recharge and discharge areas*

A logical consequence of the variable source area concept is that the catchment can be divided into two different types of areas, *recharge areas* and *discharge areas.* Those parts of the catchment where infiltration and groundwater recharge occur are called recharge areas (see Chapter 4). Here the total potential of the groundwater decreases with increasing depth, giving a vertical flow component downwards into the groundwater zone.

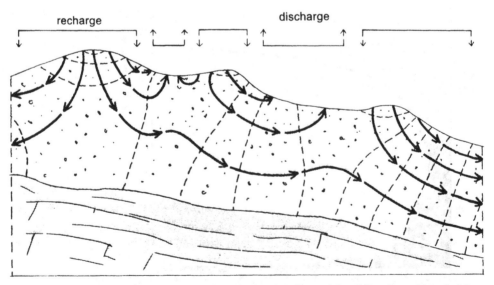

Figure 3.11. Cross section of a landscape with calculated equipotential and flow lines. Note that the horizontal and vertical scales are the same in this figure. The figure shows the fow pattern in an aquifer with undulating groundwater surface and constant hydraulic conductivity down to several hundred meters depth. (In a thin layer of till soil deposited on a less permeable bedrock, the flow would be much more concentrated to the superficial layers.) Reproduced by permission of Yngve Gustafsson.

In the discharge areas, the total potential increases with increasing depth, leading to a groundwater flow component directed up towards the soil surface. Examples of groundwater potential and flow lines are shown in Figure 3.11, demonstrating that an undulating groundwater table generates recharge and discharge areas in response to the total potential field which can be calculated from the geometry. In Figure 3.12 the total potential together with electrical conductivity is shown for a till slope in Jämtland, Sweden. The figure illustrates how the groundwater flow is directed downwards in the upper and middle part of the slope, and upwards in the lower parts, near the small stream. The electrical conductivity reveals the age and flowpaths of the water. It is low where new water is infiltrating and increases downwards and into the discharge area where old groundwater is flowing out.

The discharge areas may be *saturated* or *unsaturated*. With a saturated condition, the water table will reach the soil surface, and no infiltration can take place. From these areas both saturation overland flow from precipitation and groundwater outflow contribute to surface flow. Unsaturated discharge areas may occur if there is a very large increase in the hydraulic conductivity towards the ground surface or if the transpiration from the soil equals or is larger than the groundwater flow from below, as illustrated in Figure 3.13.

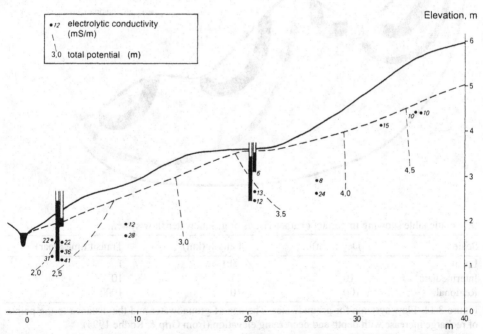

Figure 3.12. Total potential and electrolytic conductivity measured in a till hillslope in Kilmyrbäcken in Jämtland, Sweden, 1 October 1981. The total potential at different points is measured as water level in piezometric wells. The flow is directed in the direction of decreasing potential, normal to the equipotential lines. In the wells close to the small stream the potential is highest in the deepest intake showing that the total potential decreases upwards, and that the flow is directed upwards (from Grip & Rodhe 1994).

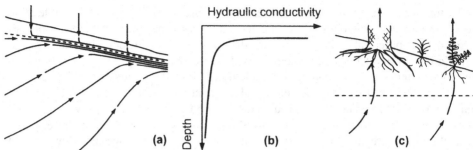

Figure 3.13. Processes in unsaturated discharge areas. a) A high hydraulic conductivity in the upper soil layer results in the lateral flow of groundwater and infiltrated water a few centimeters below the soil surface. If the conductivity near the surface is very high the soil surface may remain unsaturated even during large infiltration events; b) Saturated hydraulic conductivity as a function of soil depth. Many till soils exhibit a variation like this, with a higher conductivity near the surface, and a rapid decrease in the conductivity downwards; c) Transpiration by plants may keep the groundwater table below the ground surface in the discharge areas.

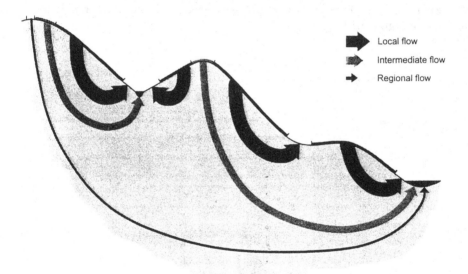

Schematic table showing important characteristics of groundwater flow systems.

Scale	Depth (m)	Length (km)	Transit time (year)
Local	1	0.1	1
Intermediate	10	1	10
Regional	100	10	1000

Figure 3.14. Flow systems for groundwater. The age of groundwater and the distance to the point of recharge increase with depth and decreasing elevation (from Grip & Rodhe 1994).

3.8.4 *Groundwater contribution in discharge areas*

Groundwater created in the recharge areas can flow towards discharge areas following different pathways depending on the topography. This is illustrated in Figure 3.14, in which different groundwater flow systems are defined. The age of the groundwater usually increases with depth and with distance from the water divide, due to the increased flowpath length and decreasing particle velocity with depth. Some typical values of depth, flow distance and mean transit time for the water (= time from recharge to discharge of the water particles) are given in the figure.

Isotope studies have shown a large and rapid response of the flow of pre-event water in the stream (water that existed in the catchment before the rainfall) to rainfall input across the catchment (see Chapters 7 and 12), see Figure 3.15 for an example. If there are no lakes or other surface water reservoirs, the pre-event water is by definition groundwater, since water can leave the ground only if the pressure is greater than that of the atmosphere, i.e. as groundwater. The discharged groundwater may, on the other hand, well be soil water that was transformed into groundwater during the event, by infiltration and percolation from above or by a rising water table from below.

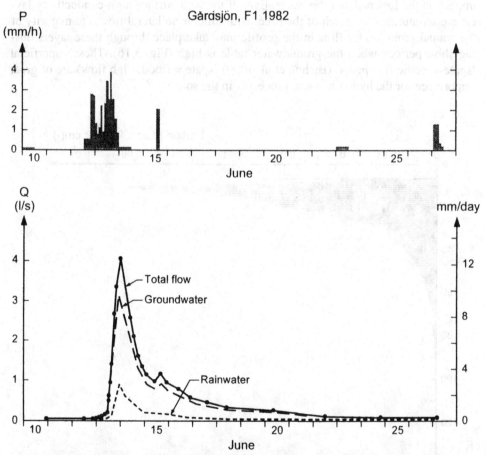

Figure 3.15. A stream hydrograph separated into flows of pre-event water and groundwater by using the stable environmental isotope oxygen-18. The runoff event (Q) is a response to a precipitation event (P), but only a small part of the streamflow actually comes from the precipitation associated with the event (event water). The largest component consists of 'old' groundwater (pre-event water), released from the catchment during the event (from Rodhe 1987).

When the water table in a hillslope rises as a result of water input, the groundwater flow increases for two reasons: increased total head gradient due to increased slope of the water table and increased transmissivity due to contribution by additional soil horizons. Since the water table largely follows the topography, the relative increase in the slope of the water table is normally small. However, the hydraulic conductivity in till soils can often increase dramatically towards the ground surface (e.g. Lundin 1982), giving a drastic increase in the transmissivity as the water table approaches the ground surface. With such an increase of the hydraulic conductivity, the groundwater flow can increase drastically following a small rise in a shallow water table, even if the slope of the water table remains essentially the same. The mechanism has been termed transmissivity feedback by Bishop (1991), who tested the concept for consistency between the isotopic, chemical and hydraulic potential signals in a hillslope and the runoff from that hillslope.

The increase in hydraulic conductivity towards the ground surface has a profound impact on the flowpaths of the water. Even if the near-surface high-conductivity layers are unsaturated for much of the time, with little or no lateral flow, the majority of the annual groundwater flow in the profile may take place through these layers during short periods when the groundwater table is high (Fig. 3.16). These superficial 'spate-specific flowpaths' (Bishop et al. 1990) (spate = flood, high flow) are of great importance for the hydrochemical processes in the soil.

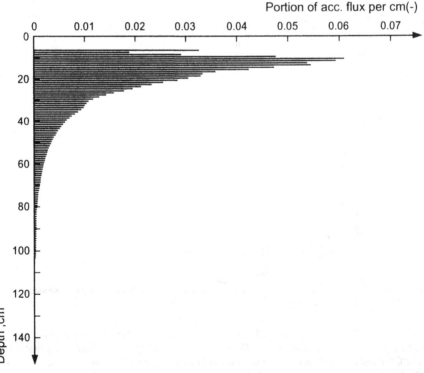

Figure 3.16. Fraction of annual groundwater flow versus depth in the lower part of a small till catchment at Lake Gårdsjön, Sweden (from Seibert 1993). Reproduced by permission of Jan Seibert.

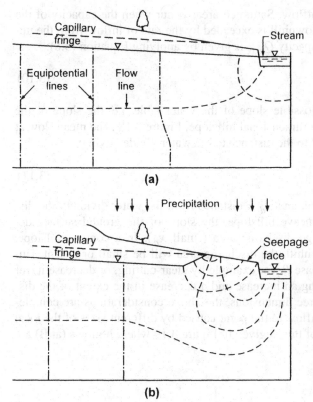

Figure 3.17. The capillary fringe model. Close to the stream the groundwater table is near the surface and the capillary fringe may reach the ground surface (from Gillham 1984). Reproduced by permission of Elsevier Science Publishers BV.

The response of the groundwater table to a certain water input depends largely on the storage coefficient of the soil (the ratio between change in groundwater storage and change in groundwater level, Equation (3.6)). Close to the stream, where the water table is shallow, most pores in the unsaturated zone may be water-filled due to a high capillary potential (the so-called capillary fringe reaches the ground surface). The storage coefficient is then very small and a small infiltrated amount of water generates a large and rapid rise in the groundwater table. A large rise in the groundwater table relative to the stream level may generate a large relative increase in the potential gradient and flow towards the stream (Fig. 3.17). This is the base for the capillary fringe model which could explain the rapid response in soils where the hydraulic conductivity does not vary vertically. The effect is probably of little importance in Nordic till soils.

3.9 THE ROLE OF TOPOGRAPHY

The wetness at a given location within a catchment, expressed as soil moisture content and groundwater level, depends on the rate of inflow to the location in relation to the ability of the ground to transport the water downhill. The catchment area of the location determines the rate of inflow, whilst the slope and transmissivity at the location determine the rate of outflow. Saturated areas occur when the capacity of the ground to transport the water downhill is exceeded by the rate of inflow from the upper part of the hillslope. The capacity Q_{max} is given by applying Equation (3.8):

$$Q_{max} = -T \cdot b \cdot \frac{dh}{dx} \tag{3.10}$$

Here dh/dx is the maximum possible slope of the water table, i.e. the slope of the ground surface. Consider a two dimensional hillslope, Figure 3.18. The mean flow at a given location is proportional to the distance to the water divide, giving:

$$R \cdot x_d \cdot b = -T \cdot b \cdot \frac{dh}{dx} \tag{3.11}$$

where $R=$ groundwater recharge, and $x_d =$ distance from the water divide to the discharge area. In an upward concave hillslope, the slope of the ground surface decreases downhill, giving a large discharge area (small x_d). In a convex hillslope, where the slope increases downhill, the discharge area will be small or absent. Human activities may either increase R, for example by clear-cutting, or decrease it, for example by afforestation, giving an increase and a decrease in the extent of the discharge area respectively. In three dimensions, the above considerations are complemented by differences in the inflow to the point caused by different sizes of the local catchment areas. An example of this is given in Figure 3.19 where hollows (at B) are wet and noses (at C) are dry.

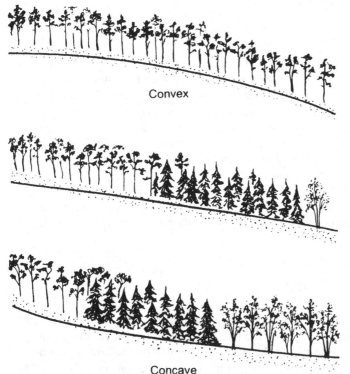

Convex

Concave

Figure 3.18. The extent of the saturated discharge areas is closely related to the topography. Three idealized situations are illustrated (from Grip & Rodhe 1994).

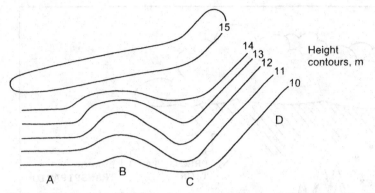

Height
contours, m

Figure 3.19. A topographic map showing various parts of a hillslope with different wetness conditions at the foot. Part B is the wettest (hollow), followed by part D (concave), part A (straight) and part C (nose), which is the driest.

3.10 THE HBV-MODEL: A PRECIPITATION/RUNOFF-MODEL

For any catchment, the relationship between precipitation and runoff is determined by the processes described above, integrated over the whole catchment. If one only wants to estimate runoff, and not the internal flowpaths and transit times of water in the soil, it is possible to use rather simple methods. One such method is the unit hydrograph method, which, as mentioned earlier, is based on the Hortonian infiltration-overland flow concept.

Another more recent method is the use of *conceptual precipitation/runoff-models*. These models are based on a simple conceptual description of the most important processes and storages in the catchment. The first well-known model of this kind was the Stanford Watershed Model (Crawford & Linsley 1966).

One of the most successful of these models is the HBV-model, which was developed in Sweden during the early 1970's (Bergström 1976). It has later been adopted for use in all the Nordic countries and in more than 30 countries worldwide (Bergström 1992).

The HBV-model, like other 'precipitation/runoff-models', is based on a simplified conceptual representation of a few main components in the land phase of the hydrological cycle as shown in Figure 3.20. Runoff from a catchment is computed from precipitation, air temperature and potential evapotranspiration. To accomplish this, the model computes the water balance for the main storage types in the catchment, and how these storages change dynamically in response to the varying meteorological inputs. The standard version of the HBV-model uses four main storage components: *Snow, soil moisture, upper zone* and *lower zone*. In addition a separate river and lake storage may be used when needed.

3.10.1 *The snow routine*

The snow routine estimates the snow accumulation and melting within each elevation zone in the catchment. The precipitation accumulates as snow when the air tempera-

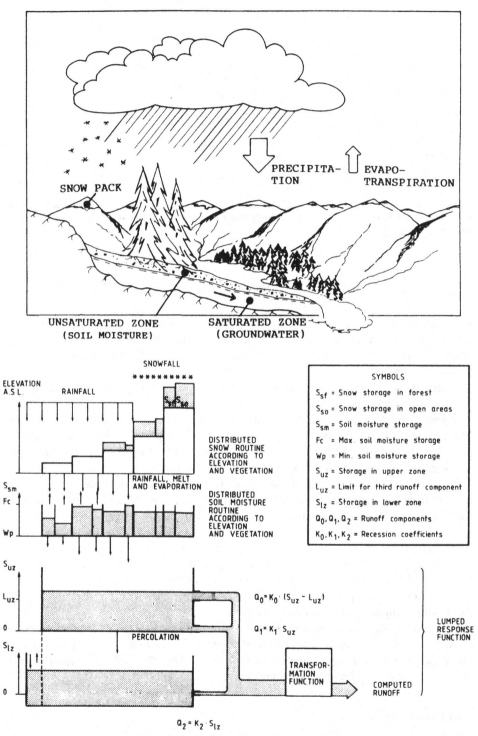

Figure 3.20. Main structure of the SMHI version of the HBV-model when applied to one subbasin (from Bergström 1992). Reproduced by permission of Sten Bergström

ture drops below a threshold value. Melting begins when the temperature exceeds the threshold. The estimation of melting is done by the degree-day model, assuming a linear relationship between air-temperature and snowmelt above the threshold, usually 0°C.

3.10.2 *The soil moisture routine*

The soil moisture routine computes an index of the degree of wetness in the catchment, and integrates storage both as interception and soil moisture. The soil moisture routine receives rainfall or snowmelt from the snow routine as input, and computes the change of storage of water in the soil moisture, actual evapotranspiration and what may be called *percolation* as output.

The field capacity is the maximum soil moisture storage in the model. If the soil moisture storage is at field capacity, no more precipitation or snowmelt can be stored as soil moisture, and all input to soil moisture storage will be transformed directly to runoff. This may lead to high runoff even from a moderate rainfall. The soil moisture storage is depleted by evapotranspiration. The *actual evapotranspiration* is computed as a linear function of potential evapotranspiration and the relative soil moisture storage deficit.

3.10.3 *The runoff response routine*

The runoff response routine transforms the *percolation* produced in the soil moisture routine into runoff. The runoff response function in the HBV-model consists of two linear tanks or reservoirs, an upper zone and a lower zone, arranged as shown in Figure 3.20. This routine also includes the effect of direct precipitation on and evaporation from rivers and lakes in the catchment.

The upper and *lower zones* delays the runoff in time, and by choosing suitable values for the parameters, the model can yield both a quick response for high flows and slow response for low flows, as normally seen in observed hydrographs. The total combined flow from the upper and lower zones can finally be filtered through a separate routine for river routing or simply for smoothing the flow. The total effect of the runoff response function is very similar to the use of a unit hydrograph, transforming a sequence of net precipitation values into a runoff hydrograph.

The upper zone conceptually represents the quick runoff components, both from overland flow and from shallow groundwater. When the percolation from the soil moisture zone exceeds the percolation capacity, the storage in the upper zone will start to fill and drainage will simultaneously start through the lower outlet. The drainage rate is determined by the recession coefficient for the lower outlet. If the upper zone storage exceeds a treshold value, an even quicker drainage will start through the upper outlet, at a rate controlled by the upper recession coefficient. In some versions of the model a third (higher) outlet with corresponding treshold and recession coefficient is used to simulate the highest runoff components. The combined effect of the upper zone is a variable response, which can be adjusted to fit the observed quickflow response in a catchment.

The lower zone conceptually represents the deeper groundwater and lake storage that contributes to base flow in the catchment. The rate of drainage is controlled by only one recession parameter. The lower zone receives water input by percolation from the upper zone and by direct precipitation on lakes and rivers. The lower zone is depleted through base flow runoff and also by evaporation from lakes and rivers. This evaporation is always equal to the potential evapotranspiration as long as there is water in the lower zone storage.

3.11 MODEL CALIBRATION AND USE

Figure 3.21 gives typical results from HBV-model simulations. Observed precipitation and air temperature are given as input to the model, which computes simulated runoff. During model calibration the simulated runoff is compared to the observed

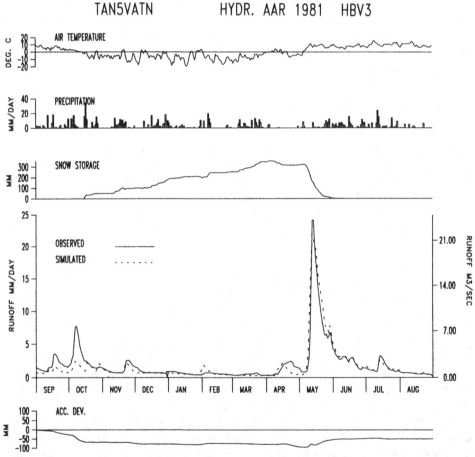

Figure 3.21. Runoff from a simulation with the HBV precipitation-runoff model in a mountainous catchment in southern Norway. The graphs shown are (from top to bottom): a) Observed air temperature in °C (daily average); b) Observed precipitation in mm/day; c) Simulated snow storage in mm; d) Observed runoff and simulated runoff in mm/day; e) Accumulated deviation between observed and simulated runoff in mm.

runoff. If a significant deviation is found, some of the model parameters are changed and a new simulation is performed. This process is repeated until the model is considered good enough for the planned application, or is found unsuitable.

The model was developed as a tool for hydrological forecasting, and this is still one of the most important uses of the model. In addition the model is used for the control of data quality, the extension of runoff records, filling in missing runoff data, the design of flood estimation systems, synoptic water balance mapping, water balance studies, simulations of the effect of climatic change and simulation of groundwater response. A successor to the HBV-model, called the PULSE-model, is used for water quality studies and simulations in ungauged catchments (Bergström 1992).

3.12 COMPONENTS OF THE WATER BUDGET IN THE NORDIC COUNTRIES

The three main water balance elements are precipitation, evaporation and runoff. Due to the variability in both climatic and physiographic factors, as described briefly in the previous chapter, the main water balance elements also show a large spatial variability within the region.

A set of water balance maps was produced after the International Hydrological Decade (1965-1974) as a result of a coordinated Nordic effort to produce consistent water balance maps, and was first published in 1976 (Forsman 1976). The maps were later redrawn and published by Otnes & Ræstad (1978) in the version shown in Figures 3.22-3.24. The maps show the distribution of *Mean annual precipitation* (P), *Mean annual runoff* (Q) and *Mean annual evaporation* (E) for the period 1931-1960, expressed in millimeters of water depth. The maps are consistent in the sense that P is intended to equal $Q + E$ at any point on the map, and they do not show any abrupt changes at the borders. The precipitation and runoff maps show the averages of areas of about 1000 km^2, while the evaporation map shows averages over areas of about 10000 km^2.

If there is no overland flow in the recharge areas (i.e. no Hortonian overland flow), the precipitation water which does not evaporate must recharge the groundwater, i.e. the rate of groundwater recharge equals $P - E = Q$. This is normally the case in the region and the map showing mean annual runoff can thus also be used to estimate the mean annual groundwater recharge to an aquifer. It should be noted, however, that such estimates are normally made for areas less than 1000 km^2. Local variations in vegetation and land use may give considerable deviations from the values given in the map.

Figure 3.22. Average annual water balance in the Nordic countries: Precipitation (from Otnes & Ræstad 1978). Reproduced by permission of Datatid A/S.

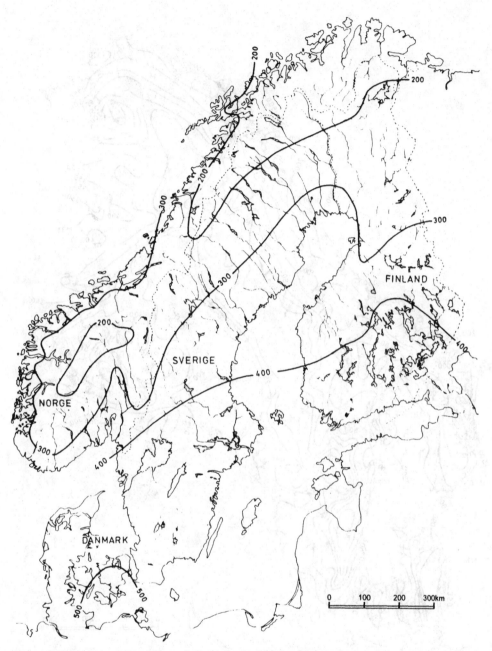

Figure 3.23. Average annual water balance in the Nordic countries: Evaporation (from Otnes & Ræstad 1978). Reproduced by permission of Datatid A/S.

Figure 3.24. Average annual water balance in the Nordic countries: Runoff (from Otnes & Ræstad 1978). Reproduced by permission of Datatid A/S.

REFERENCES

Bear, J. 1979. *Hydraulics of Groundwater*. New York: Mc Grawhill.

Bergström, S. 1976. Development and application of a conceptual runoff model for Scandinavian catchments, Swedish Meteorological and Hydrological Institute SMHI, Report No. RHO 7, Norrköping, Sweden.

Bergström, S. 1992. The HBV-model, its structure and applications, Swedish Meteorological and Hydrological Institute SMHI, Report No. RH 4, Norrköping, Sweden.

Betson, R.P. 1964. What is Watershed Runoff? *Journal of Geophysical Research*, 69(8): 1541-1551.

Bishop, K.H. 1991. Episodic increase in stream acidity, catchment flow pathways and hydrograph separation. Ph.D. thesis, Cambridge University, 246 pp.

Bishop, K.H., Grip, H. & Pigott, E. 1990. The significance of spate-specific flow pathways in an episodically acid stream. In: Mason, J. (ed.) *The Surface Water Acidification Programme*, 107-119, London: Royal Society.

Crawford, N.H. & Linsley, R.K. 1966. Digital simulation in Hydrology; Stanford Watershed Model IV, Stanford Univ., Dept. Civ. Eng. Tech. Rep. 39.

Darcy, H. 1856. *Les Fontaines Publiques de la Ville de Dijon*, Paris: V Dalmont.

Dunne, T. & Black, R.D. 1970. Partial area contribution to storm runoff in a small New England watershed , *Water Resources Research*, 6(5):1296-1311.

Eagleson, P.S. 1970. *Dynamic Hydrology,*McGraw-Hill Book Company.

Forsman A. 1976. Water balance maps of the Nordic Countries, *Vannet i Norden*, No. 4.

Germann, P. 1986. Rapid drainage response to precipitation, *Hydrological Processes*, 1: 3-13.

Gillham, R.W. 1984. The capillary fringe and its effect on water-table response, *Journal of Hydrology*, 67:307-324.

Grip, H. & Rodhe, A. 1994. *Vattnets väg från regn till bäck* (In Swedish, English title: Water Flow Pathways from Rain to Stream) Uppsala: Hallgren och Fallgren, (Earlier edition 1985, Forskningsrådens förlagstjänst, Stockholm).

Gustafsson, Y. 1968. The influence of topography on groundwater formation. In: Eriksson, E., Gustafsson, Y. & Nilsson, K. (eds), *Ground Water Problems*, 3-21, Oxford: Pergamon Press.

Hewlett, J.D. 1961. Watershed management In: Report for 1961 Southeastern Forest Experiment Station, US Forest Service, Ashville, North Carolina, USA, pp. 61-66.

Hewlett, J.D. & Hibbert, A.R. 1967. Factors affecting the response of small watersheds to precipitation in humid areas. In: Sopper, W.E. & Lull, H.W. (eds), *International Symposium on Forest Hydrology 1965*, Pennsylvania State University, 275-290, New York: Pergamon Press.

Horton, R.E. 1933. The role of infiltration in the hydrological cycle, *Transactions American Geophysical Union*, 14: 446-460.

Lundin, L. 1982. Mark- och grundvatten i moränmark och marktypens betydelse för avrinningen. (In Swedish. English title: Soil- and groundwater in till areas and the influence of soil type on discharge.) *UNGI Report 56*, Uppsala University, Dept Physical Geography, 216 pp.

Otnes, J. & Ræstad, E. 1978. *Hydrologi i praksis,* (In Norwegian. English title: Hydrology in Practice) Oslo, Ingeniørforlaget.

Rodhe, A. 1987. The origin of streamwater traced by oxygen-18. Uppsala University, Dept Physical Geography, Division Hydrology, Report Series A 41, 290 pp, Appendix 73 pp (Ph.D. thesis).

Seibert, J. 1993. Water storage and flux in a micro-catchment at Gårdsjön, Sweden, Thesis paper (Examensarbete), Department of Earth Sciences, Hydrology, Uppsala University, Sweden, 46 pp.

Shaw, E. 1988. *Hydrology in Practice*, London:Van Nostrand Reinhold (International).

Ward R.C. & Robinson, M. 1990. *Principles of Hydrology,* London: McGraw-Hill Book Company.

CHAPTER 4

Groundwater recharge

DAVID N. LERNER
Department of Civil and Environmental Engineering, University of Bradford, West Yorkshire, UK

4.1 WHAT IS RECHARGE?

4.1.1 *Recharge and related concepts*

Recharge is the water that enters the (saturated) groundwater reservoir. It can flow downwards, upwards or sideways to get there, and may only indirectly originate from precipitation.

There are alternative definitions, and terms with similar related meanings. For the purpose of this text, Figure 4.1 illustrates some of these related terms:

1. *Infiltration* is the water that enters the ground. It may come from precipitation or, for example, from leakage through a river bed. Infiltration is usually subject to soil moisture processes, such as evapotranspiration, and so only a proportion is available for downward migration.

2. *Deep percolation* is the water that escapes downwards from the soil zone and gets below the influence of roots. Some may be lost on its way to the groundwater reservoir, due to perched water tables or capillary rise. There may be substantial delays in transit through the unsaturated zone.

A distinction can sometimes be usefully made between actual recharge and potential recharge (Rushton 1988). The amount of recharge might change if conditions in the aquifer are altered even though nothing is changed at the surface. A hypothetical example is illustrated in Figure 4.2, for the cases of river infiltration with a changing water table elevation.

Assuming a simple, homogeneous, one-dimensional system, the maximum infiltration rate equals K, the bed hydraulic conductivity, for a vertical hydraulic gradient of unity for gravitational flow. The water available for recharge is the river flow per unit bed area (Q/A).

Potential recharge (PR) is the maximum water available to become recharge, and in this example is the minimum of the two quantities defined above. While the water table is deep and the river is hydraulically separate from groundwater (Fig. 4.2a), actual recharge (AR) will approximately equal PR. If the water is high (Fig. 4.2b), perhaps rising in response to recharge, and the river and groundwater are in continu-

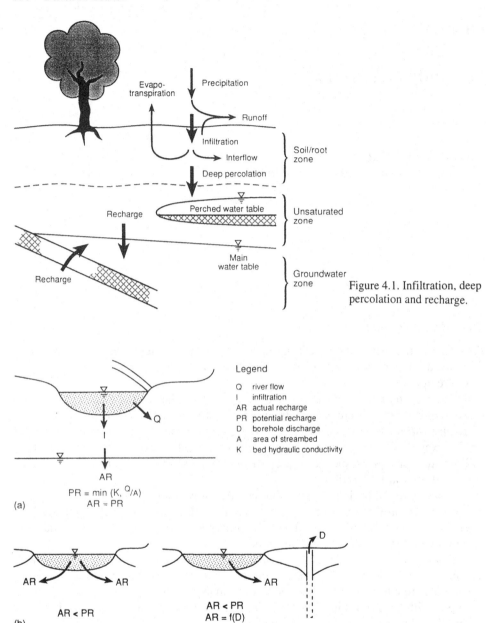

Figure 4.1. Infiltration, deep percolation and recharge.

Figure 4.2. Differences between actual and potential recharge. a) Deep water table, b) River in continuity with groundwater.

ity, then AR < PR. Other circumstances, such as a pumping well near the river, can change the relationship between AR and PR. A field example will be presented later in this chapter.

Further distinctions are conceptually useful for recharge that results from precipitation, depending on the route taken by the water. The routes are schematically illus-

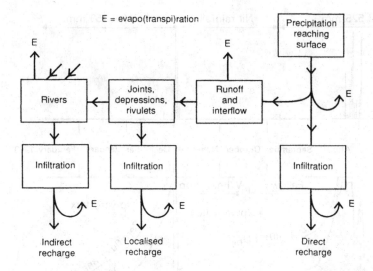

Figure 4.3. Subdivisions of precipitation derived recharge.

trated in Figure 4.3, and define three categories:

1. *Direct recharge*, which is a diffuse process, occurring beneath the point of impact of the precipitation.

2. *Localised recharge*, resulting from the horizontal movement and concentration of water into joints, rivulets and depressions. These focused points are too frequent to be mappable and measurable, but clearly involve a different mechanism from direct recharge.

3. *Indirect recharge*, also involves the concentration of water, usually into rivers, whether ephemeral or perennial. The intended distinction from localised recharge is that the water courses are sufficiently large to be mapped, counted and possibly gauged.

These distinctions are made on the basis of processes and of practicality, and cannot be rigidly adhered to. For example, preferential pathways for recharge such as fissures and root holes, which are localised recharge routes, have been documented for situations that might have been described as leading to direct recharge (e.g. Sharma & Hughes 1985). An example of transmission through fissures is given in Figure 4.4. The differences between localised and indirect recharge will depend on data availability and the methods of quantification being used; rapid studies undertaken for consultancy are likely to lump larger areas together than research studies. The latter will have the interest and resources to subdivide, monitor and interpret smaller areas.

4.1.2 *Recharge in the hydrological cycle*

Figure 3.2 of Chapter 3 shows the classical, simplified, hydrological cycle, as understood by most school children. The purpose of this section is to point out that groundwater-surface water interactions are of course more complicated than shown. In particular, there are other sources of recharge than precipitation (Table 4.1).

The earlier definition of natural recharge indicated that it could occur in the sub-

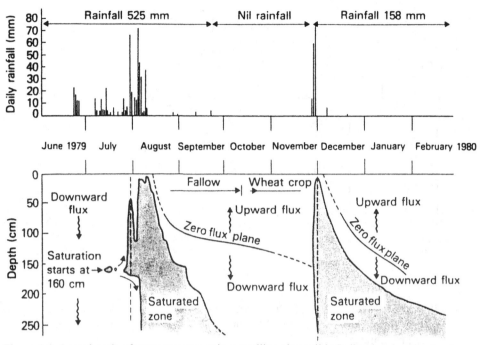

Figure 4.4. Annual cycle of water movement in a swelling clay soil in India, showing the development of saturated conditions starting at 1.6 m depth due to infiltration down shrinkage cracks (from an original diagram by Hodnett & Bell 1986).

Table 4.1. Routes of various types of water to groundwater.

Type of water	Common human interventions	Routes to groundwater
Precipitation	Agriculture Impervious surfaces (roads, etc.)	Direct infiltration Localised infiltration
Surface water	Flow regulation and abstraction Abstractions Flood control Channelisation Groundwater abstraction	Bed infiltration Overbank flooding Induced recharge (subsurface flow)
Groundwater	Abstraction	Subsurface flow (interaquifer)
Potable water Surface water Waste water	Artificial recharge: – Lagoons – River spreading – Injection boreholes	Bed infiltration Subsurface flow
Surface water Groundwater	Irrigation: – Spray – Flooding – Canals	Direct infiltration Ponded infiltration Bed infiltration
Potable water Waste water	Urbanisation: – Leaking mains and sewers – Septic tanks and soakaways	Subsurface infiltration

surface, either across the boundaries of a study area when these do not coincide with groundwater system boundaries, or from deeper aquifers (Fig. 4.1). Rivers and other surface waters are frequently a source of recharge (Fig. 4.2). Artificial recharge is widely practised and is a research topic and engineering industry in its own right, but will not be discussed here as it is less relevant to the catchment processes of interest in this book (Kivimäki & Suokko, 1996).

In many environments, the hydrological cycle has been modified by human activity, and these changes are likely to influence recharge. For example, in lowland Britain almost all rivers of any significance flow in maintained channels with regular weirs to control levels and velocities. All aquifers are exploited with consequent lowering of water tables and opportunities for increased recharge (Table 4.1). Two other human interventions deserve special mention, irrigation and urbanisation (Lerner 1990). Both modify the hydrological cycle by importing water to an area, introducing new pathways for water (canals, pipelines), and altering recharge.

Irrigation schemes are frequently a major source of recharge to aquifers, whether they use groundwater or surface water. For example, there have been many studies of groundwater in the Indus Basin where groundwater levels rose at least 20 m in the 30 years after irrigation began (Fig. 4.5). Lerner et al. (1982) estimated that the alluvial fan aquifer under Lima, Peru received 20% of its recharge from irrigation losses (Fig. 4.6). Further examples and methods of estimating recharge are discussed by Lerner et al. (1990).

An example of the impact of urbanisation is given in Figure 4.6 for Lima, Peru. Surface and groundwater are used for public water supply. The distribution system is in poor condition, losses are high, and plenty of urban recharge occurs. Sewers, where they exist, are likely to leak. Waste water, deliberately recharged in some areas, is reused for irrigation of vegetables in others. Overall, the natural system has been converted by human influence to a complex pattern of recharge sources and routes. Other aspects of urbanisation and recharge are discussed by Lerner et al. (1990).

Table 4.1 can be summarised to include four types of water that can become re-

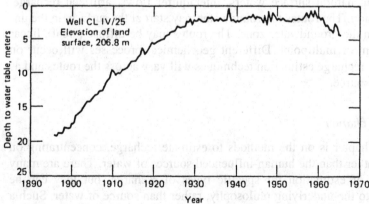

Figure 4.5. Depth to water for a well in the irrigated portion of the Indus river plain, Pakistan (Greenman et al. 1967).

Mean flows 1969 to 1978, in m³/s

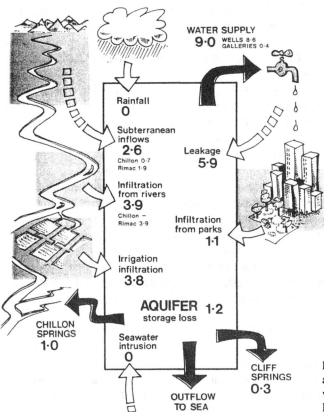

Figure 4.6. Overall water balance of the Lima (Peru) alluvial aquifer (drawn by Jon Bartlett, based on Lerner et al. 1982).

charge: Precipitation, surface water, groundwater and urban water (potable, waste). There are six situations to consider in which these waters may become recharge: natural precipitation, natural surface waters, interaquifer flows, artificial recharge, irrigation, urbanisation. The routes to groundwater may start at the surface, in the unsaturated zone, or in the groundwater zone. The routes may be diffuse (areal), linear or multi-linear, point or multipoint. Different geochemical processes will occur on the different routes; recharge estimation techniques will vary across the routes and as a function of water source.

4.1.3 *Objectives of chapter*

The focus of this chapter is on the methods to estimate recharge, concentrating on natural processes rather than the human-influenced sources of water. There are many published methods of estimation for specific types of recharge, but they can be grouped according to the underlying philosophy, rather than source of water. Such a categorisation is shown on one axis of Figure 4.7, which shows the 'space' in which any studies of recharge must be carried out.

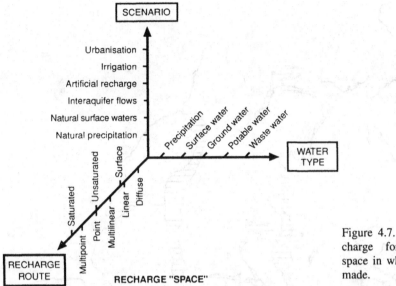

Figure 4.7. Aspects of recharge forming a 3-D space in which studies are made.

4.2 HYDROGEOLOGICAL ENVIRONMENTS

4.2.1 *Introduction*

It may well be possible to take a mechanistic approach to estimating recharge, that is going into the field, applying standard methods, and calculating results. Accuracy is more likely to be achieved when one has empathy with the environment, has a conceptual model of the processes operating, and has a prior mental estimate of the magnitude of the results. This understanding will lead to better design of measurements and enable one to spot the unusual results. The latter may be errors, requiring remeasurement, or may alert one that actual processes operation in the field are different from those expected.

This need to know the answers in advance can cause problems for the beginner! However, comparisons with similar hydrogeological environments can be very helpful, where a hydrogeological environment is defined by the combination of geology, climate, topography and human influence that gives an area its hydrogeological characteristics. Descriptions of, and case studies from, several environments are given by Lerner et al. (1990). They are geologically classified, and comprise alluvial provinces (riverbed and mountain front), sand and sandstone, limestone and dolostone, chalk, volcanic rock, and crystalline plutonic provinces. The current text does not attempt to emulate that review, and will just contrast two environments to illustrate basic aspects.

4.2.2 *Permo-Triassic sandstone of the UK*

The second most important aquifer in the UK is the Permo-Triassic sandstone (Fig. 4.8), which contains regional groundwater systems. Soils are well developed and

Figure 4.8. Outcrop of the Permo Triassic sandstone aquifers in England and Wales.

bedrock outcrops are rare; the landscape is characterised by a rolling topography. The climate is temperate, with year round rain, little snow, precipitation comfortably exceeding potential evapotranspiration, and few extreme events. There are very few areas of natural vegetation, expect for some managed forests. Agriculture is predominately arable in the drier east, with increasing livestock towards the west.

The water table is generally well below the ground, depending on topography, with an unsaturated zone up to 50 m thick. Rivers can be expected to be regional discharge points for groundwater, may be 10 km apart, and will only be perched above the water table if large boreholes are located nearby. Precipitation will lead to diffuse, direct recharge, and there is little to cause localised recharge (except as discussed below). Runoff is likely to occur by the variable source area concepts (see Chapter 3) rather than Hortonian overland flow, and so flood conditions may cause local groundwater mounds along rivers. Irrigation is confined to drier areas, sandy soils, and more valuable crops, and is always sprayed. It may lead to local increases in recharge because of the sandy soils.

The major complication to this picture is widespread presence of glacial drift cover. There is patchy superficial cover over most of England. It is very variable and unpredictable in character, from gravels to clays, and often with multiple units in any section (Fig. 4.9). The effects of drift on recharge can be large and varied, including one or more of the following:

1. Different soil types;
2. Low permeability cover for the aquifer;
3. Confined aquifer conditions;
4. Perched water tables, sometimes interacting with man-made structures;
5. Lateral movement of water in perched aquifers before becoming recharge to deep groundwater;
6. Increased surface runoff and interflow.

In resource terms, the sandstone aquifer is the one of importance, and the drift cover is only of interest from its effect on recharge. For pollution and hydrochemistry issues, the drift aquifers are much more important, where they are present.

4.2.3 *Scandinavian conditions*

Most Scandinavian groundwater environments contrast with the UK sandstone above, and have more in common with UK drift cover. Bedrock is crystalline, small outcrops are common and experience localised recharge into joints. However, bedrock is usually only of local importance as an aquifer. The important aquifers are glacial or fluvial sediments, including moraines, river valley alluvium and infills in bedrock topography. The bedrock outcrops generate localised runoff and potential recharge on their margins.

These superficial aquifers are small and contain local groundwater flow systems. Many are not exploitable, and the main interest is in their effect on runoff and its chemistry. Precipitation greatly exceeds evapotranspiration, so potential recharge is high. Snow and snowmelt are important, and can dominate the groundwater balance, occurring at times of low evapotranspiration. Water tables are frequently close to ground level and so have a controlling influence on actual recharge. High water tables and high soil organic matter can lead to anoxic conditions in many small infill deposits. Surface water is abundant, with wetlands, streams and lakes all interacting with groundwater.

Much of the land is naturally vegetated, or has managed forests. Arable agriculture is confined to accessible areas, often on the larger areas of infill between bedrock outcrops. If soils are sandy, spray irrigation is sometimes used.

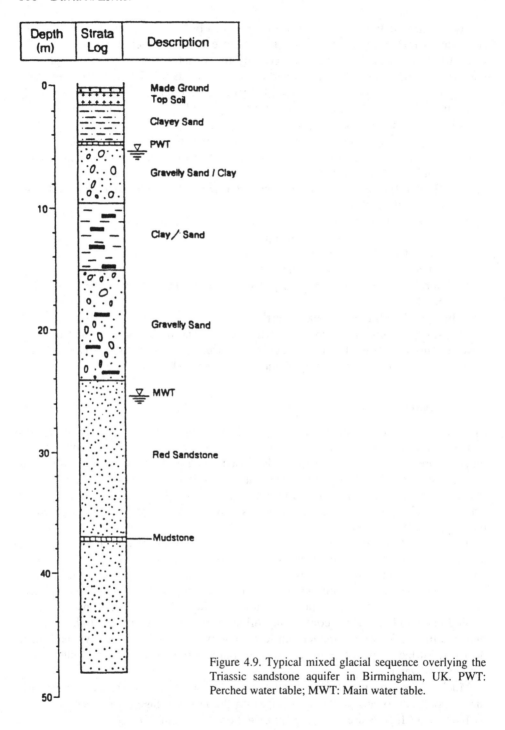

Depth (m)	Strata Log	Description

Made Ground
Top Soil

Clayey Sand

PWT

Gravelly Sand / Clay

Clay / Sand

Gravelly Sand

MWT

Red Sandstone

Mudstone

Figure 4.9. Typical mixed glacial sequence overlying the Triassic sandstone aquifer in Birmingham, UK. PWT: Perched water table; MWT: Main water table.

4.3 PRECIPITATION RECHARGE

4.3.1 *Introduction*

A simplified view of the precipitation-recharge process is shown in Figure 4.1. In essence, some of the precipitation returns to the atmosphere by various evaporation processes, some runs off laterally, and the remainder becomes direct recharge. This chapter is concerned with estimating this direct recharge, that is recharge below the point of impact of the precipitation. Including moisture storage above the water table, we can state that:

$$\text{recharge} = \text{precipitation} - \text{runoff} - \text{actual evapo-}$$
$$\text{transpiration} \pm \text{storage change} \qquad (4.1)$$

An immediate conceptual difficulty is 'How far can water move sideways and then infiltrate before being counted as indirect recharge?'. Any such movement implies variability in spatial properties, which can invalidate many of the estimation methods for direct recharge. On the other hand, small and numerous movements cannot be counted or treated individually when building a regional groundwater model. Examples include:

1. Weathered, bare, hardrock or limestone terrain where recharge is into distinct fissures;

2. Surface depressions, with sizes and spacings from centimeters upwards, where local runoff gathers and infiltrates;

3. Bare rock outcrops in permeable terrain (e.g. sand or alluvium), where runoff infiltrates at the edge of the outcrop;

4. The many minor, ephemeral drainage channels in semi-arid climates, which may only contribute to main channel flows on rare occasions; at other times all flow infiltrates into the bed.

These examples reinforce the need for a pragmatic division into three, (a) direct recharge, (b) indirect recharge, and (c) an intermediate category *localised recharge*. The last is the least well researched of the three and is discussed in the end section of this chapter.

All of the groups of methods for estimating recharge, that is direct measurement, empirical correlations, water budgets, Darcian methods and tracers, have been used for direct recharge.

The differences between potential and actual recharge have already been discussed. Actual recharge is needed when simulating historical conditions of a groundwater system. However, when modelling possible future conditions, for example when new boreholes have lowered the water table, it may be necessary to know potential recharge, and develop a method to relate actual recharge to the controlling variables, such as groundwater levels.

4.3.2 *Lysimeters: Direct measurement*

The only practicable method of measuring recharge flux is with a lysimeter. This is a block of soil instrumented so that flows through it can be measured (and not to be confused with a porous soil water sampler, also called a lysimeter!). The block is isolated from the surrounding soil but is representative because it has the same vegetation and climatic exposure. In order to minimise edge effects and average out local variations in soil and vegetation lysimeters need to be large (up to 10 m in each of the three dimensions), and so are expensive to construct. Examples of their construction and use for recharge measurement are given by Kitching et al. (1977), Kitching et al. (1980) and Kitching & Shearer (1982).

A lysimeter design for estimating recharge should fulfil the following requirements:

1. Contain undisturbed soil. Vertical flows through repacked soil will not be representative;

2. Be large enough to minimize edge effects and average small scale heterogeneities. A minimum plan area would be 1 m^2, the ideal would be 100 m^2;

3. Be large and deep enough to enclose complete root systems. This sometimes makes naturally vegetated lysimeters impractical, especially in arid and semi-arid areas where roots can be 50 m deep;

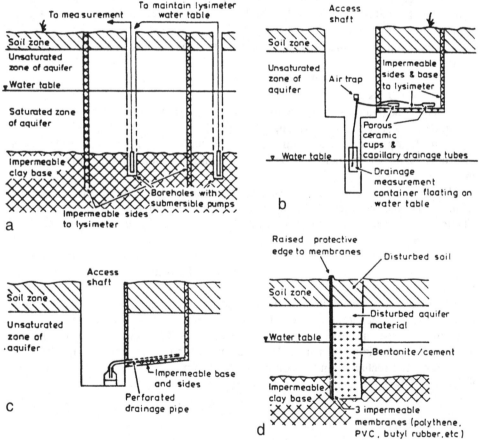

Figure 4.10. Design and construction of lysimeters. a) Lysimeter containing a water table, requires pumping to equalise conditions inside and out; b) Suction drained lysimeter; c) Gravity drained lysimeter, is suitable for coarse materials where capillary use is insignificant; d) Trench construction for lysimeter sides.

4. Be surrounded by similar vegetation to avoid oasis effects;

5. Have the same hydraulic condition at the base as found at the same depth in the surrounding soil (Fig. 4.10a-c);

6. Be watertight, except for the drainage to be measured.

Lysimeters are expensive to construct and only give 'point' measurements of recharge. Kitching et al. (1977), set up two lysimeters within 50 m of each other in a humid zone. Recharge was 159 mm/y in one and 114 mm/y in the other. The fact that the latter recharge value is only 72% of the former was not explained: Does it represent true variability in recharge or errors in measurement? In either case it shows that lysimeter methods can be as variable as other point methods.

The act of constructing a lysimeter will disturb the soil and moisture content, and time is needed for flows to settle back to a natural condition. At the extreme, this could be as long as the time for recharge to flush the lysimeter through; for a 2 m deep unit with a water content of 20% and recharge of 100 mm/y, this would be 4 years. For example, a lysimeter in Cyprus only recorded 5 mm of recharge in its first year of operation, although chloride and tritium profiles suggested average recharge was 50 mm/y (Edmunds & Walton 1980; Kitching et al. 1980). In the rather wetter conditions of Scandinavia, settling down times will be rather shorter.

4.3.3 *Empirical methods*

Many attempts have been made to find simple relationships between precipitation and recharge. Once derived by careful study, these are commonly used as 'black boxes', making recharge estimates without further consideration of hydrogeology, or whether the results are feasible.

More complex formulae often do not preserve dimensionality. The origin of many formulae are lost in the darkness of history. In general, they will have been obtained for a particular basin by correlation of precipitation with estimates of recharge obtained by other methods (water table rise, basin discharge, etc.).

There are two issues that must be resolved before an empirical formula can be used; 'How reliable is its derivation?' and 'Can it be transposed to either another catchment or another period in time?'. For example, how accurate were the recharge estimates used to derive the formula; were they checked, say by calibration of a groundwater model? When transposing, are conditions in the new catchment (or time period) the same as those in the original – depth to water table, unsaturated zone processes, land use, topographical characteristics, climate and type of rainfall. Lerner et al. (1990) give some examples of formulae and illustrations of their inaccuracy.

Given the difficulties outlined above, the occasions when empirical formulae will be most useful are for reconnaissance studies, when high margins of error are acceptable, and for catchments where groundwater conditions have little effect on recharge.

The use of empirical formulae has similarities to the use of *representative basins*. These well instrumented catchments have been set up in many parts of the world to provide accurate estimates of the components of the water balance, including recharge. The same difficulties of transposition will arise with representative basins as with empirical formulae, but at least the basic data will be of high quality.

4.3.4 *Soil moisture budgeting method*

Soil moisture budgeting methods are those that measure or estimate the items on the right hand side of Equation (4.1). At its simplest, this leads to a conceptual model like that in Figure 4.11. Water is held in a soil moisture store; precipitation adds to the store, evapotranspiration depletes it. When full, excess precipitation is routed to groundwater as recharge. The most difficult item to measure is actual evapotranspiration, and in general a conceptual quantity called 'potential evapotranspiration' is defined. A budgeting procedure involving soil moisture is used to convert potential to actual evapotranspiration. Budgeting models often measure soil water content as the deficit below field capacity (see right hand axis of Fig. 4.11). Penman (1950) introduced two constants to model the conversion of potential to actual evapotranspiration, with the *wilting point* indicating the soil moisture deficit at which plants could no longer extract water and so wilted, and the *root constant* being the lower deficit at which plants began to experience stress and so extracted water below the potential rate. There are many variations and refinements on Figure 4.11, some of which are discussed below, but at this stage it is important to note these points:

1. Such models are only simple conceptual models of the precipitation-recharge process and may not be correct for your situation;

2. The essence of the model is the relation between potential and actual evapotranspiration;

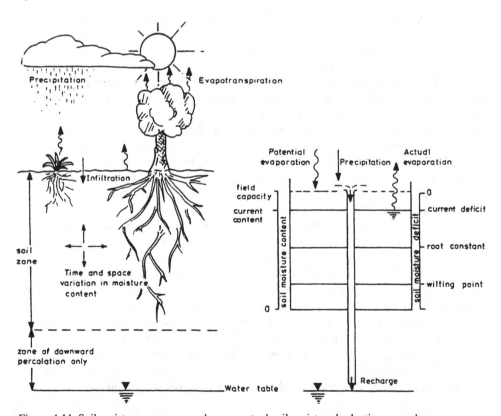

Figure 4.11. Soil moisture processes and a conceptual soil moisture budgeting procedure.

3. Estimates are for a uniform zone (see below).

Soil moisture budgeting models were developed for humid climates and have less validity in arid and semi-arid zones. They work best for seasonal patterns of recharge, well developed soils that do not dry completely, when potential and actual evapotranspiration are of similar sizes, and with precipitation that is widespread and relatively uniform. These conditions do not apply in arid and semi-arid zones, where these models normally underestimate recharge, often giving zero values. Lerner et al. (1990) give some empirical rules for deciding whether soil moisture budgeting models will be applicable.

Figure 4.11 does not describe how soil moisture behaves; there can be vertical water flows (either up or down) when a soil moisture deficit exists. Nor does it describe the recharge process, which may be dominated by fissures, root channels or topographic depressions. Many models used in real situations need empirical adjustments to make them match field conditions.

Good data on actual evapotranspiration is equally important as good precipitation data. Unfortunately, actual evapotranspiration is rarely measured except in research projects and so must be estimated from standard meteorological measurements. As mentioned above, this is usually done through a conceptual quantity called *potential evapotranspiration*, or *reference crop evapotranspiration*. This is intended to be a measure of the energy available for evaporating and transpiring water.

The Penman-Grindley model (Penman 1950; Grindley 1967) is the simplest and most widely used soil moisture budgeting model. It was originally developed to estimate soil moisture deficit and actual evaporation, recharge estimates being a byproduct. The soil moisture on day $i + 1$ (smd_{i+1}) is computed in many models on the basis of the soil moisture of the previous day (smd_i) after having calculated the intermediate variable psmd_{i+1} as shown below:

$$\text{psmd}_{i+1} = \text{smd}_i + \text{ae}_i - p_i$$
$$r_i = -\text{psmd}_{i+1}, \qquad \text{when } \text{psmd}_{i+1} < 0 \qquad (4.2a)$$
$$\text{smd}_{i+1} = \text{psmd}_{i+1} - r_i$$

Actual evapotranspiration is derived from potential as follows:

$$\text{ae}_i = \text{pe}_i, \qquad \text{when } \text{smd}_i < C \text{ or } p_i \geq \text{pe}_i$$
$$\text{ae}_i = p_i + F\,(p_i - p_i), \qquad \text{when } D > \text{smd}_i \geq C \text{ and } p_i < \text{pe}_i \qquad (4.2b)$$
$$\text{ae}_i = p_i, \qquad \text{when } \text{smd}_i = D \text{ and } p_i < \text{pe}_i$$

where smd_i = soil moisture deficit at the start of day or time period i (L), ae_i = actual evapotranspiration during day i (L), pe_i = potential evapotranspiration during day i (L), p_i = precipitation during day i (L), r_i = recharge during day i (L), psmd_i = intermediate variable (L), C = root constant (L), D = wilting point (L), F = empirical constant.

The shape of the Penman-Grindley function is shown in Figure 4.12, which shows the three parameters (C, D, F) of the model that must be calibrated or estimated. F is an empirical constant relating actual to potential evapotranspiration when deficits are greater than the root constant. All three parameters are related to the vegetation cover and, as a second order effect, to soil characteristics. In the UK, monthly values of C and D, varying with crop type, are used (Lerner et al. 1990).

A number of other models have been used. They mainly differ from the Penman-

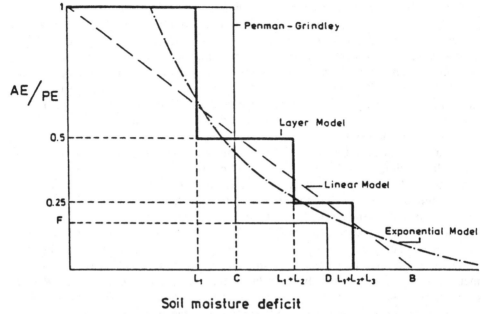

Soil moisture deficit

Figure 4.12. The form of actual over potential evapotranspiration relationships used in various soil moisture budgeting models (Based on Calder et al. 1983).

Grindley model in the shape of the actual-potential evapotranspiration relationship, as shown in Figure 4.12. They include:

1. Layer models, such as those described by Calder et al. (1983);
2. A linear model (Calder et al. 1983);
3. Exponential models (Johansson 1987).

Usually, it will be found that the difference between models will be less important than the accuracy of precipitation, potential evapotranspiration, irrigation and cropping data.

On less permeable soils or when the water table can be high, a significant proportion of precipitation may become runoff. In this case, the r_i of Equation (4.2a) is effective precipitation, and has to be split between runoff and recharge. Some workers estimate and deduct runoff before estimating the soil moisture budget, either as an empirically derived proportion of precipitation:

$$q = f(p - p_t) \tag{4.3a}$$

or related to the soil moisture deficit as well:

$$q = fp\,(1 - smd/C) \tag{4.3b}$$

where: q = runoff in time period (L), f = empirical factor (< 1), p = precipitation (L), p_t = threshold precipitation below which no runoff occurs (L), smd = soil moisture deficit (L), C = empirical constant.

The factors f, p_t and C are found by calibration against measured runoff, perhaps from an experimental catchment.

Recharge is seen to occur in some aquifers even when there is a soil moisture deficit. Rushton & Ward (1979) explored a number of ways of allowing such recharge in a soil moisture budgeting model. For a limestone aquifer in Eastern England, they finally chose a constant proportion of precipitation when it exceeded a given threshold. The remaining precipitation entered a conventional Penman-Grindley model and could give rise to additional recharge. Various other empirical estimates of such rapid or bypassing recharge have been used; a method must be chosen and calibrated to suit local conditions.

Numerous authors (e.g. Howard & Lloyd 1979) have pointed out that the time-step used in soil moisture models is critical. Longer time-steps, with the same values of the parameters, lead to lower or zero recharge estimates. All recent work recommends a daily time step for humid zones.

There is no universally correct soil moisture budgeting model. For any situation, a model should be chosen based on a conceptual model of the local recharge processes. This model should then be calibrated by one of the following methods, for point estimates:

1. Against a lysimeter;
2. Against soil moisture measured by a neutron tube or tensiometer; and for areal estimates;
3. Against other estimates of recharge, e.g. from groundwater flow modelling;
4. Against a catchment water balance.

In the UK, the Meteorological Office now provides useful data for hydrogeologists under the MORECS (Meteorological Office Rainfall and Evaporation Calculation Service) programme (Shawyer & Wescott 1987). This is primarily intended for farmers, and issues ten day, monthly and annual values of precipitation, actual and potential evapotranspiration, and effective precipitation. They are averages over 10×10 km^2 of the National Grid, and are derived from the Penman-Monteith equation, a version of the Penman equation that includes aerodynamic and surface resistance terms (Monteith 1985). These data are suitable for most groundwater studies, except for special cases, such as wetlands or where topography is very variable. Effort can then focus on hydrogeology instead of hydrology.

4.3.5 *Darcian approaches*

The flow of water in the unsaturated zone is governed by Darcy's Law, with the difference from fully saturated flow that hydraulic conductivity varies with moisture content. Moisture content and hydraulic conductivity also vary with pressure in the unsaturated zone. Borggaard (Chapter 2) explains the theory of these matters, and how observations of tensions (i.e. negative soil moisture pressures) in the unsaturated zone show whether recharge is occurring or not. Knowledge of tensions, moisture content and hydraulic conductivity can be used to estimate the vertical flow of recharge, either by numerical modelling or by interpretation of field measurements.

In transient situations, water is taken up or released from storage by changing the moisture content (degree of saturation). A differential equation to describe flow can be obtained by combining Darcy's Law with a mass conservation equation:

$$\frac{\delta \theta}{\delta t} = \frac{\delta}{\delta z}\left[k_\theta \frac{\delta h}{\delta z} \right] - s \tag{4.4a}$$

$$k_\theta = f\ (k,\theta) \tag{4.4b}$$

where: k = hydraulic conductivity (L/T), k_θ = unsaturated hydraulic conductivity (L/T), θ = moisture content, h = hydraulic head (L), and s = rate of outflow of moisture ($1/T$).

Numerical models are widely used to solve Equation (4.4) (e.g. Jansson & Halldin 1980). This is not straightforward, requiring good data on hydraulic conductivity-moisture content-pressure ($k - \theta - p$) properties and their variation, and good field observations to use for calibration. With the 3-D variability of soil properties, and the presence of often unexpected features such as macropores, models commonly have limited value for water resources and areal estimates of recharge; they can be good for point estimates, developing understanding, and research.

There are three situations where field measurements of pressure and moisture content in the unsaturated zone can potentially be used to estimate recharge:

1. When there is no input to the soil profile at the surface, and the profile is draining to the water table or to evapotranspiration;

2. When there is input into a thick unsaturated zone, but it is insufficient to saturate the soil;

3. When there is sufficient surface input to saturate the profile.

These three situations are briefly discussed below.

No input. When there is no precipitation input to a soil profile, evaporation rapidly lowers the moisture content and pressure near the surface. A zero flux plane develops and can be detected by tension measurements. Below the zero flux plane, water drains downward to become recharge (Fig. 4.13).

The zero flux plane and water table will move over the time interval (t_1-t_2). The area A represents the volume evaporated, the area B represents the volume draining to the water table. Measurements of pressure and soil moisture have been used to estimate recharge in this manner. For example, Wellings (1984) gives examples for chalk in southern England where there is continuous drainage (recharge) throughout

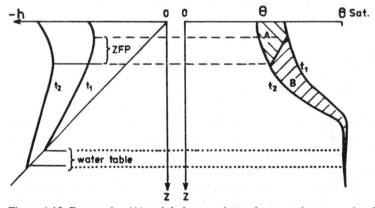

Figure 4.13. Evaporative (A) and drainage volumes between times t_1 and t_2 for a soil profile with no surface input. The division between A and B is at the zero flux plane (ZFP), which is determined from the head (h) versus depth (z) curve. Volumes are calculated from the two moisture content (q) versus depth graphs.

the summer despite the potential evapotranspiration greatly exceeding precipitation in these months. Cooper (1980) gives an example in eastern England where 43% of annual drainage occurred when there was a zero flux plane.

Small input. The inputs of water to a soil profile usually vary in time, and so there are transient changes in the soil moisture and pressure. Steady state conditions will never apply near the surface nor in the root zone. However, a thick unsaturated zone will tend to dampen and coalesce seasonal (or storm) pulses of recharge and so may reach an approximate steady state at depth. In this condition the unsaturated form of Darcy's Law, Equation (4.4), can be applied.

It is generally assumed that pressure is constant with depth under these steady state conditions, so that the hydraulic gradient is 1 and entirely due to the gravitational component of head. In this case, the recharge rate equals the unsaturated hydraulic conductivity at any depth. This implies that layers which have different saturated conductivities will be at different saturations to make their unsaturated conductivities equal.

Sammis et al. (1982) described briefly how the method was applied to a site in Arizona, where the water table was 42 m below the ground. Push tube samples were taken every 3 m during drilling of a borehole. Pressure was found by inserting a tensiometer into the samples; moisture contents were measured by the gravimetric method. They gave few details of their method for unsaturated conductivity. Their results are not impressive with k_θ estimates of 12, 790, 0.005, 226, 0.2 and 23 mm/y for various depths; under the method's assumptions, all these values should be the same.

Large input. If the infiltration rate is high enough, the ground will become saturated. Water will move downwards under gravitational forces only, that is the hydraulic gradient $dh/dz = 1$, provided that ponding on the surface is only slight so that pressure does not rise significantly above atmospheric. In this restricted circumstance, the recharge rate equals the saturated hydraulic conductivity. Although appealingly simple, this method has limited usefulness except where continuous applications of water are made, such as in irrigation schemes or river beds. Recharge will be controlled by the lowest permeability layer, which may not be at the surface.

4.3.6 *Tracer techniques*

Tracers (both isotopic and chemical) have been used widely in estimating recharge. They can be grouped into:
   – Environmental tracers, i.e. those that are already present in the geosphere;
   – Tracers applied by the researcher.

The use of tracers in catchment studies is reviewed in Chapter 7 of this book.

Environmental tracers can be assumed to be applied evenly over the whole land surface. Their movement in the unsaturated zone therefore is only in one dimension, with no lateral diffusion or dispersion. Applied tracers on the other hand are applied at a point or over a small area and so can disperse laterally as well as move vertically. This means that checks on the mass balance of an applied tracer are much more difficult to achieve.

There are several ways that tracers can be used:

1. Signature methods, in which particular parcels of water are labelled and traced, with either environmental or applied tracers;

2. Throughput methods, when fluxes of tracer and water are calculated in the unsaturated zone, usually for environmental tracers;

3. Turnover or transit time calculations, which are used for whole aquifers, usually with environmental tracers.

The signature and throughput methods generally assume a single flow system and piston displacement in the unsaturated zone. If there are two flow systems, recharge can bypass the matrix of the unsaturated zone, for example by flowing in fractures, down rootways or other macropores. Some workers have interpreted their field data in terms of two flow systems; an example is discussed below. Methods for the unsaturated zone are used where the latter is relatively deep and contains several years recharge. In Scandinavian conditions, with shallow watertables and turnover times of only a few years, these methods may be less useful.

Mass balances of tracer are very useful because, as discussed above, they can reveal if some of the tracer (and recharge) have moved by a different pathway; for example along fissures or root channels. Repetitive sampling of the same profile is also useful in this context. For example, Smith et al. (1970) showed by repetitive profiling in the English Chalk that some tritium was lost from the unsaturated zone, and estimated that 15% of recharge bypassed the matrix where piston flow was occurring.

4.3.6.1 *Signature methods*
General principles. Infiltrating water may be labelled (in time) by a particularly noticeable input of tracer. This may either be applied artificially, or be an environmental tracer, for example from an atmospheric nuclear explosion. This labelled water may be identified at depth in the soil profile at future dates, thus providing information on water movement in the unsaturated zone.

Once the water at a particular depth has been dated, two methods are available to calculate average recharge. At a time of maximum soil moisture deficit, the total water content above the dated depth is the total recharge over the intervening years, i.e.:

$$r = 1/m \sum_{i=1}^{m} \theta_i \, l_i \tag{4.5}$$

Notation below Equation (4.8).

The second method is a mass balance of tracer, and is only applied to environmental tracers, or those applied over a large enough area to prevent lateral dispersion. The amount above the dated layer is measured:

$$T_p = \sum_{i=1}^{m} T_i \, \theta_i \, l_i \tag{4.6}$$

and equated with the input in recharge after allowing for radioactive decay:

$$T_p = r \sum_{j=1}^{n} T_j \, \exp (d_j) \tag{4.7}$$

Combining Equations (4.6) and (4.7) gives average recharge as:

$$r = T_p \; / \left[\sum_{j=1}^{n} T_j \, \exp\left(d_j\right) \right] \tag{4.8}$$

where n = number of years between signature and sampling (T), r = mean annual recharge (L/T), d = depth of dated water (L), θ_i = moisture content in sampling interval i, l_i = length of sampling interval number i (L), T_i = tracer concentration in sampling interval i (M/L^3), T_p = tracer mass in profile (M), T_j = tracer concentration in rainfall in year j (M/L^3), exp (d_j) = decay of tracer input since year j, and m = number of sampling intervals.

For a conservative tracer, average recharge is estimated from Equation (4.5). A balance of tracer, Equation (4.8) can then be used to calculate the fraction of precipitation that becomes recharge.

Environmental tracers. The most commonly used environmental tracer for this method has been tritium which, as well as being produced naturally in the atmosphere, was introduced into the atmosphere in large amounts in 1952 by nuclear tests (see also Section 7.2.3). There was a significant peak, particularly in the northern hemisphere, in 1963-1964 (Fig. 4.14). Atmospheric concentrations were always much less in the southern hemisphere and, as atmospheric testing stopped in 1963, tritium levels are now falling everywhere. More sensitive analytical techniques have

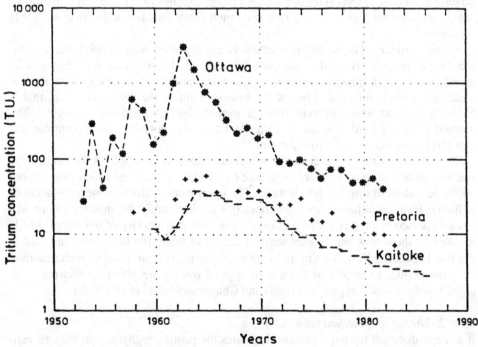

Figure 4.14. Tritium concentrations in rainfall in Ottawa (Canada), Pretoria (Republic of South Africa), and Kaitoke (New Zealand) (data courtesy of IAEA).

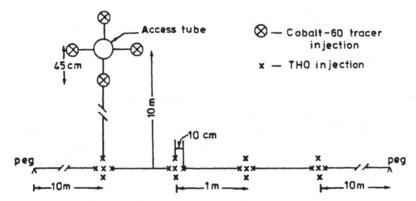

Figure 4.15. Layout of injection points for tritium and cobalt-60 (from Chandrasekharan et al. 1988, reprinted by permission of Kluwer Academic Publishers).

extended the usefulness of environmental tritium for some time, but it is now rarely of use for signature methods.

Oxygen-18 (^{18}O) and deuterium (^{2}H, or D) are fractionated during precipitation and evaporation processes, a property that has so far prevented their use in any tracer balance method of estimating recharge (see also Sections 7.2.1 and 7.2.2). They have shown useful signatures in temperate, high recharge areas. For example Thoma et al. (1979) found seasonal, markers of deuterium in sand dunes; Bath et al. (1982) and Saxena & Dressie (1984) sometimes found cyclic profiles of ^{18}O and ^{2}H corresponding to seasonal rainfall and recharge; snowmelt often contains a distinctive $^{18}O/^{2}H$ signature.

Applied tracers. The signature method is mainly used with applied tracers. Radioisotopes are usually used as they can be detected at low levels, thus they can be introduced in small quantities without large disturbances to soil or its moisture content. The tracer is introduced below the lowest depth of the zero flux plane, that is below the region where upward flow or evapotranspiration losses can occur. The method is widely used in India, and Figure 4.15 shows a typical site layout for tritium (^{3}H) and cobalt-60 (^{60}Co) injection.

The tracer position is determined at a future date, for example at the end of the wet season or one or more years later. The tracer may be detected by drilling a cored borehole and analysing the porewaters (^{3}H), or by non-destructive measurements of radiation from an adjacent open borehole (^{60}Co). Inevitably the injected tracer will have dispersed (Fig. 4.16). Conventionally the centre of gravity of the tracer profile .is taken to show how much displacement there has been; the moisture contained in the soil between injection point and centre of gravity is equivalent to recharge over the time period. Examples of the use of applied tracers are given by Sharma et al. (1985), Athavale & Rangarajan (1988) and Chandrasekharan et al. (1988).

4.3.6.2 *Throughput method (chloride)*

If a tracer does not have a signature that dates the profile, recharge can only be estimated if the concentration of the tracer is affected by the evaporation processes that reduce the precipitation to recharge, and allow the precipitation-to-recharge ratio to be determined from concentration measurements. Conservative, i.e. non-evaporated,

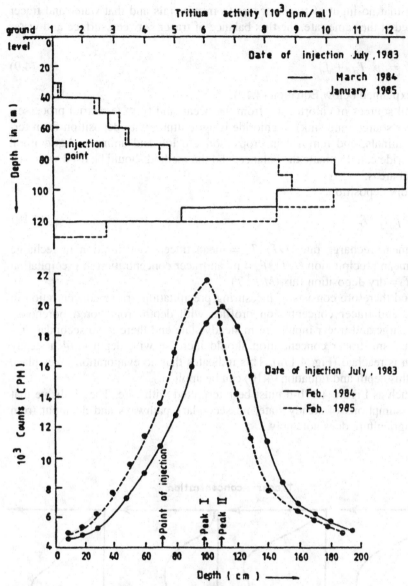

Figure 4.16. Tracer profiles for tritium and cobalt-60 at a site in Jodhpur (from Chandrasekharan et al. 1988, reprinted by permission of Kluwer Academic Publishers).

tracers are of course concentrated in recharge as evaporation proceeds, so that under steady state conditions, the flux of tracer input at the surface equals the flux of tracer reaching the water table, and the flow of tracer discharging from the aquifer. As the age of the tracer reaching the water table is unknown, it is necessary to assume a steady input of tracer at the surface. Only environmental tracers have been used, as a long enough input is needed to reach steady state; chloride is the most commonly used ion.

Assuming that no input of tracer occurs from minerals and that water and tracer are transported at the same rate, the flux balance of tracer between surface and water table is:

$$r\,T_r = p\,T_p + f_d \tag{4.9}$$

Notation is explained below Equation (4.10).

The natural sources of chloride are from the ocean and from terrestrial processes. Another major source (and sink) of chloride is agriculture, with deposition from fertilisers and animals, and removal in crops, soil erosion and animals. Agricultural fluxes of chloride can dominate the balance, and the method should be used with care in agricultural areas.

Omitting dry deposition:

$$r = p\,T_p/T_r \tag{4.10}$$

where: r = mean recharge rate (L/T), T_r = mean tracer concentration in recharge (M/L^3), p = mean precipitation (L/T), T_p = mean tracer concentration in precipitation (M/L^3), and f_d = dry deposition flux ($M/L^2\,T$).

The method therefore consists of measuring precipitation, tracer concentration in precipitation, and tracer concentration profiles with depth from cored boreholes. Provided recharge and tracer inputs are in steady state and there is no secondary recharge mechanism, tracer concentration should increase with depth until a steady concentration is reached (Fig. 4.17a). This indicates that no evaporation takes place from below this depth and Equation (4.10) can be applied.

Profiles such as Figure 4.17b-d must be interpreted with care. They indicate that one of the assumptions of steady state, no secondary pathways and no input from minerals or agriculture does not apply.

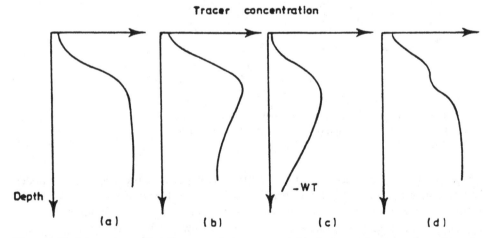

Figure 4.17. Schematic profiles of a conservative tracer. a) Ideal case, recharge can be estimated; b) Probable change in recharge history; c) Secondary pathways for recharge; d) Addition of tracer from soil.

4.3.7 *Variability of recharge across catchments*

Recharge varies across catchments because the controlling factors vary, both in their nature and size. These factors include:
1. Precipitation and other water supplies;
2. Geology and soil;
3. Topography and landform; and
4. Groundwater condition.

Because recharge is a non-linear process, it is not possible to use average values of each controlling factor to derive an average recharge. Recharge should be estimated separately for each homogeneous zone; the spatially varying values are of course essential for groundwater modelling studies.

In general, the more detailed the subdivision into zones, the more accurate will be the recharge values, but the more expensive and time consuming will be their estimation. The amount of data available (maps, climate stations, Landsat imagery, etc.) will partly determine the detail that can sensibly be achieved. Sensitivity analyses will help decide how important variations in the controlling factors are.

Table 4.2 lists a set of factors that should be used to zone a catchment for recharge calculations. Initially, each factor should be mapped on a separate overlay, then the overlays combined to produce a master map of homogeneous zones. Recharge is then estimated for each zone by one of the methods discussed above.

Some authors have observed important changes in the factors controlling recharge over very short distances (< 1 to 100 m) in apparently homogeneous terrain (e.g. Nielsen et al. 1973; Berndtsson & Larson 1987). Others have reported significantly varying recharge over similarly small distances (e.g. Kitching et al. 1977; Sharma & Hughes 1985). The importance of these variations has not yet been established, and

Table 4.2. Factors for classifying recharge zones.

Factor	Example values or classification scheme
Precipitation type	Monsoon/thunderstorm/winter/summer etc.
Precipitation amount	20% increase in precipitation between zones
Irrigation type	Sprinkler/furrow/paddy/flood/canal etc.
Irrigation amount	As precipitation
Evaporation potential	20% increase in Penman potential evapotranspiration between zones. Solar or net radiation can be used in arid and semi-arid areas
General lithology	Alluvium/chalk/limestone/sand etc.
Soil classification	Important factors are infiltration capacity, moisture storage and depth. Natural vegetation or drainage density may be adequate substitutes if no soil data available
Land cover	Grass/arable (crop type)/natural forest/plantation/phreatophyte/urban etc.
Landform	Floodplain/rolling hills/plateau/etc.
Depth to groundwater or capillary fringe	> root depth/ within reach of plants/ within soil zone/direct evaporation possible

Notes: This classification is not intended to be exhaustive. On the other hand, it will often be unnecessary to use all classifying factors. Data sources: Topographic maps, geological and soils maps, Landsat and other satellite imagery, aerial photography, climate maps and meteorological stations, precipitation records, irrigation scheme maps and records, field visits!

there is certainly no practical way to take account of them for engineering studies. Three to ten repeat samples within each zone (as defined by the factors in Table 4.2) would certainly be worthwhile if they can be afforded, but first priority should go to obtaining one value from each zone.

4.3.8 *Localised recharge*

The introduction to this chapter pointed out that there is a category of recharge intermediate between direct (at the spot where the precipitation falls) and indirect (along main river channels). This intermediate category is called localised, implying some horizontal movement of water before recharging groundwater. This movement is on a scale too detailed to map for engineering studies, and therefore causes great difficulty in estimating recharge. Unfortunately this type of recharge is often the largest in arid and semi-arid areas, and can be important in humid zones such as the UK (drift cover) and Scandinavia (bedrock outcrops).

In weathered, bare hardrock or limestone terrain, recharge is into distinct fissures. There are few satisfactory ways of estimating the localised recharge component in such terrain. Tracers, Darcy's law below the water table, and water balances of project or representative basins offer the best possibilities.

Topographical depressions can range in size from centimeters to enclosed catchments of several square kilometers. Studies in a wide range of terrains have shown that recharge is focused in such depressions, where they exist (e.g. Freeze & Banner 1970; Rehm et al. 1982). This is true even for highly permeable materials. The most common approach to estimating recharge in these cases in practice is a mixture of water balance and empirical formula. Rainfall-runoff relationships are estimated or transposed from other areas, and the proportions of runoff going to each destination (evaporation, soil moisture storage, irrigation, etc.) is measured, estimated or guessed; the residual is recharge.

Many alluvial aquifers can be described as mountain front systems, including alluvial fans, piedmont plains and subsidence basins. The major sources of recharge to these systems are often along the mountain boundary, consisting of two components, subsurface inflow from the mountain mass to the basin infill, and infiltration of runoff. These are illustrated in Figure 4.18.

The two components of mountain front recharge are difficult to separate in practice. The runoff component can be estimated by water balance methods, as discussed above, but the subsurface flow is difficult to isolate. One solution is to estimate the total by a Darcy throughflow calculation within the main aquifer. The section over which flow is estimated should be well within the aquifer, away from the mountain boundary, for two reasons. All of the mountain front recharge will occur upslope of the section as all the ephemeral stream (wadi) flows will have infiltrated. Secondly, it is probable that more data on groundwater gradients, hydraulic conductivities and aquifer cross-section will be available in the main body of the aquifer, so giving greater accuracy and confidence in the results.

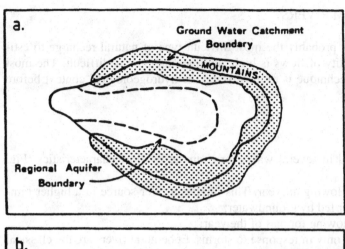

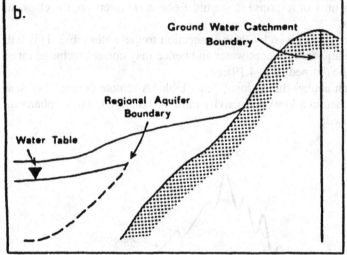

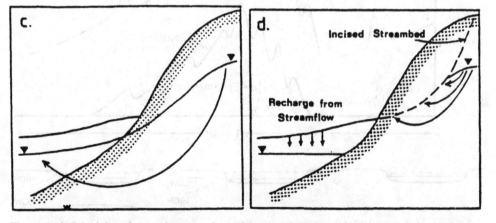

Figure 4.18. Principles of groundwater recharge in mountain front basins. a) Plan view of boundaries of groundwater catchment and regional aquifer; b) Cross-section of boundaries; c) Mountain front recharge by subsurface inflow; d) Mountain front recharge through streamflow infiltration (WRRC 1980).

4.4 RECHARGE FROM RIVERS

Recharge from rivers is probably the most difficult type of natural recharge to estimate. Temporal variability of flows is high, and measurement is difficult. The most
important part of any technique is to understand the hydrogeological context before
making any calculations.

4.4.1 *River types*

Rivers can be classified in several ways, for example by flow characteristics (Fig.
4.19):
   1. Perennial, that is flowing all year. This implies a river source in a higher rainfall catchment, or a river fed by groundwater;
   2. Seasonal, that is flowing for part of the year;
   3. Ephemeral, that is only in response to storms. Ephemeral rivers are the classical
type found in arid areas.
   An additional dimension is given by their connection to the water table. This will
affect the ability of the aquifer to accept water and hence may control recharge rates.
Three broad types can be defined (Fig. 4.19):
   1. Remote from (high above) the regional water table. A remote perennial or seasonal river will be perched in a low conductivity (alluvial) material, while ephemeral
rivers need not be perched;

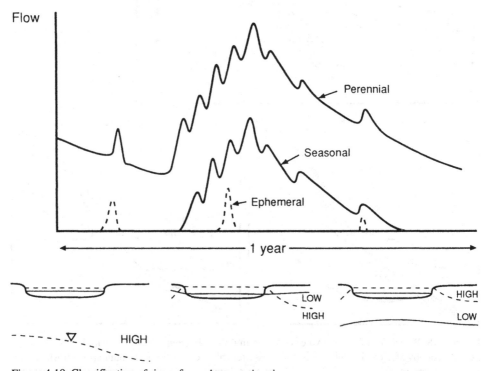

Figure 4.19. Classification of rivers for recharge estimation.

2. Connected to groundwater;

3. Above the water table, but close enough for the water table to rise to the river in response to recharge. For the purposes of this chapter, this intermediate category has been included with category 2 above (connected rivers).

In humid climates, rivers are normally in the discharge zone for groundwater, and are perennial or seasonal. In arid and semi-arid climates, perennial rivers are either discharge zones, or originate in more humid areas upstream. Perennial and seasonal rivers are most likely to contribute to groundwater recharge when they travel across lithological contacts (e.g. from bedrock to alluvium), or when the water table had been artificially lowered. In Scandinavian conditions, rivers are unlikely to be re-charging groundwater except over short lengths and under untypical conditions.

Different methods of recharge estimation apply to perennial/seasonal rivers on the one hand and ephemeral rivers on the other. Remote rivers require differences of approach from those connected to the water table.

4.4.2 *Rivers in contact with the water table*

Recharge from rivers that contact the water table is affected by conditions below that water table. In these cases, the empirical methods outlined above rarely work. Figure 4.20 shows data for a perennial river crossing a major alluvial aquifer in North Africa. There is no clear relation between transmission loss and river flow because of the effect of the groundwater system. The data can be interpreted to show an upper envelope to transmission loss as 100% of flow, with many months having lower losses. The high loss months are in late summer when groundwater levels are low, which suggests that the state of groundwater is the second control on recharge. No general relationship can be devised for these rivers and recharge for each time period must be explicitly calculated from measurements, or assessed through modelling.

4.4.3 *River recharge estimation methods*

Direct measurement of recharge from river is not possible. A number of empirical methods have been devised and are discussed by Lerner et al. (1990); they are not recommended. Some combined empirical/budgeting methods based on groundwater response are outlined below. Water balance methods are common. Darcian approaches, that is based on the groundwater flow equation, are possible in theory but are difficult to use in practice without important simplifications. Tracer techniques have very limited use for quantifying river recharge. Models of river hydraulics and of the complete hydrological cycle in catchments have been used for river recharge estimation.

4.4.4 *Groundwater response under ephemeral rivers*

There are a number of empirical techniques based upon analysing groundwater response. They are typified by the convolution approach, and include a spectral analysis method (Gelhar et al. 1979) and a linear transformation of the annual distribution of rainfall (Flug et al. 1980). The latter two require more simplifying assumptions,

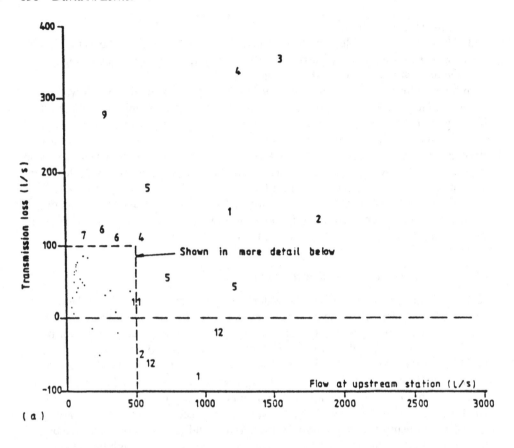

(a)

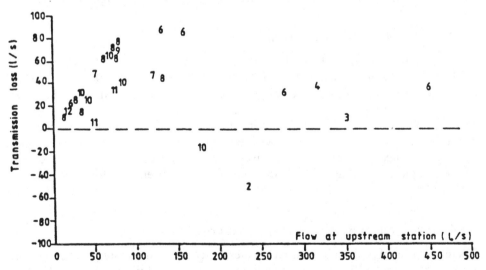

(b)

Figure 4.20. Transmission loss – vs – flow for a river in contact with the water table showing little correlation. Each point is one month, identified by its month number (e.g. 1 = January). a) Complete flow range; b) Low flows.

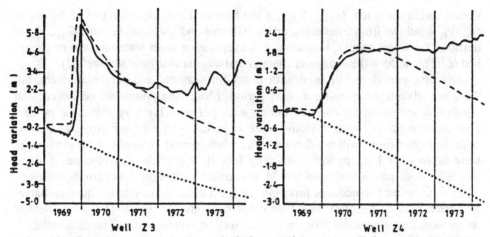

Figure 4.21. Groundwater responses to a single large recharge event, Kairowan, Tunisia. Solid lines: Observed heads. Dotted lines: Heads computed by groundwater model assuming no recharge. Dashed lines: Heads computed with estimated recharge (Besbes et al. 1978, copyright by the American Geophysical Union.)

but are more mathematically complex than the convolution method described by Besbes et al. (1978) and others.

All of the groundwater response methods require a good independent estimate of recharge for at least one event. Without this, the parameters of the method cannot be found. Therefore any results for other events will be less accurate than the one independent estimate.

Figure 4.21 shows some examples from Besbes et al. (1978) who give a thorough comparative study of several methods in a semi-arid area of Tunisia. They observed the initial and peak response and constructed the remainder using a model. The volume of the recharge event was initially estimated by a water table rise calculation using a guessed specific yield, then refined by calibrating the model. Other authors have used transmission losses to estimate recharge volume.

4.4.5 *Water balances*

4.4.5.1 *Channel water balance*
A water balance of the flows along a river reach is the most straightforward way to estimate transmission losses. By careful consideration of processes, this may be extended to deep percolation or recharge. Using river flow rates:

$$R = Q_{up} - Q_{down} + \sum Q_{in} - \sum Q_{out} - E_a - \frac{\delta S}{\delta t} \qquad (4.11)$$

or in volume terms,

$$R\,\delta t = V_{up} - V_{down} + \sum V_{in} - \sum V_{out} - E_a\,\delta t - \delta S \qquad (4.12)$$

where Q = flow rate (L^3/T), V = flow volume (L^3), Q_{up}, V_{up} = the flow at the upstream end of the reach, Q_{down}, V_{down} = the flow at the downstream end of the reach, Q_{in}, V_{in} = inflows from tributaries, urban effluents and irrigation returns, Q_{out}, V_{out} = outflows for water supply, irrigation, E = evaporation from water surface or stream bed (L^3/T), and δS = the change in channel and unsaturated zone storage (L^3).

Any time period can be used with flows for seasonal and perennial rivers, but there are advantages in using a short period. More data points are generated, and conditions are relatively constant during each period. This may make the process controlling recharge more obvious. More importantly, the higher errors associated with the higher flows will not dominate the whole period as would occur if averages were taken over long periods, including low flow periods. Travel time of flood waves between gauging stations should be allowed for if short time periods are used.

Channel water balances are probably the most accurate way to estimate recharge, and often provide the data on which other methods are based. They are essential when modelling rivers that may receive as well as donate water. The data collected may have many other uses, including rainfall-runoff modelling for transposition to ungauged catchments.

The method is prone to inaccuracy as recharge is calculated as the difference between large numbers. Measurement errors on high river flows are commonly ± 25%, and leading to higher errors in recharge estimates.

4.4.5.2 *Water table rise*

An alternative to a surface water balance for transmission loss is a groundwater balance for recharge from remote rivers. Observation wells perpendicular to the axis of the river will show the profile of a recharge mound. Under ideal conditions the volume of water in the mound is the amount of recharge. Several points should be noted:

1. The groundwater flowing away from the mound may be significant in relation to the change in storage;

2. Storage in the unsaturated zone is common below ephemeral rivers on alluvial aquifers. Wilson & De Cook (1968) found that 33% of recharge showed up immediately as a water table rise; the remainder took several months to arrive;

3. Specific yield is a critical parameter, but difficult to estimate.

4.4.6 *Darcian approaches*

There are three ways to use Darcy's law in estimating recharge from rivers as follows:

1. Using infiltration equations and flow nets, that is desk studies of homogeneous and isotropic aquifers with assumed boundary conditions;

2. Gathering field data on aquifer properties, moisture content and pressure in the unsaturated zone, and piezometric heads in the saturated zone, and using these data in the equations of flow to analyse real events;

3. Using numerical models.

The infiltration approach is for remote rivers, and assumes that no preferential pathways and no lateral flow occur. Field data on moisture content and pressure in the unsaturated zone beneath river beds have been collected for research projects on recharge estimation. I know of no instances where such data have been collected for resources studies, presumably because of the expense and difficulty of collecting enough data to describe a transient process with three dimensional variability. Head and hydraulic conductivity data from connected rivers could be used to estimate flow away from the river; velocity measurements using tracers could also be used. The difficulty in using them for rivers is the transient, three dimensional nature of the flow pattern.

Numerical models may be surface or groundwater based (with the other water type being taken as a boundary condition) or may be integrated stream-aquifer models. In all cases, substantial quantities of field data are needed to calibrate and validate the models.

4.4.7 *Tracer techniques for groundwater recharge from river*

Environmental tracers in groundwater seem to have little role in quantifying river recharge. Almost all of the groundwater in the vicinity of a recharging river will have come from that river. Therefore, the local groundwater will carry the same environmental tracers as the recharging water, and the two cannot be separated.

Of course isotopes and other tracers are very valuable in identifying recharge sources and in delineating zones of river-recharged groundwater. Some discussion of methods is given in the Guidebook on Nuclear Techniques in Hydrology (IAEA 1983). See also Chapter 7.

4.5 INTERAQUIFER FLOWS

Flows to and from other aquifers can be important parts of the water balance of an aquifer, and must be taken into account when modelling water and solute fluxes. Usually, seepage is from an underlying aquifer with low flows and reasonably steady conditions. Although seepage may be small, when expressed per unit area, large areas are often involved.

Using Darcy's Law can give good estimates and is an approach easily linked to modelling. It is usually possible to measure hydraulic heads in both aquifers; the major uncertainty is knowing the hydraulic conductivity of the intervening material, and whether there are any preferential flow zones, such as fractures or windows in the aquitard.

Water balances of both the donating and receiving aquifers should always be used to set limits on the likely interaquifer flow. More accurate results will be obtained when the interaquifer flow is a large proportion of the water balance. The direction of flow can be determined by head measurements, hydrochemical or isotopic data, or from preliminary numerical modelling of the system.

Tracer techniques, particularly isotopes, are useful in detecting flow between aquifers. The main condition is, of course, that the isotopic composition of each aquifer's water is significantly different, say by several parts per mill in the case of oxygen-18. However, tracer methods rarely give firm estimates of the amount of interaquifer flow, because of the unknown mixing of the two waters in any sample. There may be a role for transit time calculations in steady state groundwater systems.

4.6 NET RECHARGE OVER A REGION

4.6.1 *Introduction*

The previous chapters have dealt with the various components of recharge to an aquifer, based on the origin of the water (precipitation, irrigation, etc.) and its route to groundwater. Alternative approaches are to estimate the total recharge. Such approaches seek to avoid describing any of the complex processes controlling recharge and groundwater movement, such as evaporative losses from the unsaturated zone, rivers with both influent and effluent reaches, and phreatophyte evapotranspiration. There are four types of methods described in this section:

1. Storage change or water table rise, i.e.

$$\text{total recharge} = \text{change in saturated volume} \times \text{specific} \atop \text{yield} + \text{outflows} \tag{4.13}$$

2. Discharge from the aquifer, that is the sum of the measured discharge of groundwater across the boundary of a study area plus internal discharges by wells, etc. is equal to the net recharge inside the study area.

3. Inverse techniques, that is inverting the groundwater flow equation to solve for recharge instead of groundwater heads.

4. Tracer techniques, either estimating residence time of water in an aquifer, or modelling the distribution of several tracers throughout an aquifer.

All the methods have advantages and disadvantages as discussed below. Numbers 1, 2 and 4 are useful as checks on a water balance for the aquifer calculated by considering all the components of recharge individually by the methods outlined above. The methods are only accurate for groundwater systems which are in hydraulic equilibrium.

4.6.2 *Water table rise*

The total recharge to an aquifer is often estimated from the volume of water stored as the water table rises during the wet season. The method clearly only applies to aquifers with a well defined recharge season. Allowances are made for pumping wells and other discharges as required, and recharge is estimated as:

$$r = (\delta s + \sum Q_A \, \delta t + V_D) \, / \, A \, \delta t \tag{4.14}$$

where: r = total recharge per unit area (L), δs = volume of water stored between lowest and highest water table positions (L^3), Q_A = abstraction rate from wells pumping during the water table rise (L^3/T), δt = the time interval between high and low water table positions (T), A = area of the aquifer (L^2), and V_D = the volume of water discharged to springs, river bed seeps, etc. (L^3).

The method is appealing because of its simplicity and because it appears to be a direct measurement of total recharge. There are plenty of reasons why it may not give a good answer. These need to be identified, because the method is in such widespread use. Minor problems include:

1. Seasonal abstractions inducing recharge from perennial water courses at some times of year (A in Fig. 4.22);

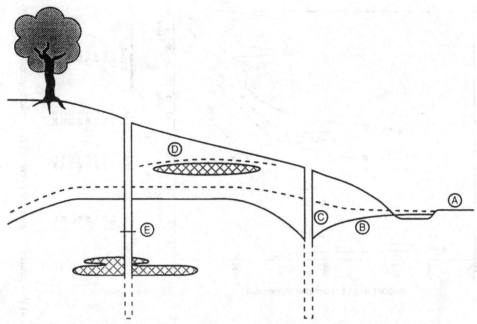

Figure 4.22. Problems that may arise when estimating recharge by water table rise – see text for explanation.

2. The common association of observation wells with abstraction wells, overestimating the degree of regional drawdown (B);

3. Delayed recharge, perhaps due to perched water tables (C). Recent regional studies of recharge in the Chalk aquifer in the Thames area of the UK suggest that up to 50% of recharge is still in transit in the unsaturated zone when the water table peaks (J. Barker, British Geological Survey, pers. comm.). This water drains to the saturated zone over the summer at a slower rate than groundwater discharges to rivers.

4. Observation wells in deeper parts of the system showing confined responses but being interpreted as phreatic (D).

4.6.3 *Hydrograph analysis for groundwater discharge*

One method to estimate total recharge over a whole catchment relies on analysis of river hydrographs. The concept can be expressed as:

$$\text{recharge} = \text{baseflow} + \text{withdrawals and discharges} + \text{rate of storage, depletion} \tag{4.15}$$

where all terms are averaged over a substantial time. Baseflow is taken to mean groundwater discharges to springs and rivers, and is estimated by hydrograph separation, as discussed below. Withdrawals by wells, discharges to lakes, seas and sabkahs, and evapotranspiration from the water table must be estimated and allowed for. Significant storage depletion will only occur in over-exploited aquifers or during

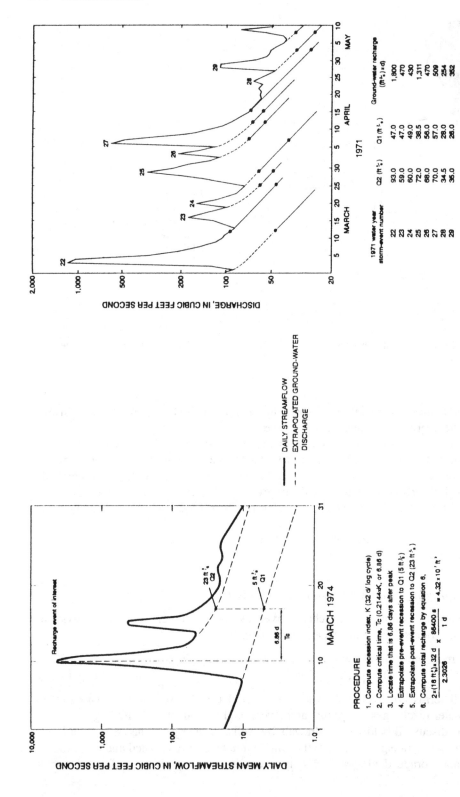

Figure 4.23. Recession curve-displacement method for estimating groundwater discharge. a) Procedures for a single event; b) Example application to Big Alamance Creek, North Carolina (Rutledge & Daniel 1994).

long droughts, but is more likely to be significant in arid lands.

Runoff processes, and stream-aquifer interactions, are not simple processes. A lengthy discussion is possible on the definition of baseflow and its relation to groundwater discharge – but it will not be conducted here. The following assumes that baseflow separation is a useful exercise, and discusses how it might be done.

It is straightforward to identify baseflow during the recession season, when soil moisture deficits are high enough to prevent further recharge, surface runoff is negligible, and all river flow is groundwater derived. The difficult part is during the winter recharge season, when surface runoff must be separated, and multiple and overlapping precipitation events follow each other. There are three ways to approach the problem:

1. An empirical approach, sketching in a baseflow separation subjectively or according to some simple rules. The Institute of Hydrology (1980) developed an automatic procedure based on five day minima which has the advantage of being routine and consistent, and the disadvantage of having no underlying conceptual model;

2. Developing a groundwater rating curve, that is a graphical relationship between a representative groundwater level and baseflow. The rating curve is derived for recession periods when all flow is groundwater, and then used to estimate the groundwater component at other times of the year;

3. A recession curve-displacement method developed by Rutledge & Daniel (1994) from earlier analytical solutions by Roraborough (1964). In essence, the upward displacement of the recession by a recharge event is calculated and used to estimate the increase in groundwater volume (Fig. 4.23). The method requires extrapolation of the recession and assumes, in common with most approaches, that delays in the unsaturated zone are unimportant.

4.6.4 *Inverse techniques*

The flow of groundwater can be represented by an equation that includes: 1) Groundwater heads; 2) Aquifer properties; 3) Recharges and boundary fluxes. Knowing the boundary conditions and two of these sets of data allows solution of the equation for the third. Conventional groundwater models solve for groundwater heads, inverse models solve for aquifer properties or, more unusually, recharge.

Inverse techniques to find aquifer properties were a popular research field in the late 1970's. Three problems occur with most methods; great sensitivity to errors in water level and recharge data, low sensitivity in unstressed areas of an aquifer, and non-uniqueness of solutions. These problems have meant that inverse techniques have had little practical application.

Few authors have attempted inverse techniques to estimate recharge. In general, they can be expected to suffer from the same problems identified for aquifer property solutions above. Inverse models are probably best left to experienced modellers; anyone insisting on trying should use the MODFLOWP package (Hill 1992).

4.6.5 *Aquifer-wide tracers*

There are a number of ways to estimate net recharge from the concentrations of environmental tracers throughout an aquifer. Although the methods are interrelated, it is

useful to subdivide them into groups, as follows:
1. Analytical transit time models;
2. Distributed numerical models of single or multiple tracers.

Comprehensive reviews of mathematical transit time models are given by Maloszewski & Zuber (1982) and Zuber (1986). The models relate the mean age of water leaving a groundwater system to the mean transit time of a tracer, via alternative mathematical models of the way the tracer moves through the system.

In order to use data resulting from a continuous, and possibly varying, input concentration, we need to derive the weighting function or impulse response of the system and use the convolution integral. The impulse response theoretically could be obtained by a tracer experiment, but it is not normally possible to apply an impulse of tracer and wait long enough to observe the complete output response. Therefore a number of conceptual models have been developed, and their impulse responses determined theoretically. Details can be obtained in the references cited above.

The method consists of the following:
1. Measuring input and output concentrations of tracer over a sufficiently long period, preferably at least as long as the expected turnover time;
2. Choosing an appropriate groundwater flow and tracer sampling model, and so the form of impulse response;
3. Solving the convolution equation for the numerical values of the parameters of the impulse response, usually by trial and error comparison of measured and predicted output values;
4. Estimating the volume of mobile groundwater (V) and hence the recharge rate (R) on the basis that mean age = V/R.

The limitations of the mathematical transit time models can sometimes be overcome by using distributed, numerical models of tracer movement. The aquifer is divided into cells and the movement of tracer and water between inputs, cells and outputs is modelled using mass conservation and assuming each cell is fully mixed. The various fluxes in the model are calibrated until the model predicts the same distribution of tracer as observed in the aquifer. All models that I know of use a mixing cell approach instead of a numerical solution to the solute transport equation. Examples include Campana & Simpson (1984) and Campana & Mahin (1985), who looked at a single tracer, and Adar & Neuman (1988) and Adar et al. (1988) who use a quadratic programming algorithm to find a weighted least squares solution for recharge to a semi-arid alluvial aquifer with multiple tracers.

4.7 CONCLUDING REMARKS

Recharge is the water that enters the saturated groundwater reservoir. It is related to, but not necessarily the same as, infiltration and deep percolation. Precipitation (rain, snow, etc.) is the principal direct source of recharge, but there are many exceptions, that is groundwater systems that receive significant amounts of recharge from other sources. These may be confined (with significant subsurface inflows), urbanised or irrigated, river-fed alluvial systems, or in arid climates. Man's influence on the hydrological cycle is often profound, with changes in vegetation, surface and groundwater regimes, and other effects leading to changes in recharge.

Recharge is one of the two water links between surface and groundwater (the other being discharge). It would be convenient if recharge were a one way link, that is with no feedback from groundwater. Unfortunately this is frequently not the case, and changes in the groundwater system, such as the water table elevation, can alter the amount of recharge. Feedback is especially common in shallow systems, such as those common in Scandinavia, where groundwater does not conform to the classical model of an exploitable resource. Rather it is just a slightly slower responding, sub-surface part of the runoff process. Hence it is difficult, and probably unwise, to try to separate the hydrological process into independent surface and groundwater components.

A complicating factor in recharge studies is the inherent spatial variability in recharge, at all scales from the influence of single plants up to basin wide. Direct recharge is the residual of precipitation, runoff and evapotranspiration, and the variability of these components of the water balance will accumulate in the residual, i.e. recharge. Another source of variability is less well known, probably as important, and much more difficult to estimate. Recharge can be focused into preferred zones by several mechanisms. The simplest example is recharge from a river. More relevant causes for Scandinavia include fissures in hard rocks, outcrops of hard rock within superficial aquifers, and topographical depressions in ground surface and sub-surface layers. Variability, and focusing, are particularly important for geochemical studies, where kinetics of reactions may mean that knowledge of the distribution of recharge water velocities are needed to complete chemical mass balances.

The bulk of this chapter has been about methods of estimating recharge, and some concluding remarks would be appropriate. Direct measurement is really only feasible for precipitation recharge, using lysimeters. Most other methods, particularly in engineering studies, turn out to be water balances, estimating recharge as the residual of more easily estimated fluxes. Balances have the advantage that they are balances, that is they account for all the water and there is little risk of a gross error. They have the severe disadvantage of accumulating all the errors of measurement into the most important item, the recharge estimate, and so can generate wide error bands. Tracers have a long and respectable history of use for recharge estimation, particularly in research studies and the more arid climates. They often have the advantage of revealing processes, for example picking up the existence of preferential flow paths. They don't always identify a mass and water balance. Measurements of groundwater heads (and tensions in the unsaturated zone) can be very useful process studies, identifying sources and timings of recharge. They have less value for numerical estimation. In general, more than one method should be used for any project, to provide checks.

The editors have asked for some comments on future research needs. A wish list of techniques would include new tracers that are cheap, safe and conservative, one to be unaffected by evaporation processes, and another to mimic water. A water flux meter to be buried in soil or under rivers would be useful. For broader research on processes, two topics worth addressing in a variety of hydrogeological environments are:

1. Scaling up from point measurements of recharge to areal estimates, and
2. The importance of topography in focusing recharge at a range of scales.

REFERENCES

Adar, E.M. & Neuman, S.P. 1988. Estimation of spatial recharge distribution using environmental isotopes and hydrochemical data. II. Application to Aravaipa Valley in southern Arizona, USA. *J. Hydrol.* 97: 279-302.

Adar, E.M., Neuman, S.P. & Woolhiser, D.A. 1988. Estimation of spatial recharge distribution using environmental isotopes and hydrochemical data. I. Mathematical model and application to synthetic data. *J. Hydrol.* 97: 251-277.

Athavale, R.N. & Rangarajan, R. 1988. Natural recharge measurements in the hard-rock regions of semi-arid india using tritium injection – a review. In: I. Simmers (ed.), Estimation of natural groundwater recharge. NATO ASI Series C, Vol. 222 (*Proc. of the NATO Advanced Research Workshop, Antalya, Turkey*, March 1987). D. Reidel Publ. Co., Dordrecht: 175-194.

Bath, A.H., Darling, W.G. & Brunsden, A.P. 1982. The stable isotopic composition of infiltration moisture in the unsaturated zone of the English Chalk. *Stable Isotopes* (H.L. Schmidt et al. (eds), Elsevier, Amsterdam: 161-166.

Berndtsson, R. & Larson, M. 1987. Spatial variability of infiltration in a semi-arid environment. *J. Hydrol.* 90: 117-133.

Besbes, M., Delhomme, J.P. & De Marsily, G. 1978. Estimating recharge from ephemeral streams in arid regions: a case study of Kairouan, Tunisia. *Water Resour. Res.* 14: 281-290.

Calder, I.R., Harding, R.H. & Rosier, P.T.W. 1983. An objective assessment of soil-moisture deficit models. *J. Hydrol.* 60: 329-355.

Campana, M.E. & Mahin, D.A. 1985. Model-derived estimates of groundwater mean ages, recharge rates, effective porosities and storage in a limestone aquifer. *J. Hydrol.* 76: 247-264.

Campana, M.E. & Simpson, E.S. 1984. Groundwater residence times and recharge rates using a discrete-state compartment model and 14C data. *J. Hydrol.* 72: 171-185.

Chandrasekharan, H., Nevada, S.V., Jain, S.K., Rao, S.M. & Singh, Y.P. 1988. Studies on natural recharge to groundwater by isotope techniques in arid Western Rajasthan, India. In: I. Simmers (ed.), Estimation of natural groundwater recharge. NATO ASI Series C, Vol. 222 (*Proc. of the NATO Advanced Research Workshop, Antalya, Turkey*, March 1987). D. Reidel Publ. Co., Dordrecht: 205-220.

Cooper, J.D. 1980. Measurement of moisture fluxes in unsaturated soil in Thetford Forest. Report No. 6: Inst. Hydrol., Wallingford.

Edmunds, W.M. & Walton, N.R.G. 1980. A geological and isotopic approach to recharge evaluation in semi-arid zones – past and present. In: Arid Zone Hydrology: Investigations with Isotope Techniques, 47-68. IAEA, Vienna.

Flug, N., Abi-Ghanem, G.V. & Duckstein, L. 1980. An event-based model of recharge from an ephemeral stream. *Water Resour. Res.* 16: 685-690.

Freeze, R.A. & Banner, J. 1970. The mechanism of natural groundwater recharge and discharge, 2. Laboratory column experiments and field measurements. *Water Resour. Res.* 6: 1388-155.

Gelhar, L.W., Gross, G.W. & Duffy, C.J. 1979. Stochastic methods of analysing groundwater recharge. In: Hydrology of areas of low precipitation. *Proc. of the Canberra Symp.*, December 1979. IAHS-AISH Publ. No. 128: 313-321.

Greenman, D.W., Swarzenski, W.V. & Bennett, C.D. 1967. Groundwater hydrology of the Punjab, West Pakistan, with emphasis on problems caused by canal irrigation. USGS Water Supply Paper 1608-H, 66 pp.

Grindley, J. 1967. The estimation of soil moisture deficits. *Meteorol. Mag.* 96: 97-108.

Hill, M.C. 1992. A computer program (MODFLOWP) for estimating parameters of a transient, three dimensional, groundwater flow model using non-linear regression. USGS Open-File Report, 91-484.

Hodnett, M.G. & Bell, J.P. 1986. Soil moisture investigations of groundwater recharge through black cotton soils in Madhya Pradesh, India. *Hydrological Sciences Journal*, 31: 361-381.

Howard, K.W.F. & Lloyd, J.W. 1979. The sensitivity of parameters in the Penman evaporation equations and direct recharge balance. *J. Hydrol.* 41: 329-344.

IAEA (International Atomic Energy Agency) 1983. Guidebook on nuclear techniques in Hydrology. Technical Report Ser. No. 91, IAEA, Vienna. 439 pp.

Institute of Hydrology 1980. Low flow studies. NERC, Wallingford.

Jansson, P-E. & Halldin, S. 1980. Soil water and heat model. Technical description. Swedish Coniferous Forest Project, Uppsala. Tech. Rep. 26, 81 pp.

Johansson, P.-O. 1987. Estimation of groundwater recharge in sandy till with two different methods using groundwater level fluctuations. *J. Hydrol.* 90: 183-198.

Kitching, R. & Shearer, T.R. 1982. Construction and operation of a large undisturbed lysimeter to measure recharge to the chalk aquifer, England. *J. Hydrol.* 58: 267-277.

Kitching, R., Shearer, T.R. & Shedlock, S.L. 1977. Recharge to the Bunter Sandstone determined from lysimeters. *J. Hydrol.* 33: 217-232.

Kitching, R., Edmunds, W.M., Shearer, T.R., Walton, N.R.G. & Jacovides, J.,1980. Assessment of recharge to aquifers. *Hydrol. Sci. Bull.* 25: 217-235.

Kivimäki A.L. & Suokko, T., (eds) 1996. Artificial rechange of groundwater. *Nordic Hydrological Programme* NHP Report, No 38, 309 pp.

Lerner, D.N. 1990. Groundwater recharge in urban areas. Atmospheric Environment, 248: 29-33.

Lerner, D.N., Issar, S.A. & Simmers, I. 1990. Groundwater recharge – a guide to understanding and estimating natural recharge. Heise, Hannover, 345 pp.

Lerner, D.N., Mansell-Moullin, M., Dellow, D.J. & Lloyd, J.W. 1982. Groundwater studies in Lima, Peru. From: Optimal application of water resources, *Proc. Exeter Symp.* July 1982. IAHS Publ. No. 135: 17-30.

Maloszewski, P. & Zuber, A. 1982. Determining the turnover time of groundwater systems with the aid of environmental tracers. 1. Models and their applicability. *J. Hydrol.* 57: 207-231.

Monteith, J.L. 1985. Evaporation from land surfaces: progress in analysis and prediction since 1948. *Advances in Evaporation*, American Society of Agricultural Engineers, 4-12.

Nielsen, D.R., Biggar, J.W. & Erh, K.T. 1973. Spatial variability of field measured soil-water properties. Hilgardia 42: 215-259.

Penman, H.L. 1950. The water balance of the Stour catchment area. *J. Inst. Water Eng.* 4: 457-469.

Rehm, B.W., Moran, S.R. & Groenewold, G.H. 1982. Natural groundwater recharge in an upland area of central North Dakota, USA. *J. Hydrol.* 59: 293-314.

Rorabaugh, M.I. 1964. Estimating changes in bank storage and groundwater contribution to streamflows. IAHS Publ. 63: 432-441.

Rushton, K.R. & Ward, C.J. 1979. The estimation of groundwater recharge. *J. Hydrol.* 41: 345-361.

Rushton, K.R. 1988. Numerical and conceptual models for recharge estimation in arid and semi-arid zones. In: I. Simmers (ed.), Estimation of natural groundwater recharge. NATO ASI Series C, Vol. 222 (*Proc. of the NATO Advanced Research Workshop, Antalya, Turkey*, March 1987). D. Reidel Publ. Co., Dordrecht: 223-238.

Rutledge, A.T. & Daniel, C.C. 1994. Testing an automated method to estimate groundwater recharge from streamflow records. *Ground Water*, 32: 180-189.

Sammis, T.W, Evans, D.D. & Warwick, A.W. 1982. Comparison of methods to estimate deep percolation rates. *Water Resour. Bull.* 18: 465-470.

Saxena, R.K. & Dressie, Z. 1984. Estimation of groundwater recharge and moisture movement in sandy formations by tracing natural oxygen-18 and injected tritium profiles in the unsaturated zone. Isotope Hydrology 1983, *Proc. Symp. Vienna*, IAEA, 139-150.

Sharma, M.L. & Hughes, M.W. 1985. Groundwater recharge estimation using chloride, deuterium and oxygen-18 profiles in the deep coastal sands of Western Australia. *J. Hydrol.* 81: 93-109.

Sharma, M.L., Cresswell, I.D. & Watson, J.D. 1985. Estimates of natural groundwater recharge from the depth distribution of an applied tracer. *Proc. 21st Int. Assoc. Hydraulic Res.*, Melbourne, 65-70.

Shawyer, M.S. & Wescott, P. 1987. The MORECS climatological data set – A history of water-balance variables over Great Britain since 1961. *Meteorological Magazine*, 116: 205-211.

Smith, D.N., Wearn, P.L., Richards, H.J. & Rowe, P.C. 1970. Water movement in the unsaturated zone of high and low permeability strata by measuring natural tritium. In: *Isotope Hydrology*, IAEA, Vienna, 73-87.

Thoma, G., Esser, N., Sonntag, C., Weiss, W., Rudolph, J. & Leveque, P. 1979. New technique of in-situ soil-moisture sampling for environmental isotope analysis applied at Pilat sand dune near Bordeaux. *Isotope Hydrology* 1978. Proc. Symp. Neuherberg, IAEA, Vol. 2: 753-766.

Wellings, S.R. 1984. Recharge of the upper chalk aquifer at a site in Hampshire, England. 1. Water balance and unsaturated flow. *J. Hydrol.* 69: 259-273.

Wilson, L.G. & De Cook, K.J. 1968. Field observations on changes in the subsurface water regime during influent seepage in the Santa Cruz River. *Water Resour. Res.* 4: 1219-1234.

WRRC (Water Resources Research Center) 1980. Regional recharge research for southwest alluvial basins. Final report on USGS contract 14-08-0001-18257. Dept. of Hydrology and Water Resources, Univ. of Arizona, Tuscon. 417 pp.

Zuber, A. 1986. Mathematical models for the interpretation of environmental radio-isotopes in groundwater systems. In: Fritz, P. & Fontes, J.-Ch. (eds), *Handbook of Environmental Geochemistry*. Vol. 2. The Terrestrial Environment, B.

2. Techniques for catchment studies

CHAPTER 5

Chemical analysis of rocks and soils

MAGNE ØDEGÅRD
Laboratories Division, Geological Survey of Norway, Trondheim, Norway

5.1 INTRODUCTION

Information on the chemical composition of geological materials is often of fundamental importance within many branches of modern geoscience, and applies to geomaterials in their widest sense, as solid rocks and ores, sediments, water and air. Analytical chemistry is therefore an important element within all fields of geology and related activities. The ever increasing demand for geoanalyses is mainly concentrated within three areas, namely:
1. Basic research within geology and geochemistry;
2. Exploitation and use of georesources;
3. Environmental surveillance.
Basic research within many fields of geology and geochemistry has created an extended interest for chemical data, not only with regard to element coverage and concentration level, but also concerning the type of geological material. The same applies to the vast group of activities designated exploitation and use of georesources, a group which also includes the rapidly developing field of material science. Even if the increasing demand for chemical analysis of geomaterials primarily is concentrated within these two groups, considerable geoanalytical service is also attached to environmental surveillance and control, an activity of vital importance and with high political priority.

Today's extensive use of analytical data on geological materials at all concentration levels would not have been possible without intensive research in analytical inorganic chemistry, research which has led to the development of rapid and accurate instrumental analytical methods with good element coverage and detection capabilities.

5.2 HISTORICAL DEVELOPMENT

In the early days of the development of mineralogy and petrology, geological materials like rocks, ores and minerals were exclusively analysed with gravimetric and titrimetric wet chemical methods. These classical methods were largely suitable only for major elements, i.e. concentration levels above 0.1%, and used mainly for the characterisation and description of minerals and rocks. If minor and trace elements were to be determined, often complicated and time-consuming separation and/or enrichment procedures had to be used. Many wet chemical methods are very accurate, and are even today important supplements to modern instrumental methods.

The great breakthrough in the field of inorganic chemical analysis came with the development of the spectrochemical analytical methods X-ray fluorescence (XRF) and optical emission spectrometry (OES). From the time around 1930 and up to now, these techniques have developed to be indispensable analytical methods with high precision and great capacity. Within the field of OES the so-called inductively coupled plasma (ICP) represented a significant breakthrough as excitation source in the analysis of solutions (ICP-AES).

A major part of the inorganic chemical analyses being done today on geological materials are done by either of these two methods, or by atomic absorption spectrometry (AAS).

Mass spectrometry (MS), which is of particular significance within geology, is also a relatively old analytical technique originally applied on solid samples. The latest development in this field is the so-called plasma mass spectrometer, where an inductively coupled plasma is used as ionization source (ICP-MS). This technique is therefore mostly used on solutions, and is a very sensitive and valuable method for many purposes, e.g. environmental analyses at low concentration levels.

Another recognized analytical technique with a fairly long history is neutron activation analysis (NAA), but since this technique normally is dependent on an atomic reactor for irradiation, its use is relatively limited.

Determination of inorganic anions were originally done by gravimetric and titrimetric methods, but these methods are largely abandoned. The most used methods for anions today are ion chromatography (IC) and ion selective electrodes (ISE).

Within the field of inorganic chemical analysis there exist today several refined combinations of analytical techniques with excellent detection capabilities, e.g. combination of chromatography and plasma emission spectrometry or plasma mass spectrometry. By this technique elements or groups of elements are chromatographically isolated and enriched before they are determined by either of the two plasma techniques. Useful references on many of these techniques are found in Johnson & Maxwell (1981), Augustitis (1983), Potts (1987) and Skoog & Leary (1992).

5.3 TOTAL ANALYSIS VERSUS PARTIAL ANALYSIS

There are, in principle, two possible modes of analysis of geological materials, namely:
1. Analysis of total content;
2. Partial analysis based on some form of extraction.

Analysis of the total element content of a sample material is a well defined task which does not need further comments. Partial analysis, on the other hand, is a way of analysis which needs several comments. This way of analysis is mainly used in two cases, namely:

1. Environmental studies;
2. Geochemical prospecting for ores and minerals.

Both environmental pollution studies and geochemical prospecting for ores and minerals are activities which have three basic elements in common, namely:

1. A primary source for supply of chemical elements;
2. A mechanism for dispersing the elements;
3. A medium for picking up or collecting the elements.

The main aim both in environmental studies and geochemical prospecting will be to determine quantitatively and selectively the elements or element compounds which secondarily have been supplied to the geological medium from an external source. This selective task can never be totally fulfilled.

There are two factors related to the extraction technique which are of particular importance, namely:

1. The extraction attack must be strong enough to bring the secondarily supplied elements or element compounds to be studied in solution;
2. The extraction attack should influence the primary sample minerals as little as possible to avoid introduction of unwanted elements.

Unwanted elements in this context are first of all potential analyte elements primarily present in the geological medium, and which at the same time are completely or partly released during the extraction. In extreme cases this can disturb or even overshadow effects which are to be studied, the danger being especially great when working at low concentration levels and with marginal effects. This is often the situation when studying the effects of moderate or marginal environmental loads.

Unwanted elements are also all other matrix elements which go into solution during extraction. This increases the content of dissolved solids in the analysis solutions, which in most cases complicates the subsequent analysis.

The degree to which unwanted elements of the two categories are introduced from geological materials during extraction is dependent on the mineralogical composition of the material. Information on general solubility of rock-forming silicate minerals is very scarce in the literature, probably because this group of minerals traditionally have been considered as largely resistant against attacks from mineral acids, except HF. However, systematic research at the Geological Survey of Norway (Graff & Røste 1985) has shown that a lot of silicate minerals have substantial solubility in mineral acids as HCl and HNO_3, in some cases up to 100%. Studies by Faye (1982) show that many common accessory minerals of rocks, and which also have been regarded as very resistant against acid attacks, have considerable solubility.

The most important aspects of these studies, especially in relation to environmental studies and geochemical prospecting, will be the release of analyte elements bound in the lattice of the primary silicate minerals. The possibility of introducing such elements can in many cases be evaluated, but this requires:

1. Knowledge of the composition of the rock material;
2. Knowledge of the basic geochemical distribution laws;
3. Knowledge of the general solubility of rock-forming silicate minerals.

It is obvious that a total analysis and a partial analysis in most cases give entirely different information, and the choice of method must therefore be carefully based on the purpose of the analysis. This choice only influences on the sample preparation stage, and does not in principle influence the subsequent instrumental determination.

5.4 ANALYTICAL METHODS

The analytical methods which will be treated in this paper are confined to the methods used at the Geological Survey of Norway (NGU) (Faye & Ødegård 1975, Ødegård 1979, Ødegård 1981, Løbersli et al. 1990, Ødegård 1990, Saether et al. 1992 and Steinnes et al. 1993).

Some basic features are common to all spectrochemical analytical techniques. All techniques are based on measurement of electromagnetic radiation, either emitted or absorbed, and all techniques are based on a calibration system where a relationship between radiation intensity and concentration is established. The techniques are further dependent on an excitation source, a device for isolating the used wavelength, and a device to detect the radiant energy. These main elements of a spectrochemical analytical method are outlined in Figure 5.1. Emission spectroscopy is chosen as example, and the figure also shows the historic development which has taken place within this technique, especially on the excitation and registration sides.

5.4.1 *X-ray fluorescence (XRF)*

XRF represents one of the few instrumental analytical methods which normally uses solid samples. The principle of XRF is in short that a representative surface of the sample material is irradiated by X-rays from an X-ray tube, whereby the atoms of the sample are excited and emit fluorescence radiation. This radiation is normally isolated according to wavelength by means of an analysing crystal and finally recorded by a counter (wavelength dispersive XRF). In some cases the radiation is isolated according to energy (energy dispersive XRF). The basis for both techniques is that the intensity of the fluorescence radiation is a function of concentration.

XRF is one of the most important and accurate analytical methods for geological materials, both for major elements and for many elements at the minor and trace level. This is first of all due to the very high stability of XRF instruments, which gives good analytical precision.

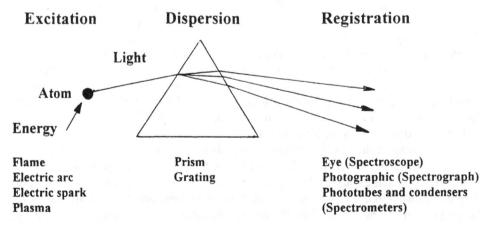

Figure 5.1. Simplified outline of emission spectroscopy.

There are two major problems related to XRF, which have to be overcome. These are matrix effects and grain size effects. Matrix effects are due mainly to absorption of primary and secondary X-rays by the elements of the sample. Less important are enhancement effects. Grain size effects are physical effects which are related to the size of the sample particles in relation to the penetration depth of the X-rays.

Major elements
A major rock element analysis includes the elements Si, Al, Fe, Ti, Mg, Ca, Na, K, Mn and P, and is today most often done by XRF. The characteristic fluorescence radiation from the light rock elements is relatively long, and therefore sensitive to grain size effects. Most geological materials are crystalline, and even after prolonged grinding grain size effects will normally be considerable. Analysis of major rock elements are therefore mostly based on fusion of the samples by a flux material. This is a sample preparation technique which completely eliminates grain size effects, and a technique which also in other respects is nearly ideal. By fusion of the crystalline rock sample with a flux the crystal structure is broken down, and the sample is transformed to an amorphous and homogeneous material. This isoformation technique also has the great advantage that synthetical standards of high accuracy can be prepared from oxides with known stoichiometry, and used in the same way as natural standard reference materials for instrumental calibration. With this nearly ideal sample preparation technique, combined with mathematical correction for matrix effects, the XRF method is today without competition in major rock element analysis, both as regards quality and speed.

The sample preparation method in use at NGU is based on a sample weight of 0.8 g and a flux weight of 5.6 g $Li_2B_4O_7$. The material is fused in Au/Pt-crucibles in an automatic fusion machine (Claisse fluxer), and finally cast to beads with a flat surface. The X-ray analyses are carried out on a Philips X-ray spectrometer model PW 1480. Table 5.1 gives an overview of the analytical range of the different components together with analytical accuracy.

Minor and trace elements
XRF-analysis of minor and trace elements in geological materials is normally based on finely ground samples which are mixed with a binder and finally pressed to pellets. The sample preparation technique used for major elements gives a dilution of the sample by a factor of 8, a dilution which one normally cannot afford for trace elements at low concentration levels. Grain size effects are less dominant for the heavier elements due to their relatively short wavelengths.

The preparation technique used at NGU uses a sample weight of 5.4 g which is mixed with 1.2 g of wax. The samples are pressed to pellets with a hydraulic press with a pressure of 20 kN.

Instrumental calibration is based on international reference materials which are prepared in the same way as the samples. Correction for matrix effects are done either by referring the net line intensities to a Compton peak or to a chosen background position.

Table 5.2 shows the minor and trace elements which are determined by XRF at NGU, together with their detection limits in ppm (3σ-level). Shown are also precision obtained by repeated measurements over a time period of 6 months of a natural rock sample.

Table 5.1. X-ray fluorescence analysis of major rock elements based on fusion with $Li_2B_4O_7$.

Component	Concentration range (% wt.)	Accuracy
SiO_2	0.5-5	± 15% rel.
	5-25	± 10% rel.
	25-100	≤ 5% rel.
Al_2O_3	0.05-0.5	± 0.05% wt.
	0.5-1.5	± 0.1% wt.
	1.5-5	± 0.2% wt.
	5-10	± 0.4% wt.
	>10	≤ ±0.8% wt.
Fe_2O_3	0.03-0.1	± 0.03% wt.
	0.1-7.5	± 15% rel.
	7.5-25	± 10% rel.
	>25	≤ ± 5% rel.
TiO_2	0.005-0.05	±0.005% rel.
	0.05-0.3	± 15% rel.
	0.3-0.75	± 10% rel.
	0.75-2.5	± 5% rel.
	>2.5	≤ ± 2.5% rel.
MgO	0.03-0.5	± 0.03% wt.
	0.5-5	± 10% rel.
	5-25	± 5% rel.
	>25	≤ ± 2.5% rel.
CaO	0.007-0.05	± 0.007% wt.
	0.05-1	± 15% rel.
	1-5	± 10% rel.
	5-20	± 5% rel.
	>2.5	≤ ± 2% rel.
Na_2O	0.03-0.1	± 0.03% wt.
	0.1-0.5	± 0.07% wt.
	0.5-2	± 15% rel.
	2-5	± 10% rel.
	>5	≤ ± 5% rel.
K_2O	0.005-0.02	± 0.005% wt.
	0.02-0.5	± 15% rel.
	0.5-2.5	± 10% rel.
	2.5-7.5	± 5% rel.
	>7.5	≤ ± 2.5% rel.
MnO	0.004-0.02	± 0.004% wt.
	0.02-0.2	± 15% rel.
	0.2-0.5	± 10% rel.
	>0.5	≤ ± 5% rel.
P_2O_5	0.03-0.1	± 0.03% wt.
	0.1-0.5	± 15% rel.
	0.5-1	± 10% rel.
	>1	≤ ± 5% rel.

Table 5.2. Lower limits of detection (LLD) together with precision for minor and trace elements determined by XRF.

Element	LLD in ppm	Precision test based on natural sample	
		Conc. (ppm)	σ (ppm)
Pb	9	32.8	4.3
U	7	19.6	4.0
Th	10	<LLD	1.6
Rb	4	176	2
Y	3	46.9	1.5
Nb	4	20.9	1.0
Sr	4	86.7	1.0
Zr	3	291	3
Mo	4	<LLD	1.5
As	6	<LLD	1.7
V	3	173	2.8
Cr	4	88.8	3.1
Sc	8	15.9	2.4
S	6	<LLD	180
Cl	13	<LLD	10
F	300	800	80
Ba	9	611	11
Sb	8	<LLD	2.2
Sn	8	6.8	2.8
Cd	6	<LLD	2.4
Ag	8	<LLD	2.4
Ga	5	20.4	2.5
Zn	3	117	2
Cu	6	52.0	2.5
Ni	5	47.9	1.1
Yb	16	<LLD	4.7
Co	10	20.9	1.9
Ce	12	131	10
Nd	8	48.6	3.0
La	11	67.1	4.1
W	10	<LLD	4.6

Analytical accuracy, defined as the difference between correct concentration and the found value, is a function of concentration. The absolute accuracy in ppm at concentrations near the detection limit is of the order of 5 to 20 ppm for the element group Nb, Y, Rb, Pb, Th, U, Sr, Mo, Sc, As, Sr, V, Zr, Sn, La, Ni, Ga, Sb, Cd, Nd, Ce, Ag, Co, Yb, Zn and W. The accuracy for Cr and Ba is correspondingly in the range 30 to 50 ppm.

Even if X-ray fluorescence uses solid samples without any dilution its detection capabilities are often insufficient for many purposes, e.g. environmental analyses and geochemical studies. For concentrations above around 10 ppm X-ray fluorescence is a valuable alternative with high grade of accuracy.

5.4.2 *Inductively coupled plasma atomic emission spectrometry (ICP-AES)*

Emission techniques with flame, electric arc or spark excitation have long traditions in the analysis of geological materials, especially for minor and trace elements. This is mainly due to low detection limits and excellent multielement capability. Electric arc and spark excitation were mostly used on solid samples, but for some purposes high voltage spark excitation was used on solutions. Today, superior excitation properties for solutions are obtained with the ICP. The principle for this excitation source is in short that argon gas is heated to a very high temperature, up to 10 000°C, by an induction coil carrying an alternating current of high frequency, mostly of 27.12 MHz. The sample to be analysed passes through a nebulizer where the solution is transformed to a mist of small droplets before entering the plasma, where the atoms are excited.

The development of ICP-AES represented in many ways an important break-through in the emission analysis of solutions, and this technique is today well established as a multielement technique with low detection limits and high precision. At NGU the method has been in use since 1978 for analyses related to geochemical prospecting and mapping, and in later years for environmental investigations. The instrumentation is a Thermo Jarrell Ash ICAP 61.

Since the method works with solutions, solid samples must be brought into solution prior to analysis. Depending on the purpose of the analysis this can be done in different ways, either as a total dissolution or as a partial dissolution with some extrant.

A standard procedure used at NGU is autoclave extraction in 7N HNO_3 in accordance with Norwegian Standard NS 4770. By this technique a sample weight of 1 g and an analysis volume of 100 ml are mostly used. This means that the samples are diluted by a factor of 100, which must be born in mind when establishing the lower limits of determination. The determination of these limits is based on acid blanks which have undergone the same procedure in autoclave as real samples. Table 5.3 shows the lower limits of determination, which are obtained by multiplying the standard deviation (σ) by a factor of at least 4 to 5. This has been done to take into consideration that most solutions are very complex in composition and that many interelement corrections are in use in the analytical program. As can be seen from the table many of the major rock elements have relatively high limits of determination. This is due to leaching of elements from the glass material together with element contributions from contaminations in the acid. Table 5.3 also shows, for comparison, lower limits of determination in pure water samples. In this case there will of course be no dilution connected to sample preparation.

Precision

Determination of precision of an analytical method must include all steps related to the analysis. In the case of extraction analysis both the extraction and the subsequent instrumental analysis must be included. Table 5.4 shows both instrumental precision and precision of complete method. The precision tests are based on extraction and analysis of the reference sample PACS-1. Precision of the complete method is based on 10 extractions of PACS-1, while instrumental precision is based on 10 consecutive analyses of one of these solutions.

Table 5.3. Lower limits of determination of solid samples with ICP-AES, based on autoclave extraction in 7N HNO_3 according to NS 4770, and a standard dilution of 1 g sample in 100 ml analysis solution. Last column shows correspondingly lower limits of determination in water samples.

Element	Lower limits of determination with ICP-AES	
	Extraction analysis (ppm)	Water analysis (ppm)
Si	100	0.02
Al	20	0.02
Fe	5	0.01
Ti	1	0.005
Mg	100	0.05
Ca	200	0.02
Na	200	0.05
K	100	0.50
Mn	0.2	0.001
P	10	0.10
Cu	1	0.005
Zn	2	0.002
Pb	5	0.05
Ni	2	0.02
Co	1	0.01
V	1	0.005
Mo	1	0.01
Cd	1	0.005
Cr	1	0.01
Ba	1	0.002
Sr	2	0.001
Zr	1	0.005
Ag	1	0.01
B	5	0.01
Be	0.2	0.001
Li	1	0.005
Sc	0.2	0.001
Ce	10	0.05
La	1	0.01
Y	0.2	0.001
Sb	5	0.05
As	10	0.10

Analytical accuracy

Analytical accuracy is dependent both on the instrument and on the extraction, and both steps must therefore be included in the tests on analytical accuracy. Since the ICP-AES technique is extensively used also for other solutions than extracts, e.g. water samples, it is important to document expected analytical accuracy in these cases as well. For this purpose commercially available quality control standards have been used. Table 5.5 shows analytical accuracy obtained on two quality control standards, QC-24 and QC-7.

Accuracy of complete procedure involving extraction and analysis is based on international reference materials. Table 5.6 shows the results obtained on respectively PACS-1 and CRM-277. It is very important to be aware of the fact that the reference

Table 5.4. Instrumental precision together with precision of complete analytical method, which is based on autoclave extraction in 7N HNO_3 according to NS 4770, a standard dilution of 1 g sample in 100 ml analysis solution, and subsequent ICP-AES analysis.

Element	Instrumental precision (based on 10 consecutive runs of one extract of PACS-1)		Precision of complete method (based on analysis of 10 extracts of PACS-1)	
	Average	St. dev. (σ)	Average	St. dev. (σ)
Si	190.7	1.7	98.0	17.0
Al	2.33%	138.0	2.42%	0.10%
Fe	4.53%	324.0	4.49%	0.11%
Ti	0.13%	10.0	0.13%	23.9
Mg	1.06%	73.0	1.04%	186.6
Ca	1.04%	64.0	1.01%	186.8
Na	1.83%	138.0	1.81%	353.0
K	0.39%	24.0	0.39%	56.8
Mn	302.1	1.9	298.5	5.8
P	887.9	11.7	887.4	17.4
Cu	442.6	3.2	454.1	7.7
Zn	772.8	4.4	765.9	3.2
Pb	377.1	3.9	375.2	6.0
Ni	36.5	0.8	38.3	2.5
Co	16.0	0.2	16.2	0.4
V	89.4	0.6	88.2	1.6
Mo	9.1	0.3	9.0	0.4
Cd	2.8	0.2	3.0	0.3
Cr	55.9	0.4	55.4	0.9
Ba	376.3	3.0	371.2	3.2
Sr	85.4	0.8	79.6	1.5
Zr	18.8	0.1	18.6	0.3
Ag	1.3	0.1	0.4	0.1
B	47.1	0.8	44.2	1.6
Be	6.3	0.1	5.9	0.1
Li	26.6	0.3	26.1	0.5
Sc	7.1	0.0	7.0	0.1
Ce	21.1	1.0	20.4	0.9
La	13.1	0.3	14.5	0.3
Y	8.9	0.1	8.7	0.2
Sb	26.3	0.9	27.6	1.6
As	167.1	3.3	155.5	4.7

Note: All value are in ppm except those marked %.

materials are certified for total element content, while the ICP-AES analyses are based on partial extraction. The agreement is therefore expected to be good only for easily extractable elements. Analytical accuracy is in such cases best controlled by cross-testing based on identical extraction procedures.

5.4.3 *Atomic absorption spectrometry (AAS)*

AAS is a well established analytical technique which in many cases can be an alternative to ICP-AES, since detection capabilities with some exceptions are comparable, and both techniques most often work with solutions.

Table 5.5. Accuracy obtained by ICP-AES analysis on quality control standards QC-24 og QC-7.

Element	Obtained results (µg/ml)	Correct values (µg/ml)	Relative deviation (%)
Si	–	–	–
Al	25.60	25.00	2.4
Fe	25.25	25.00	1.0
Ti	5.26	5.00	5.2
Mg	24.79	25.00	0.8
Ca	24.82	25.00	0.7
Na	25.42	25.00	1.7
K	50.08	50.00	0.2
Mn	5.00	5.00	0.0
P	5.06	5.00	1.2
Cu	5.08	5.00	1.6
Zn	4.82	5.00	3.6
Pb	4.87	5.00	2.6
Ni	5.06	5.00	1.2
Co	4.93	5.00	1.4
V	4.86	5.00	2.8
Mo	5.01	5.00	0.2
Cd	0.495	0.500	1.0
Cr	4.98	5.00	0.4
Ba	4.90	5.00	2.0
Sr	4.45	5.00	11.0
Zr	5.20	5.00	4.0
Ag	–	–	–
B	5.03	5.00	0.6
Be	5.11	5.00	2.2
Li	5.03	5.00	0.6
Sc	4.97	5.00	0.6
Ce	5.05	5.00	1.0
La	5.00	5.00	0.0
Y	4.98	5.00	0.4
Sb	5.18	5.00	3.6
As	5.06	5.00	1.2

The main principle for AAS is that free atoms in the ground state absorb resonance radiation which is emitted from a lamp and isolated in a monochromator. Absorption of light is a function of concentration. Because the temperature in a flame is much lower than in a plasma, the detection limits for easily ionizable elements are often better by AAS than by ICP-AES. Elements forming refractory compounds in the flame have as a rule far lower detection limits by ICP-AES.

The detection capabilities of ICP-AES for the elements Cd, Pb and Hg are not sufficient for many geochemical and environmental purposes, and at NGU AAS is therefore more or less routinely used for these elements. Cd and Pb are analysed by the graphite furnace AAS technique, and Hg by the cold vapour AAS technique. The standard extraction procedure used for ICP-AES is also used for the AAS analyses. Since Hg routinely is determined in most samples, the material is dried at a temperature as low as 50°C for 48 hours to avoid loss of Hg by evaporation.

Table 5.6. Results obtained by ICP-AES analysis of PACS-1 and CRM-277, based on autoclave extraction in 7N HNO_3 according to NS 4770 and a standard dilution of 1 g sample in 100 ml analysis solution.

Element	PACS-1		CRM-277	
	Obtained values	Certified values (total)	Obtained values	Certified values (total)*
Si	98.0 ppm	55.7 ± 0.5% SiO_2	102.2 ppm	(23.0%)**
Al	2.42%	12.23 ± 0.22% Al_2O_3	2.60%	(4.8%)
Fe	4.49%	6.96 ± 0.12% Fe_2O_3	4.67%	(4.55%) [4.17%]
Ti	0.13%	0.703 ± 0.011% TiO_2	262.9 ppm	(0.30%)
Mg	1.04%	2.41 ± 0.09% MgO	0.88%	(1.0%)
Ca	1.01%	2.92 ± 0.13% CaO	3.63%	(6.0%)
Na	1.81%	4.40 ± 0.11% Na_2O	0.81%	(1.2%)
K	0.39%	1.50 ± 0.09% K_2O	0.64%	(1.6%)
Mn	298.5 ppm	470 ± 12 ppm	0.14%	(0.16%) [0.162]
P	887.4 ppm	0.253 ± 0.018% P_2O_5	0.40%	(0.41%)
Cu	454.1 ppm	452 ± 16 ppm	101.9 ppm	101.7 ± 1.6 ppm [97.2 ppm]
Zn	765.9 ppm	824 ± 22 ppm	508.0 ppm	547 ± 12 ppm [557 ppm]
Pb	375.2 ppm	404 ± 20 ppm	129.8 ppm	146 ± 3 ppm [137.5 ppm]
Ni	38.3 ppm	44.1 ± 2.0 ppm	40.3 ppm	43.4 ± 1.6 ppm [34.9 ppm]
Co	16.2 ppm	17.5 ± 1.1 ppm	14.4 ppm	(17 ppm)
V	88.2 ppm	127 ± 5 ppm	72.1 ppm	(102 ppm)
Mo	9.0 ppm	12.3 ± – ppm	1.2 ppm	(1.5 ppm)
Cd	3.0 ppm	2.38 ± 0.20 ppm	11.0 ppm	11.9 ± 0.4 ppm [10.8 ppm]
Cr	55.4 ppm	113 ± 8 ppm	138.2 ppm	192 ± 7 ppm [145.6 ppm]
Ba	371.2 ppm	–	117.9 ppm	(329 ppm)
Sr	79.6 ppm	277 ± – ppm	241.8 ppm	–
Zr	18.6 ppm	–	18.7 ppm	–
Ag	0.4 ppm	–	0.1 ppm	(3.3 ppm)
B	44.2 ppm	–	47.5 ppm	–
Be	5.9 ppm	–	6.9 ppm	(1.6 ppm)
Li	26.1 ppm	–	30.4 ppm	–
Sc	7.0 ppm	–	5.9 ppm	9.0 ± 0.12 ppm
Ce	20.4 ppm	–	78.0 ppm	–
La	14.5 ppm	–	36.6 ppm	(45 ppm)
Y	8.7 ppm	–	29.2 ppm	–
Sb	27.6 ppm	171 ± 14 ppm	19.9 ppm	(4.0 ppm)
As	155.5 ppm	211 ± 11 ppm	11.9 ppm	47.3 ± 1.6 ppm

*Values in brackets are not certified. **Figures are uncertified values based on aqua regia dissolution.

The AAS instrumentation used at NGU is of Perkin Elmer make.

The lower limits of determination of Cd and Pb in solid geological materials, based on autoclave extraction in 7N HNO_3 according to NS 4770 and graphite furnace AAS, are respectively:

Cd: 5.0 ppb
Pb: 50 ppb

Precision of the Cd and Pb analyses by graphite furnace AAS has been determined to be of the order of 10%.

Tests on accuracy are based on reference materials certified for their total element contents. It is expected that most Cd and Pb go into solution during extraction, and analytical accuracy seems in most cases to be within ± 10% rel. for both elements.

The lower limit of determination of Hg in geological materials based on autoclave extraction in 7N HNO_3 according to NS 4770 is:

Hg: 10 ppb

Precision has been tested to be around 10% rel.

Accuracy has been found to be in the order of ± 10% rel., with the same reservation that the reference materials are certified for their total element contents.

5.5 QUALITY CONTROL

The quality system being used in the accredited laboratories of NGU is in accordance with the requirements in NS-EN 45001.

One of the most important elements in the quality control is the use of so-called control charts (X-charts). This is a system where a homogeneous natural sample is analysed with regular intervals and used for controlling analytical quality. Based on a certain number of determinations, for instance 10, over a certain time period, average values and standard deviations are calculated for the actual elements. This information is used as criterion for accepting future analytical values of this control sample. The system operates with two limits, the one being a warning limit which is average value ± 2σ, and the other being an action limit which is average value ± 3σ.

Figure 5.2 shows as example the control chart for Y in reference sample CRM-277, which is being used as control for extraction analyses. The topic of quality control (QC) is discussed further in e.g. Kateman & Pijpers (1981) and Montgomery (1985).

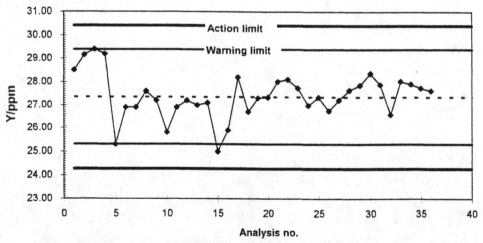

Figure 5.2. Example of control chart, showing obtained values for Y in reference sample CRM-277, together with warning limits and action limits.

REFERENCES

Augustitis, S.S. (ed.) 1983. *Leaching and diffusion in rocks and their weathering products.* Theophrastus Publications S.A. Athens 1983.

Faye, G.C. & Ødegård, M. 1975. Determination of major and trace elements in rocks employing optical emission spectroscopy and X-ray fluorescence. *Norges geol. unders. Report* 322, 35-53.

Faye, G.C. 1982. Metodestudier i geokjemi – HNO_3-ekstraksjon av geokjemiske prøver. (Method evaluation - HNO_3 - extraction of geochemical samples). *Norges geol. unders. Report* 1687 C.

Graff, P.R. & Røste, J.R. 1985. Utluting av silikatmineraler med mineralsyrer. (Leaching of silicate minerals with inorganic acids). *Norges geol. unders. Report* 85.105.

Johnson, W.M. & Maxwell, J.A. 1981. *Rock and Mineral Analysis*, John Wiley & Sons, 489 pp.

Kateman, G. & Pijpers, F.W. 1981. *Quality control in analytical chemistry*, Chemical Analysis, 60. John Wiley, New York, 276 pp.

Løbersli, E.M., Steinnes, E. & Ødegård, M. 1990. A Historical Study of Mineral Elements in Forest Plants from South Norway. *Environm. Monitoring and Assessment.* Vol.15, pp. 111-129.

Montgomery, D.C. 1985. *Introduction to Statistical Quality Control.* John Wiley, New York, 520 pp.

Potts, P.J. 1987. *A Handbood of Silicate Rock Analysis.* Blackie & Sons, Glasgow and London, 622 pp.

Skoog, D.A. & Leary, J.J. 1992, *Principles of Instrumental Analysis,* 4th ed., Philadelphia: Sanders College Publishing, 700 pp.

Steinnes, E., Johansen, O., Røyset, O. & Ødegård, M. 1993. Comparison of different multielement techniques for analysis of mosses used as biomonitors. *Environm. Monitoring and Assessment.* Vol.25, pp 87-97.

Saether, O.M., Misund, A., Ødegård, M., Andreassen, B.T. & Voss, A. 1992. Groundwater contamination at Trandum landfill, Southeastern Norway. *Nor. geol. unders. Bull.* 422, 83-95.

Ødegård, M. 1979. Determination of Major Elements in Geological Materials by ICAP Spectroscopy. *Jarrell-Ash Plasma Newsletter*, Vol. 2. No. 1.

Ødegård, M. 1981. The use of inductively coupled argon plasma (ICAP) atomic emission spectroscopy in the analysis of stream sediments. *J. Geochem. Explor.*, Vol.14, 119-130.

Ødegård, M. 1990. Den analytiske kjemis plass i geoinstitusjonen. (The role of analytical chemistry in the geo-institution). *Geonytt. Årgang* 17. Nr. 2.

CHAPTER 6

Collection and analysis of groundwater samples

JOHN MATHER
Royal Holloway, University of London, Egham, Surrey, UK

6.1 INTRODUCTION

Procedures are available for the analysis of waters of a wide range of quality, including water suitable for domestic supply, surface water, groundwater, cooling or circulating water, boiler water, treated or untreated effluents and sea water. This presentation will concentrate on the analysis of groundwaters, but even here the total range of constituents that could be determined is very large. Each constituent may give some information about the nature of the aquifer, residence time, amount of contamination and so on. It will generally be necessary for reasons of economy to restrict the number of parameters measured.

Historically, the type of chemical analysis carried out by the water industry has not been adequate for geochemical or hydrogeological analysis, giving only potability criteria. The minimum set of data useful for geochemical study is the mineral analysis for the major ions, namely Ca^{2+}, Mg^{2+}, Na^+, K^+, SO_4^{2-}, Cl^-, HCO_3^- and NO_3^-, as well as pH value. Using only these parameters together with temperature, it is possible to infer a lot about the history of the water and the rocks through which it has passed. This should form the basic unit of analysis and interpretation; minor elements, trace elements, organic species and isotopic studies can provide useful additional information. The main inorganic species and their expected concentrations in potable groundwaters are given in Table 6.1.

6.2 DATA QUALITY

The principal indicators of data quality are its bias and precision which, when combined, express its accuracy. Bias can be defined as 'consistent deviation of measured values from the true value, caused by systematic errors in procedure' and precision as 'measure of the degree of agreement among replicate analyses of a sample, usually expressed as the standard deviation'. Accuracy combines bias and precision to reflect the closeness of a measured value to a true value (Clesceri et al. 1989). The

Table 6.1. Concentration range of major and minor ionic species in dilute, oxygenated groundwaters at pH = 7 (after Edmunds et al. 1989).

Trace elements						Major elements		
1 ng/l			1 µg /l			1 mg/l		
Less than 0.00001 mg /l	0.00001-0.0001 mg/l	0.0001-0.001 mg/l	0.001-0.01 mg/l	0.01-0.1 mg/l	0.1-1.0 mg/l	1.0-10 mg/l	10-100 mg/l	> 100 mg/l
Nb	Be	As	U	Zn			NO_3^-	
Pd	Cs	Rb	Li	P	Sr	Mg	Na	HCO_3^-
Pt	Zv	La	Ba	B	F	K	Ca	
Po	Mo	V	Cu	Br		Si	SO_4^{2-}	
Au	Ag	Se	Mn	Fe			Cl	
	Hg	Cd	I					
	Sb	Co						
	Sn	Ni						
	Te	Cr						
	Bi	Pb						
	W	Al						

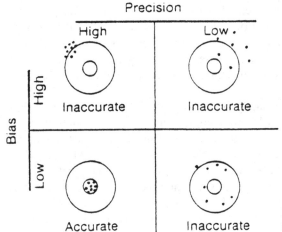

Figure 6.1. Definition of accuracy (from Clesceri et al. 1989).

relationship between these terms is shown in Figure 6.1. Of the four possible outcomes only the condition of low bias and high precision is accurate.

Other data quality indicators are method detection limit and representativeness. The detection limit is the smallest amount that can be detected above the noise in a procedure and within a stated confidence limit (usually 3σ). Representativeness can relate both to the sample itself and to the sampled population. A method may be very accurate, but if the results do not represent the population, the data are useless. These problems are likely to be less significant in the analysis of water samples than in analyses of rocks and soils.

The correctness of a groundwater analysis can be checked by carrying out a cation-anion balance. The sums of cations and anions, when expressed as millequivalents per litre (meq/l), must balance because all groundwaters are electrically neutral. The criterion for acceptance is that:

$$\frac{\sum cations - \sum anions}{\sum cations + \sum anions} \times 100\% \text{ is within} \pm 5\% \tag{6.1}$$

Most laboratories have also established a set of operating principles that, if strictly followed during sample collection and analysis, will produce data of known quality. Such quality assurance (QA) includes both quality control and quality assessment. Quality control will include such things as certifying operator competence, analysis of externally supplied standards and reagent blanks, calibration with standards, analysis of duplicates etc. (Clesceri et al. 1989). Quality assessment is the process of using both internal and external quality control measures to determine the quality of data produced by a laboratory. It includes such things as samples for performance evaluation, samples for laboratory intercomparison and performance audits, as well as internal quality control. These are applied to test the recovery, bias, precision, detection limit and adherence to standard operating procedures. Adequate quality assurance procedures are now a prerequisite before a laboratory can tender for much contract work in the UK, particularly contract work related to the radioactive waste programme.

6.3 SAMPLE COLLECTION AND ANALYSIS

Analysing a sample of groundwater should not represent a serious challenge given that the species to be determined are already in solution. Unfortunately, the situation is not quite as simple as it appears and there are many problems to frustrate the analyst.

The first difficulty is that the analysis should represent as closely as possible the in situ aquifer conditions. Some species, especially those involved in carbonate and redox reactions, will undergo changes upon reaching the atmosphere and need to be analysed in the field at the well head. Redox changes lead to other difficulties, such as the precipitation of metals, and thus samples need to be stabilised to keep metal ions in solution. A protocol for the collection and analysis of groundwater samples is given in Figure 6.2. The remainder of this section will focus on the different parts of this protocol.

6.3.1 *Field parameters*

The field parameters normally measured are pH, electrical conductance (EC), alkalinity (HCO_3^-), temperature (T), redox potential (e.g. Eh) and dissolved oxygen (DO). Some of these require in–line measurements, whereas others may be analysed on site, but not necessarily in-line.

pH
When groundwater is discharged from an aquifer, the physical controls governing H^+ ion activity are changed, and it is very important that pH is measured at the well

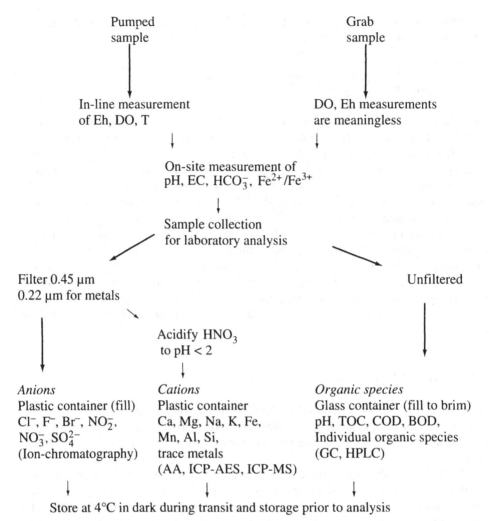

Figure 6.2. Flowchart for the collection and analysis of groundwater samples. BOD = biochemical oxygen demand, COD = chemical oxygen demand, DO = dissolved oxygen, EC = electrical conductance, T = Temperature, TOC = total organic carbon. Abbreviations of instrumental methods as in text.

head. The pH is determined using a glass electrode compared with a reference electrode of a known potential, by means of a pH-meter or other potential-measuring device. Using modern equipment, pH measurements in the field can realistically be made to ± 0.01 of a unit.

Electrical conductance

The electrical conductance (EC) of a water sample provides a rapid estimation of its total mineralisation and reliable equipment is readily available. Conductances in groundwater are normally reported as micromhos per centimetre or in SI units, microsiemens per centimetre (μS/cm). As ionic activity is affected by temperature, electrical conductance increases with temperature, and most measurements are reported at 25°C. This is automatically calibrated on modern instruments.

Alkalinity

Measured by titration with acid to a fixed end-point or to the inflection point on the titration curve, alkalinity is a measure of the acid neutralising capacity of a water. In groundwaters, and most other natural waters, the carbonate equilibria predominate and the end-point at pH 4.5 essentially measures the HCO_3^- concentration (Fig. 6.3). In some waters, other weak acids (e.g. borate, silicate and organic acids) contribute to the alkalinity and must be subtracted for more accurate interpretation of the carbonate system. Computer programmes are available to perform this function. It should be noted that alkalinity is not affected by gain or loss of $CO_{2(gas)}$. However, the pH will change, it will increase with $CO_{2(gas)}$ loss and decrease with $CO_{2(gas)}$ gain.

Temperature

All geochemical reactions are temperature dependent so that measurements are essential for assessing the extent to which equilibrium has been attained. Measurements can be made with a standard mercury thermometer or using the probe attached to the pH-meter for temperature compensation.

Redox potential

Eh, the oxidation-reduction potential, is a qualitative measurement and does not provide information on the specific behaviour of a single redox couple, or on the oxidising or reducing condition of the water. However, it provides a convenient theo-

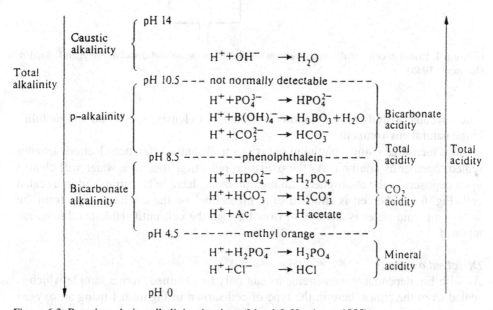

Figure 6.3. Reactions during alkalinity titrations (Lloyd & Heathcote 1985).

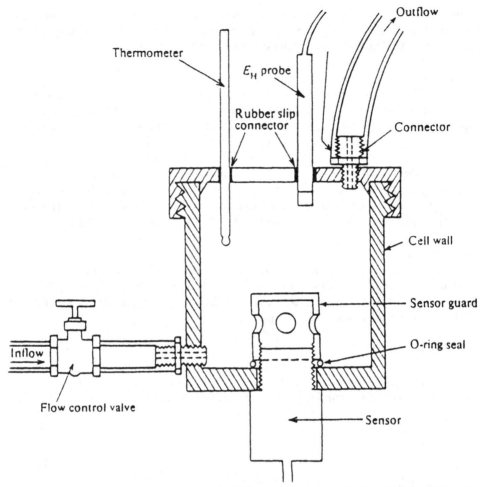

Figure 6.4. Isolation cell for the measurement of Eh, temperature and dissolved oxygen (Lloyd & Heathcote 1985).

retical means of studying the limiting conditions of element speciation and mobility in the natural environment.

Eh is measured using a platinum redox electrode and a reference electrode usually sealed together as one unit. As the oxidation-reduction state of a water will change upon exposure to the atmosphere, Eh measurements have to be carried out in a sealed cell (Fig. 6.4). The cell is attached to a 'bleed line' on the discharge pipe from the well pump, and water is allowed to flow through the cell until stable conditions are attained.

Dissolved oxygen

As with Eh, dependable measurements can only be obtained from a sample which is sealed from the atmosphere in the type of cell shown in Figure 6.4 using an oxygen electrode.

6.3.2 *Laboratory measurements*

Samples for laboratory analysis need to be collected with the species to be determined firmly in mind. Figure 6.2 shows three distinct routes as follows:

– Filtration through a 0.45 μm filter into a plastic container filled to the brim for anion analysis;

– Filtration through a 0.45 μm filter into a plastic container acidified to pH < 2 by HNO_3 for cation analysis;

– Collection in a glass bottle of an unfiltered sample for the analysis of organic species.

Filtration is necessary as most groundwaters contain particulate and colloidal material. Samples for cation analysis are normally filtered through a 0.45 μm filter and acidified with concentrated suprapure nitric acid. This should remove particulate matter, prevent sorption of trace ions on the container wall, and maintain the main cations in stable solution. Samples for anion analysis require filtration but not acidification. Samples for organic analysis need to be collected in glass bottles without filtration. Special techniques may be required to collect such samples because of loss of volatile components. Consideration of these is outside the scope of this chapter.

Cation analysis

Atomic absorption (AA) spectrometry has been the mainstay of elemental analysis of waters for many years now, and is still one of the standard instrumental methods (Walsh 1993). An atomic absorption spectrometer consists of an atomic line source (normally a hollow cathode lamp coated with the element of interest), a means of atomising a sample solution, a monochromator, a photomultiplier detector and a readout system (Fig. 6.5). Radiation from the lamp is passed through the flame and the degree of atomic absorption detected using the monochromator tuned to the wavelength of the atomic emission line.

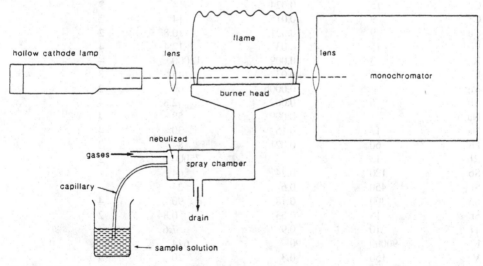

Figure 6.5. Typical instrumental configuration for atomic absorption spectrometry (after Potts 1987).

Despite the introduction of more versatile simultaneous multielement techniques, AA retains many advantages. It is above all simple to calibrate and operate, it is comparatively inexpensive, rapid and generally free from interferences. Flame atomic absorption is still one of the best methods for routine analysis of some of the easily volatilised metals such as Zn, Cd, Cu, Ag, Pb, Rb etc. For many of these elements, it will give excellent detection limits during routine analysis of water samples (Table 6.2). However, there are numerous occasions when the detection limits attainable through flame atomic absorption analysis are inadequate. This applies even to the 'good' flame AA elements noted above. Several options are available to improve detection limits. Flame AA can be replaced by flameless AA (electrothermal atomisation), the samples can be preconcentrated or some other analytical instrumental method can be used.

Flameless or graphite furnace AA has become a popular method in water analysis and certainly offers dramatic improvements in detection limits in comparison with flame AA (Table 6.2). In this technique the flame is replaced with a graphite furnace as an atom cell. The furnace normally consists of a hollow graphite tube that can be

Table 6.2. Detection limits for a representative series of elements in ideal solution (taken to represent elements in dilute groundwaters) measured using flame AA, flameless AA, ICP-AES and ICP-MS (after Potts 1987).

Element	Detection limit (ppb)			
	Flame AA	Flameless AA	ICP/AES	ICP/MS
Al	60	0.03	46	3
B	2100	45	9.6	4
Ba	60	0.12	2.6	2
Ca	3	0.03	20	–
Cd	3	0.0006	6.8	2
Co	15	0.024	14	2
Cr	9	0.015	14	2
Cu	9	0.021	10.8	2
Fe	15	0.03	12.4	4
K	3	0.009	12000	–
Li	3	0.03	45	8
Mg	0.6	0.0006	60	2
Mn	6	0.002	2.8	4
Na	1	0.009	58	4
Ni	15	0.15	30	4
Pb	60	0.021	84	2
Rb	15	–	75000	2
Sb	120	0.24	64	–
Si	450	0.6	24	–
Sn	300	0.18	90	4
Sr	15	0.03	0.84	2
Ti	210	0.9	7.6	4
U	90000	90	674	2
V	150	0.3	20	4
Zn	3	0.006	8	6
Zr	4500	–	14.2	2

heated by passing a low voltage DC current through the graphite. The sample is deposited as a solution on to the inner wall of the graphite tube (Potts 1987).

A flameless attachment can be added to the basic flame AA at a relatively small cost, however, it does have several disadvantages. It is comparatively slow, although automation of sample input reduces this problem. A more serious disadvantage is the matrix interferences that are found. This is especially troublesome where the analyte element is at trace levels and the matrix elements are high levels of refractory elements (Si, Al, etc.). This has severely limited the use of flameless AA in analysis of refractory samples (notably geological and ceramic materials). For water analysis, the problem is not so severe, but high (and very variable) concentrations of Ca and Na have restricted the application of the technique (Walsh 1993). Flameless AA is likely to remain the most popular method for certain trace elements. The determination of Pb and Cd are examples of elements where the detection limits required are beyond the capabilities of most alternative techniques (Table 6.2). However, in cases where an alternative technique is available, this is likely to be used in preference to flameless AA.

Preconcentration of water samples prior to analysis is an obvious method for enhancing detection limits, and it is appropriate for many water samples. The total dissolved solids (TDS) content of potable groundwater is low and preconcentration by simple evaporation can be used. A typical level of preconcentration will be 10 or 20 fold. This technique cannot be used for waters that already contain high dissolved solids and it is a somewhat tedious procedure to implement for routine analysis of large numbers of samples.

A technique that has gained widespread acceptance in recent years for elemental analysis of water samples is inductively coupled plasma atomic emission spectrometry (ICP-AES). This is a conventional spectrometric technique whose unique properties derive from the particular excitation source used. Power is transferred into the plasma gas by inductive heating. Argon gas carries the sample material through the plasma torch which is located in an induction coil carrying a very high frequency alternating current (Fig. 6.6). This heats the gas to high temperatures, up to around 10,000 K.

The light emitted by the atoms or ions in the ICP is converted to an electrical signal by a photomultiplier in the spectrometer, and by comparison to a previous measured intensity of known concentration of the element, the concentration can be computed. ICP-AES has the capability to measure a very wide range of elemental concentrations simultaneously on one small (~ 1 ml) aliquot of sample. The method gives good results for a far greater range of elements than flame AA. It is especially useful for the determination of many of the more refractory elements; notable examples are Si, Al, Ti, etc. (See Table 6.2).

Although the cost of instrumentation is high, the costs of analysis per sample where a range of elements are sought, can be dramatically reduced. ICP-AES is capable of measuring element concentrations at both high levels and low levels. Thus, for a water analysis programme the traditional major cations Ca, Mg, Na K, Si, can be determined directly at the 5-500 ppm level. In addition, trace element concentrations can be determined at the same time at the parts per billion level (Table 6.2).

However, it is important also to consider the limitations of ICP-AES. It is not entirely without interferences and these have to be carefully documented and detection

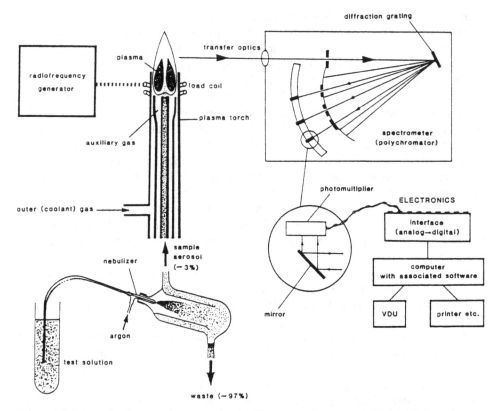

Figure 6.6. Schematic diagram of a conventional ICP system (Thompson & Walsh 1989).

limits, which are low, may still not be low enough for groundwaters. This has lead to the introduction of another ICP based technique, inductively coupled plasma mass spectrometry (ICP-MS). The linking of a conventional inductively coupled plasma source unit to a mass spectrometer gives an analytical technique capable of measuring elemental abundance levels 1000-fold lower than ICP-AES (or the best elements in flame AA). ICP-MS is not as rapid as ICP-AES; in analysis sample throughput is significantly slower. It is also not really suitable for the analysis of the major cation constituents of waters. A water analysis laboratory will almost certainly require both ICP-AES and ICP-MS to cover the range of elements required. However, it does offer outstanding performance at the ultra-trace level. Many elements of environmental concern can be measured down to their natural abundance levels in groundwaters with ICP-MS (Edmunds et al. 1989).

Anion analysis
Although determination of the common anions can be made by conventional colorimetric or titrimetric techniques, there is one technique, ion chromatography (IC), that provides a single instrumental method for their rapid, sequential measurement. IC eliminates the need to use hazardous reagents and it also distinguishes between the various halides and the oxides (e.g. NO_2^- and NO_3^-).

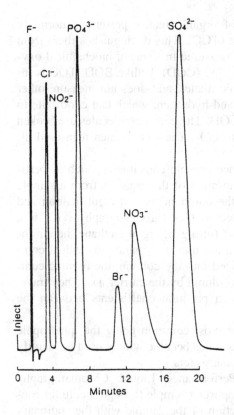

Figure 6.7. Ion chromatography: Typical inorganic anion separation (Clesceri et al. 1989).

In ion chromotography, a water sample is injected into a stream of carbonate/bicarbonate eluant and passed through a series of ion exchangers. The anions of interest are separated on the basis of their relative affinities for a strongly basic anion exchanger. The separated anions in their acid forms are measured by conductivity, and identified on the basis of retention times as compared to standards. Any substance that has a retention time coinciding with that of any anion to be determined will interfere. For example, high concentrations of low-molecular-weight organic acids interfere with the determination of chloride. Sample dilution overcomes many interferences. A typical inorganic anion separation is shown in Figure 6.7.

Another technique used for the rapid estimation of anion concentrations is the ion-selective electrode. This carries out a similar function to a pH-meter, except that instead of responding to changes in hydrogen ion concentration, the electrode is designed to respond to changes in the activity of specific ions in solution. Such electrodes are now available for a wide range of ions, but for groundwaters the outstanding applications are in the determination of fluoride and ammonium ions.

Organic analysis
Analyses for organic matter in water can be classified into two general types of measurements (Clesceri et al. 1989), those that quantify an aggregate amount of organic matter with a common characteristic and those that quantify individual organic compounds.

In groundwaters, the aggregate amount of organic matter present is normally measured in terms of the total organic carbon (TOC). This distinguishes them from wastewaters, where organic matter is usually measured in terms of biochemical oxygen demand (BOD) and chemical oxygen demand (COD). Unlike BOD, TOC is independent of the oxidation state of the organic matter, and does not measure other organically based elements such as nitrogen and hydrogen, which can contribute to the oxygen demand measured by BOD and COD. The carbon molecules are broken down into single carbon units and oxidised to CO_2. The CO_2 is then measured directly with an infra-red analyser.

The identification and quantification of trace organic constituents, such as pesticides, generally involves isolation and concentration of the organics from a sample by solvent or gas extraction, separation of the components and identification and quantification of the compounds with a detector. Gas chromatography (GC) is a commonly used method in which a carrier gas (nitrogen, argon-methane, helium or hydrogen) and a stationary phase in a column are used to separate individual compounds. When the sample solution is introduced into the column, the organic compounds are vaporised and moved through the column by the carrier gas. They travel at different rates, depending on differences in partition coefficients between the mobile and stationary phases.

Various detectors can be used, one of the most common being the quadropole mass spectrometer (MS). The GC-MS technique has become widely used to identify unknown organic compounds occurring in groundwaters.

Another technique available is High-Performance Liquid Chromatography (HPLC), in which a liquid mobile phase transports a sample through a column containing a stationary liquid phase. The interaction of the sample with the stationary phase selectively retains individual compounds and permits separation of sample components. Detection is by absorbance detectors, where the detector measures the absorbance of a sample from an incident light source.

6.3.3 *Pore water analysis*

So far this discussion has assumed that samples are derived from pumping wells and springs. Unfortunately, hydrochemical stratification is a common feature of aquifers so that the integrated samples taken from a borehole pump will inevitable have mixed chemistries. Mixing of oxygenated and reducing waters may precipitate iron and if pumping is not continuous, stagnation of water in the well will result in chemical changes.

The large volumes of stored intergranular fluids will exert a major influence on water quality over a long period, and it is desirable to sample and analyse such matrix fluids. Various drilling techniques can be used to recover uncontaminated core material for investigation, including augering, percussion drilling and airflush or waterflush rotary. If waterflush drilling is used, a suitable tracer, such as Li, may be added to quantify formation invasion; otherwise air flush (dry) sampling is to be preferred (Edmunds 1986).

Extraction of interstitial fluids may be carried out by centrifugation, squeezing or fluid displacement. For detailed investigation, such interstitial fluid sampling is far the best method of examining the water profile with depth.

6.4 REPRESENTATION OF DATA

The units chosen for hydrochemical species are generally mg/l, equivalent in dilute waters to parts per million (ppm, or mg/kg). Above concentrations of about 7000 mg/l, however, solution density must be taken into account and values converted according to the equation:

$$mg/kg = \frac{mg/l}{\text{solution density}} \qquad (6.2)$$

Trace constituents are often reported as µg/l or parts per billion (ppb, or µg/kg).

Whilst the analysis in mg/l is most readily understood, analyses expressed in millimoles per litre (mmol/l) or in millequivalents per litre (meq/l) are generally more useful in geochemical interpretation where chemically equivalent quantities involved in reactions must be compared:

$$mmol/l = \frac{mg/l}{\text{molecular weight}} \qquad (6.3)$$

and

$$meq/l = mmol/l \times \text{ionic charge} \qquad (6.4)$$

Conversion factors are given in Table 6.3 (after Edmunds 1986).

Individual analyses can then be treated to reveal a large amount of additional information using a variety of computer software. This includes information on speciation, ionic balance, and the extent of saturation with respect to various minerals.

Table 6.3. Conversion factors from mg/l to (a) meq/l and (b) mmol/l (Edmunds 1986).

Element of species	Conversion Factors (a)	(b)	Element and species	Conversion Factors (a)	(b)
Aluminium (Al^{3+})	0.11119	0.03715	Lead (Pb)	–	0.00483
Ammonium (NH_4^+)	0.05544	0.05544	Lithium (Li^+)	0.14411	0.14411
Barium (Ba^{2+})	0.01456	0.00728	Magnesium (Mg^{2+})	0.08226	0.04113
Beryllium (Be^{3+})	0.33288	0.11096	Manganese (Mn^{2+})	0.03640	0.01820
Bicarbonate (HCO_3^-)	0.01639	0.01639	Molybdenum (Mo)	–	0.01042
Boron (B)	–	0.09250	Nickel (Ni^+)	–	0.01703
Bromide (Br^-)	0.01251	0.01251	Nitrate(NO_3^-)	0.01613	0.01613
Cadmium (Cd^{2+})	0.01779	0.00890	Nitrite (NO_2^-)	0.02174	0.02174
Calcium (Ca^{2+})	0.04990	0.02495	Phosphate (HPO_4^{2-})	0.02084	0.01042
Carbonate (Co_1^{2-})	0.03333	0.01666	Potassium (K^+)	0.02557	0.02557
Caesium (Cs^+)	0.00752	0.00752	Rubidium (Rb^+)	0.01170	0.01170
Chloride (Cl^-)	0.02821	0.02821	Silica (SiO_2)	–	0.01664
Cobalt (Co^{2+})	0.03394	0.01697	Silver (Ag^{2+})	–	0.00927
Fluoride (F^-)	0.05264	0.05264	Sodium (Na^+)	0.04350	0.04350
Hyrogen (H^+)	0.99209	0.99209	Strontium (Sr^{2+})	0.02283	0.01141
Hydroxide (OH^-)	0.05880	0.05880	Sulphate (SO_4^{2-})	0.02082	0.01041
Iodide (I^-)	0.00788	0.00788	Sulphide (S^{2-})	0.06238	0.03119
Iron (Fe^{2+})	0.03581	0.01791	Zinc (Zn^{2+})	0.03060	0.01530
Iron (Fe^{3+})	0.05372	0.01791			

Note: Multiply concentration in mg/l by conversion factor.

The problem with all such modelling of groundwater chemistry is that it is based on the assumption that inorganic chemical equilibrium exists in solution. Unfortunately, such equilibrium is often not attained by redox-sensitive species, such as NO_3^-/N_2 and H_2S/SO_4^{2-} and the models need to be interpreted with caution.

Describing the concentration or relative abundance of major and minor constituents and the pattern of variability is part of nearly every groundwater investigation. Over the years a number of different graphical and statistical techniques have been developed, particularly for the presentation of major ion patterns. Examples are:

1. Bar (or Collins) diagrams;
2. Pie diagrams;
3. Pattern (or Stiff) diagrams;
4. Vertical scale (or Schoeller) diagrams;
5. Bivariate or scatter plots;
6. Trilinear (or Piper) diagrams; and
7. Durov diagrams (development of Piper).

Examples of some of these graphical representations are shown in Figure 6.8.

The most useful plot for comparison of large numbers of groundwater analyses is the Piper diagram. By classifying samples on such a diagram one can identify geological units with chemically similar water and define the evolution of groundwater chemistry along a flow system. Drawbacks are that, because concentrations are expressed as percentages, variations in absolute concentrations are not shown clearly, and analyses of similar mixtures but different total concentrations are not differentiated.

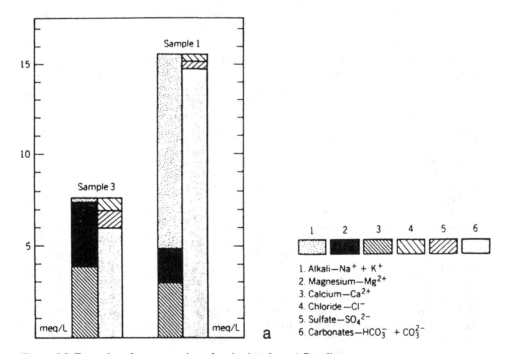

Figure 6.8. Examples of representation of major ion data. a) Bar diagram.

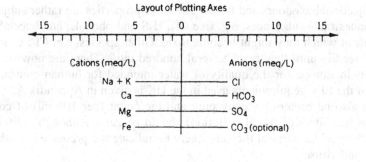

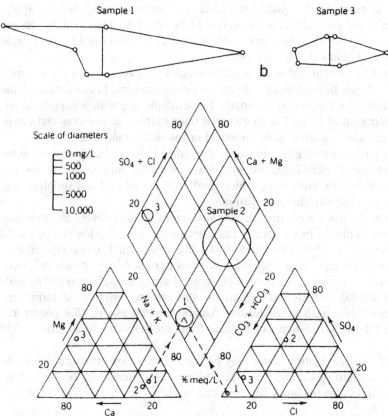

Figure 6.8. Continued. b) Stiff diagram; c) Piper diagram (from Domenico & Schwartz 1990).

6.5 WATER QUALITY STANDARDS

For use as drinking water and for irrigation, quality standards have been set by many national and intergovernmental agencies. Water supplied for drinking must be wholesome, in other words it must meet high standards of physical, chemical and biological purity. It must be free from undesirable physical properties such as colour, turbidity, and objectionable odours and tastes. The latter properties are rather subjective and many undesirable substances, e.g. free Cl_2, H_2S and phenols, are detectable at or below levels at which they might become toxic. Other species, e.g. Cl^-, cannot be distinguished readily until they reach several hundred mg/l. There are now statutory requirements in Europe for the quality of water intended for human consumption. A version of the EEC requirements used in the UK is given in Appendix A.

Water should also be bacteriologically pure and the count (per 100 ml) of coliform bacteria (as indicators of faecal pollution) should be zero. Although coliform bacteria are not themselves harmful they are likely to indicate the presence of pathogenic organisms and viruses.

For livestock, many of the quality standards for domestic supplies also apply. However, animals are generally able to tolerate higher total dissolved solids concentrations that humans. For example, up to 10,000 mg/l can be tolerated by beef cattle although levels well below this are desirable.

It is important to recognise that a particular species may be toxic above a certain concentration, yet may be essential to life at low concentrations. In certain cases, the optimum concentration range may be small. For example Se, with a maximum admissible concentration of 10 μg/l is an essential trace nutrient at low concentrations, and Se deficiency diseases are well documented in veterinary medicine.

The most extensive use of groundwater in the world is for the irrigation of crops. The main difference between irrigation waters and domestic supplies is the absence of pretreatment and anaerobic waters, waters high in iron and polluted supplies (e.g. nitrate-rich) may be used for direct irrigation.

For irrigation, the main concerns are the build-up of salinity levels and presence of elements toxic to plants. In the former case, geochemical evaluation can be useful in predicting minerals likely to precipitate following irrigation. The drainage characteristics of the soil are equally as important as the water chemistry. It may be possible to irrigate with quite saline water on well-drained sandy soils, but on clayey soils saline build-up and the formation of gypsum or other minerals may occur during irrigation with reasonably fresh groundwater. Although some plants, like cotton and sugar beet, are relatively resistant to Cl, some fruits are sensitive to as little as 350 mg/l Cl.

An excess of Na in irrigation water may be damaging to soil structure due to the replacement by sodium of Ca and Mg. The sodium adsorption ratio (SAR) is commonly used to express the tendency for these cation exchange reactions to occur:

$$SAR = \frac{Na^+}{\sqrt{(Ca^{2+} + Mg^{2+})/2}} \quad \text{(cations expressed as meq/l} \tag{6.5}$$

Other elements are essential for plant growth at low concentrations but extremely toxic at concentrations slightly above the optimum. For example B is an essential element but in excess of 2 ppm in irrigation water it can become toxic and some plants can be adversely affected by concentrations as low as 1 ppm particularly in a commercial greenhouse environment.

APPENDIX A

Prescribed concentrations or values specified in the UK Water Supply (Water Quality) regulations 1989.

Coulour	mg/l Pt/Co scale	20
Turbidity	Formazin Turbidity Units FTU	4
Odour	Dilution number	3 at 25°C
Taste	Dilution number	3 at 25°C
Temperature	°C	25
Hydrogen ion (pH)	pH value	9.5 max. 5.5 min.
Sulphate	mg SO_4/l	250
Magnesium	mg Mg/l	50
Sodium	mg Na/l	150
Potassium	mg K/l	12
Dry residues	mg/l	15000
Nitrate	mg NO_3/l	50
Nitrite	mg NO_2/l	0.1
Ammonium	mg NH_4/l	0.5
Kjeldahl nitrogen	mg N/l	1
Oxidizability	mg O_2/l	5
Dissolved or emulsified hydrocarbons	µg/l	10
Phenols	µg C_6H_5OH/l	0.5
Surfactants	µg/l as lauryl sulphate	200
Aluminium	µg Al/l	200
Iron	µg Fe/l	200
Manganese	µg Mn/l	50
Copper	µg Cu/l	3000
Zinc	µg Zn/l	5000
Phosphorus	µg P/l	2200
Fluoride	µg F/l	1500
Silver	µg Ag/l	10
Arsenic	µg As/l	50
Cadmium	µg Cd/l	5
Cyanide	µg CN/l	50
Chromium	µg Cr/l	50
Mercury	µg Hg/l	1
Nickel	µg Ni/l	50
Lead	µg Pb/l	50
Antimony	µg Sb/l	10
Selenium	µg Se/l	10
Individual pesticides	µg/l	0.1
Total pesticides	µg/l	0.5
Polycyclic aromatic hydrocarbons	µg/l	0.2
Total coliforms	number/100 ml	0
Faecal coloforms	number/100 ml	0
Faecal streptococci	number/100 ml	0
Sulphite-reducing clostridia	number/20 ml	<1
Conductivity	µg S/cm	1500 at 20°C
Chloride	mg Cl/l	400
Calcium	mg Ca/l	250

Substances extractable in chloroform	mg/l dry residues	1
Boron	μg B/l	2000
Barium	μg Ba/l	1000
Benzo 3,4 pyrene	ng/l	10
Tetrachloromethane	μg/l	3
Tricholoethane	μg/l	30
Tetrachloroethane	μg/l	10
Total hardness	mg Ca/l	60 minimum if supply is softened
Alkalinity	mg HCO3/l	30 minimum if supply is softened

REFERENCES

Clesceri, L.S., Greenberg, A.E. & Trussell, R.R. (eds) 1989. *Standard Methods for the Examination of Water and Wastewater*, 17th edition. American Public Health Association Washington USA.

Domenico, P.A. & Schwartz, F.W. 1990. *Physical and Chemical Hydrogeology*. John Wiley, New York.

Edmunds, W.M. 1986. Groundwater chemistry. In: Brandon, T.W. (ed.), *Groundwater: Occurrence, Development and Protection*. Inst. Water Engineers and Scientists, London, 49-107.

Edmunds, W.M., Cook, J.M., Kinniburgh, D.G., Miles, D.L. & Trafford, J.M. 1989. *Trace Element Occurrence in British Groundwaters*. British Geological Survey Research Report SD/89/3.

Lloyd, J.W. & Heathcote, J.A. 1985. *Natural Inorganic Hydrochemisty in Relation to Groundwater: An introduction*. Clarendon Press, Oxford.

Potts, P.J. 1987. *A Handbook of Silicate Rock Analysis*. Blackie, Glasgow and London, 622 pp.

Thompson, M. & Walsh, J.N. 1989. *Handbook of Inductively Coupled Plasma Spectrometry*, 2nd Ed. Blackie, Glasgow and London, 316 pp.

Walsh, J.N. 1993. Modern analytical methods applied to groundwater monitoring. In: *Conference Documentation E0136 Conference on Groundwater Pollution 16/17 March 1993*, London IBC Technical Services Ltd.

CHAPTER 7

Environmental isotopes as tracers in catchments

SYLVI HALDORSEN, GUNNHILD RIISE & BERIT SWENSEN
Department of Soil and Water Sciences, Agricultural University of Norway, Ås, Norway

RONALD S. SLETTEN
Quaternary Research Center, University of Washington, Seattle, USA

7.1 INTRODUCTION

7.1.1 *Environmental isotopes*

Individual forms of an element, characterised by the number of neutrons they contain, are called isotopes of an element. Isotopes of a given element differ slightly in mass due to different number of neutrons in relation to proton number (atomic number). The relative abundance of the different isotopes of an element will vary, one of them being dominant. 'Environmental isotopes' are those isotopes that are widespread in nature and are part of the hydrological and/or geological cycle. They may be stable or radioactive (Table 7.1) and may partly be introduced into nature by human pollution (e.g. by thermonuclear atmospheric experiments). Isotopes of oxygen (^{16}O, ^{18}O) and hydrogen (^1H, ^2H and ^3H) are by far the most frequently used isotopes in water studies. Their applicability was pointed out more than thirty years ago (Dansgaard 1964), and their use in hydrological studies has increased gradually since that time and they currently are some of the most commonly applied and most useful parameters in many catchment studies. In addition, other isotopes are also used in the study of catchment processes, among which carbon (^{12}C, ^{13}C, ^{14}C), nitrogen (^{14}N, ^{15}N), strontium (^{86}Sr, ^{87}Sr), sulphur (^{32}S, ^{34}S), uranium (^{234}U, ^{238}U), thorium (^{232}Th), radium (^{226}Ra), radon (^{222}Rn) are of special importance. The applicability of isotopes for a large variety of natural environments is documented in a number of symposium books and other publications (e.g. IAEA 1974; Fritz & Fontes 1980; IAEA 1983; Fontes 1985; Faure 1986; Fritz & Fontes 1986; Hoefs 1987; Kendall 1990; Peters et al. 1993; Lajtha & Michener 1994; McDonnell & Kendall 1996), and the reader is referred to these publications for more detailed descriptions of isotope studies in catchments. In this chapter the use of oxygen, hydrogen, carbon, nitrogen and sulphur isotopes is described. The selection is based on the fact that these isotopes are the ones most frequently studied in catchments, and very applicable for the study of catchments where the storage time of water is relatively short.

Why are environmental isotopes so useful in catchment studies? Many of them are introduced with precipitation and thereby form a natural part of the water system. In contrast to artificially injected tracers, they are distributed evenly over the whole catchment. Other isotopes are mainly related to geology and thereby give valuable information about chemical processes taking place between water, biological material and minerals in a catchment.

Normally, a given isotope of an element (e.g. ^{16}O) is overwhelmingly abundant compared to other isotopes of that same element (see Table 7.1). Natural changes in the relation between dominant and rarer isotopes due to fractionation are often too small to be measured accurately. Therefore isotopic abundances are reported as positive or negative deviations away from a standard. This convention, yielding the δ-values, was established in the following general equation (Fritz & Fontes 1980):

$$\delta\ (\text{‰}) = \frac{R_{\text{sample}} - R_{\text{standard}}}{R_{\text{standard}}} \times 1000 \tag{7.1}$$

where R is the isotope ratio given by the relation between the number of atoms for two isotopes, e.g. $^{18}O/^{16}O$, $^2H/^1H$, $^{13}C/^{12}C$, $^{15}N/^{14}N$, $^{34}S/^{32}S$.

Radioactive isotopes are characterised by decay, which occurs mainly by emission of α or β particles and γ-radiation. The decrease in the abundance or activity of a radioactive substance is commonly expressed in terms of radioactive half-life ($t_{1/2}$), which is the time required to reduce the number of parent atoms by one-half. The half-life time varies considerably for the radionuclides of interest (see Table 7.1).

Table 7.1. Average terrestrial abundance of the isotopes of major elements used in environmental studies (based on Fritz & Fontes 1980).

Element	Isotopes	Average terrestrial abundance (%)	Comments
Hydrogen	^1H	99.984	
	^2H	0.015	
	^3H	10^{-14} to 10^{-16}	Radioactive $t_{1/2} = 12.35$ years
Carbon	^{12}C	98.89	
	^{13}C	1.11	
	^{14}C	10^{-10}	Radioactive $t_{1/2} = 5730$ years
Oxygen	^{16}O	99.76	
	^{17}O	0.04	
	^{18}O	0.1	
Nitrogen	^{14}N	99.64	
	^{15}N	0.366	
Sulphur	^{32}S	95.02	
	^{33}S	0.76	
	^{34}S	4.21	
	^{36}S	0.02	
Strontium	^{86}Sr	9.86	
	^{87}Sr	7.02	
	^{88}Sr	82.56	
Uranium	^{234}U	0.006	Radioactive $t_{1/2} = 2.47 \cdot 10^5$ years
	^{238}U	99.27	Radioactive $t_{1/2} = 4.51 \cdot 10^9$ years
Thorium	^{230}Th		Radioactive $t_{1/2} = 1.4 \cdot 10^{10}$ years
Radium	^{226}Ra		Radioactive $t_{1/2} = 1.6 \cdot 10^3$ days
	^{222}Ra		Radioactive $t_{1/2} = 3.8$ days

The groundwater age (t) can be described mathematically as:

$$t = -\frac{t_{1/2}}{\ln 2} \cdot \ln \frac{A_{obs}}{A_o} \tag{7.2}$$

where A_o is the assumed activity before any decay took place and A_{obs} is the activity at a given subsequent time.

7.1.2 *The use of environmental isotopes in small catchments*

Northern catchments such as in Scandinavia often are characterised by rather rugged topography, and well-defined catchment boundaries. River discharges have strong seasonal variations with snow melt as the main flood event. Rather thin unconsolidated glacial sediments, mostly tills, cover the lithified bedrock. The unconsolidated sediments usually have a storativity and permeability considerably higher than the underlying fractured bedrock (Knutsson & Morfeldt 1993). Most of the groundwater flow thus occurs in the Quaternary sediments (Fig. 7.1). Due to the limited thickness of the sediment overburden, saturated zones are thin and shallow groundwater often dominates the stream base flows (Anderson & Burt 1990; Haldorsen et al. 1992; Grip & Rodhe 1994). Deep groundwater related to fractured bedrock normally forms a smaller part of the base flow (Haldorsen et al. 1993).

Environmental isotopes may be useful in connection with groundwater flow and weathering studies in such catchments and can assist in solving the following problems:

1. Hydrograph separation (i.e. determining the origin of water in streams): Surface runoff ($\delta^{18}O$, δ^2H), shallow groundwater ($\delta^{18}O$, δ^2H, ^3H), groundwater components with transit times of years (^3H);

2. The annual variation in the hydrogeological regime: When and where does the groundwater recharge occur ($\delta^{18}O$);

3. Aquifer residence time: Groundwater discharging into rivers ($\delta^{18}O$, ^3H), and, more rarely, groundwater components with transit times of thousands of years (^{14}C);

4. Weathering processes in the catchment ($\delta^{13}C$);

5. Biochemical processes influencing upon weathering ($\delta^{15}N$, $\delta^{34}S$).

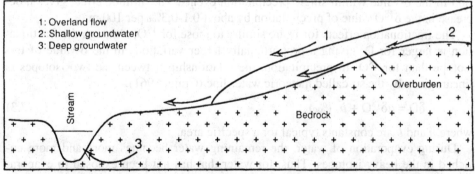

Figure 7.1. Main water components of a northern catchment: 1. Overland flow, 2. Shallow groundwater in unconsolidated sediments, 3. Deep groundwater.

7.2 OXYGEN AND HYDROGEN ISOTOPES

7.2.1 *Stable isotopes:* ^{18}O *and* ^2H *(D)*

The ^{18}O and ^2H (D = deuterium) isotopes are integrated parts of the water molecule and are therefore well-suited for hydrological tracing and budget studies. The ^{18}O and D contents are normally expressed as $\delta^{18}O$ and δD, related to the standard SMOW (Standard Mean Ocean Water, Craig 1961). The general principles of fractionation of these isotopes and their use in catchment studies are discussed by, e.g. Fontes (1980), Gat & Gonfiantini (1981), and Rodhe (1987).

The use of ^{18}O and D in catchment hydrology is mostly based on temporal variations in the isotope content of the water input (rainfall or snowmelt). Some studies rely on areal variations (variations with altitude) of the isotope input and some may rely on isotope changes of the water due to fractionation processes within the catchment. Temporal and geographical variations in the $\delta^{18}O$ and δD of precipitation are caused by fractionation during condensation of the atmospheric water vapour. Water molecules containing the heavy isotopes are less likely to evaporate and more likely to condensate than those containing the lighter isotopes. During the poleward atmospheric transport from the vapour supply regions around the equator, the vapour becomes more and more depleted in the heavy isotopes, giving a progressively more depleted precipitation. The lower the temperature is at a certain location, or at a certain time of the year, the larger is the effect of this preferential condensation and the more depleted is the precipitation. From an analysis of global precipitation data and the effects of fractionation processes, Dansgaard (1964) showed that the $\delta^{18}O$ of precipitation decreases with:
– Decreasing surface air temperature, shown as a geographical as well as a seasonal variation;
 – Increasing latitude;
 – Increasing altitude;
 – Increasing distance from the oceans in the direction of vapour transport.

At high latitudes, the winter precipitation may be depleted in ^{18}O by about 10‰ in the monthly mean values as compared to the monthly mean values during summer (Fig. 7.2). Superimposed on a sinusoidal variation in the monthly mean values, there is a large scatter in the daily δ-values and there may also be a large temporal variation in the δ-value within single precipitation events. The altitude effect gives a decrease in the $\delta^{18}O$ value of precipitation by about 0.1-0.3‰ per 100 m.

The fractionation effects for D are similar to those for ^{18}O, although the fractionation is larger for D, giving a proportionally larger variations in the δ-value of this isotope than for ^{18}O. In precipitation, the relationship between the two isotopes is linear, given by the so-called meteoric water line (Craig, 1961):

$$\delta D = a\delta^{18}O + b \quad (‰) \tag{7.3}$$

where a and b are constants typical for a specific area.

During evaporation of water, the remaining water becomes more and more enriched in the heavy isotopes. Thus, for water that has not been exposed to evaporation, ^{18}O and D will give the same information since they are uniquely related. However, for tracing water that has been exposed to evaporation, e.g. groundwater

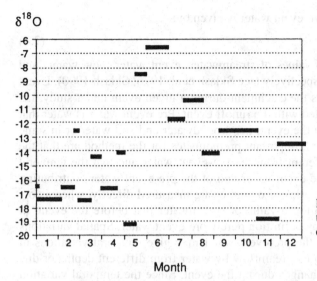

Figure 7.2. Variation of $\delta^{18}O$ in the precipitation, Venabu, southeastern Norway 1992.

recharged by lake water, simultaneous measurements of both isotopes may give valuable information.

In catchment studies, ^{18}O and D are normally treated as conservative tracers, i.e. concentration changes are assumed to be caused only by mixing of water with different δ-values. This is a safe assumption for non-thermal groundwater and a relatively safe assumption for soil water in vegetated areas, since transpiration is non-fractionating (in contrast to the direct evaporation of soil water). It should to be noted, however, that any exposure of water to the atmosphere, such as storage in permanent or temporary surface reservoirs, may cause an enrichment which must be considered during date interpretation.

There is also a fractionation taking place during melting/freezing of water. The heavy isotopes are slightly more inclined to freeze and less inclined to melt than the lighter isotopes, making the meltwater depleted in ^{18}O and D compared to the snowpack, whose δ-value largely reflects the δ-values of the winter snowfalls. For this reason, and also due to a possible enrichment of the snowpack by molecular exchange with the atmosphere, the δ-values of meltwater must be measured directly in order to obtain a good accuracy of the isotopic input to the ground during snowmelt.

7.2.2 *Application of $\delta^{18}O$ and δD in catchment studies*

7.2.2.1 *Hydrograph separation*
In isotopic hydrograph separation the streamwater is normally separated into flows of new water (event water) and old water (pre-event) water. Old water is water that was already in the catchment before the runoff event. The technique is based on the dampening of the temporal variations of the isotope input signal taking place during water flow through a catchment. During some events, the isotopic composition of the rainfall or meltwater differs from that of the water in the catchment, making separa-

tion possible. The fraction of pre-event water is given by:

$$X = (\delta_s - \delta_e)/(\delta_{pe} - \delta_e) \qquad (7.4)$$

where δ_s, δ_e and δ_{pe} are the δ-values of streamwater, event water (new water) and pre-event (old water) water, respectively (see Sklash et al. 1976; Sklash 1990). Event water is liquid water that enters the catchment during a given event under study, i.e. rainwater or meltwater associated with the runoff event. Pre-event water is water that existed in the catchment before the event, i.e. groundwater and soil water or in some cases water in surface reservoirs. Important prerequisites for the method are that the event water is isotopically different from the pre-event water and that the temporal variations in the δ-values of the event water and of the pre-event water contributing to streamflow are small. Time series of δ_s and δ_e are measured directly, whereas δ_{pe} often is assumed equivalent to the δ-value of streamwater just before the event, at which time the streamwater is by definition purely pre-event water. Spatial variations in the isotopic composition of the pre-event water may give temporal variations in δ_{pe}, as the relative contribution to streamflow by water from different depths or different parts of the catchment changes during the event. Since the temporal variation of δ_{pe} cannot be measured directly, the uncertainty in δ_{pe} constitutes an important problem in isotopic hydrograph separation.

The assumption of a constant δ_{pe}, equal to the δ-value of streamflow before the event, may make the total fraction of groundwater during the event larger than that of pre-event water. This can occur where a proportion of the new water reaches the stream as groundwater after a short residence time in the groundwater zone.

Apart from the ^{18}O and D isotopes, the radioactive hydrogen isotope ^3H (tritium = T) has been used in hydrograph separations. As with the stable isotopes, these studies rely on the seasonal variation or daily scatter of the isotope content of precipitation, and, since the three isotopes are mainly conservative during water flow through a catchment, they give largely the same information. Equation (7.4) has also been used with chemical compounds as tracers, i.e. electrical conductivity and Si. In contrast to the isotopes, these tracers are not conservative and the tracer concentration may change along the flowpaths due to chemical reactions with the surrounding minerals. As pointed out by Wels et al. (1990), separation by isotopes gives information on the origin of water (event/pre-event water) whereas separation by non-conservative chemicals may give information on the flowpaths (e.g. flow through mineral soil or organic layer only).

Early isotope studies of catchment response to melting of seasonal snowpacks are those by Dincer et al. (1970) and Martinec et al. (1974). These studies were carried out in the mountaineous areas of Central Europe. These studies gave important contributions to the development of the modern view of streamflow generation in humid catchments in that they indicated a dominance of pre-event water in snowmelt runoff. Indications of large contributions by groundwater to stream runoff during events generated by rainfall also have been given by other studies using electrical conductivity or major ions (Newbury et al. 1969; Pinder & Jones 1969) or tritium (Crouzet et al. 1970). However, the snowmelt studies by Dincer et al. (1970) and Martinec et al. (1974) gave the first and more striking evidence of the indirect mechanism of streamflow generation. Their findings have been supported by several later isotopic studies of snowmelt as well as events generated by rainfall, e.g. Fritz et al. (1976),

Sklash & Farvolden (1979), Wels et al. (1990) and Harum & Frank (1992) in Canada, Hooper & Shoemaker (1986) and Kennedy et al. (1986) in USA, Lindström & Rodhe (1986), Rodhe (1987) in Sweden, McDonnell (1990) in New Zealand and Gunyakti & Altinbilek (1993) in Turkey.

A large number of studies of old and new water components have been carried out, and the Gwy catchment studied by Sklash et al. (in print) is very illustrative for what is now a commonly observed situation in many catchments. The study was based on δD-analysis, and verified by $\delta^{18}O$ and electrical conductivity measurements. Water samples were taken of precipitation of the Gwy stream, of a shallow pipe (diameter > 4 cm) and of groundwater. Even for the pipe flow, the water during a storm runoff event consisted mainly of old water, with a storage time of more than forty days.

The study by Sklash et al. (in press), as well as many other hydrograph separation studies which involve isotope studies, are based on a two component approach, where the new water and the old water are assumed to be spatially and temporarily homogeneous. The old water composition may be estimated from the base flow composition of the stream. However, this approach has been questioned by several authors (e.g. Bazemore et al. 1994), as each of the water components may have a significant isotopic variation during a certain runoff event. McDonnell et al. (1990) and Kendall & McDonnell (1993) found that the isotopic composition may vary considerably within the same precipitation event, and suggested that a varying isotope composition for the new water component is required in many hydrograph separations. Hooper & Shoemaker (1986) suggested that the old water component may also vary considerably during a specific event. The water in the unsaturated zone may also be important for the composition of the old water component (Kennedy et al. 1986; De Walle 1988; Mulholland 1993). This composition may be different from that of the groundwater, which forms the base flow of many streams. Old water in the unsaturated zone may be mixed with groundwater during a storm event, and can form an important part of the old water component at least during the last part of the runoff event.

7.2.2.2 *Transit times for shallow groundwater*
The seasonal variation of stable oxygen and hydrogen isotopes can be traced in the soil moisture and the movement of the depleted or enriched moisture layers can be monitored in time, giving a direct measure of water particle velocity, fluxes, and transit time in the unsaturated zone (see Dincer & Davis 1984).

Fontes (1980) concluded that oxygen isotopes and deuterium are of limited value for calculation of residence times of groundwater compared with ^3H and ^{14}C. For groundwater with a long residence time and for confined aquifers this is clearly the case. However, in mountain catchments, where small phreatic groundwater systems dominate, there is a distinct variation in the chemistry and groundwater level during the year. In such aquifers the average transit time may be less than one year and stable isotopes may give information on residence times (Rodhe 1987; Haldorsen et al. 1993; Kendall & McDonnell 1993; Nyberg 1995).

An example of this can be given from south-eastern Norway, where a groundwater spring had the lowest $\delta^{18}O$ values in early May 1989 (Fig. 7.3; Haldorsen 1994). This reflects the discharge of meltwater from the cold winter precipitation (see Fig.

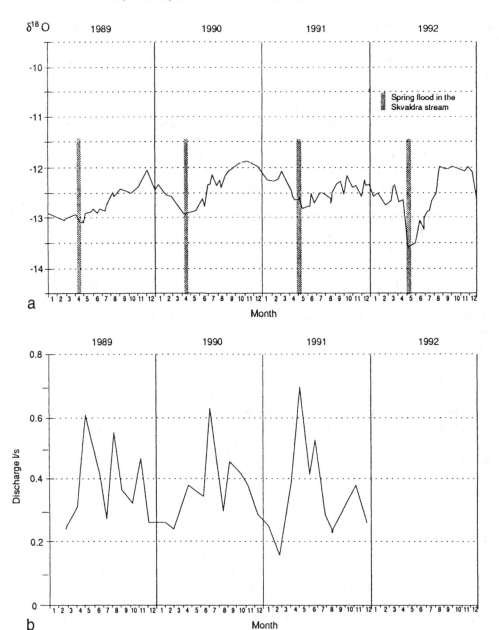

Figure 7.3. The δ^{18}O-variation in a groundwater spring in Åstdalen, southeastern Norway. The snowmelt period (springflood) is indicated by the thick bars.

7.2) reaching the spring outlet. At that time the discharge was at a maximum (Fig. 7.3B). The snow melt took place less than a month earlier (Fig. 7.3A). The average residence time for the groundwater was thus only a few weeks. The δ^{18}O of the groundwater spring increased during the summer and reached a maximum in late autumn. This reflects the influence of a mildly ^{18}O-enriched summer precipitation component within the groundwater. The maximum $\delta^{18}O$-content in groundwater (−12‰) reflects the relative maximum of summer precipitation (−8‰), which occurred at a time when the spring discharge was low. The $\delta^{18}O$ of precipitation was quite high in August 1989, which is normally a wet month in inland southern Norway. From September to November 1989, the $\delta^{18}O$ value of the precipitation is normally very low (on average −17‰). The precipitation that caused the high $\delta^{18}O$ values of the groundwater spring in November must date from August or earlier. Thus, the transit time during summer must be at least two months, i.e. longer than during the spring.

7.2.2.3 *Recharge of groundwater: Area and time*
The first attempt to use oxygen isotopes for estimating groundwater recharge zones was made in France by Fontes et al. (1967). The general basis for this method is that the decrease of $\delta^{18}O$ in precipitation with increasing altitude is also reflected in the groundwater (Payne & Yurtsever 1974; Fontes 1980; Kovac & Drost 1992).

Yasuhara et al. (1993) studied spring outlets at different altitudes along the same slopes, and concluded that the recharge zones were mainly restricted to the areas directly upstream of each spring level. A parallel study was done for three groundwater springs in a mountain catchment in south-eastern Norway (Fig. 7.4; Haldorsen 1994). Spring A at altitude 700 m a.s.l. has on average $\delta^{18}O$ values 0.2‰ higher than spring B at 760 m a.s.l. Spring C at 920 m a.s.l. has on average $\delta^{18}O$ values 0.3‰ lower than spring B. This is in accordance with the expected altitude-related difference in precipitation, indicating that recharge areas are located not far from the spring outlets.

Calculations of recharge periods for groundwater within a catchment can be based on monthly $\delta^{18}O$ values of precipitation, precipitation rates, $\delta^{18}O$ of the groundwater and groundwater discharge measurements. A groundwater spring in Norway had an annual average $\delta^{18}O$ value of −12.5‰ for the period October 1990 to October 1991. The average $\delta^{18}O$ value of the precipitation in winter (October 1990 to April 1991) was −12.6‰, and −10.5‰ in summer (May 1991 to September 1991). It can be concluded that winter precipitation dominated groundwater recharge.

7.2.2.4 *Oxygen isotopes in carbonate minerals*
The stable isotopes ^{13}C and ^{18}O in dissolved CO_2 and H_2O, respectively, may be reflected in the isotopes of carbonate minerals that are precipitated from water. The discussion below is for ^{18}O; carbon isotopes are discussed separately.

The fractionation factor, α, is defined as the increase or decrease of a given isotope in a substance that undergoes a phase change. The values of α at 0°C for the fractionation of ^{18}O in $CaCO_3$ (calcite) in isotopic equilibrium with $H_2O_{(l)}$ is given by:

$$\alpha\,(^{18}O) = \frac{1 + \delta^{18}O_{CaCO_3}\,/\,1000}{1 + \delta^{18}O_{H_2O}\,/\,1000} = 1.0349 \tag{7.5}$$

Relative to the water from which it precipitates, calcite will be enriched in ^{18}O by approximately 35‰. Fractionation factors at other temperatures are given in Freidman and O'Neil (1977). Precipitation of $CaCO_3$ can occur when the activity of Ca^{2+} and CO_3^{2-} exceeds the solubility product, and this may be induced by pH changes or

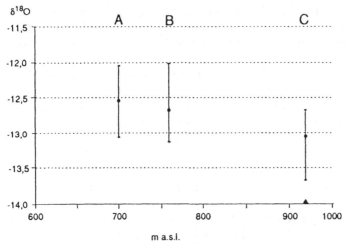

Figure 7.4. The relation between the $\delta^{18}O$ and altitude of three springs in Åstdalen, 1989, southeastern Norway.

by concentrations of ions during freezing or evaporation. If precipitation occurs due to a pH change, $\delta^{18}O$ of water will not change and Equation (7.5) can be used to estimate the ^{18}O in the water from the calcite or the expected value of the ^{18}O in calcite. In the case of ionic enrichment due to freezing or evaporation, the fractionation of ^{18}O in the water must be accounted for.

Fractionation factors during water condensation and freezing are given by:

$$\alpha\,(^{18}O) = \frac{1 + \delta^{18}O_{(H_2O_{(l)})} / 1000}{1 + \delta^{18}O_{(H_2O_{(g)})} / 1000} = 1.003 \tag{7.6}$$

$$\alpha\,(^{18}O) = \frac{1 + \delta^{18}O_{(H_2O_{(s)})} / 1000}{1 + \delta^{18}O_{(H_2O_{(l)})} / 1000} = 1.0111 \tag{7.7}$$

Liquid water is enriched in ^{18}O during evaporation since it is the heavier isotope, and it is depleted during freezing (note that $H_2O_{(l)}$ appears in the denominator for the equation describing freezing). Carbonates precipitated from solutions that have become supersaturated due to evaporation will tend to be enriched in ^{18}O. This is typically the case for carbonates in hot deserts (Schlesinger 1985) and cold, dry environments such as the Arctic (Sletten 1988). If supersaturation is induced by freezing, the $\delta^{18}O$ of precipitating carbonates will be lower (^{18}O accumulates preferentially in the ice). However, little evidence is available at present to support this theory.

A correlation between $\delta^{13}C$ and $\delta^{18}O$ in carbonates has been noted in several studies. This has been attributed to the fact that the abundance of both these isotopes may be correlated with temperature (Cerling & Quade 1993). The ^{18}O fraction in precipitation can be correlated with freezing in the atmosphere and ^{13}C can be correlated with biological activity; both of these are affected by temperature.

The $\delta^{18}O$ values of the pedogenic carbonates may indicate both freezing and evaporation as mechanisms for generating supersaturation. Both of these processes may occur in a given soil and therefore it may not be possible to quantitatively assess the relative significance of each mechanism. It appears that both of these processes may be important and qualitative estimates of the intensity may be made by comparing the $\delta^{18}O$ levels in the carbonates.

7.2.3 *Tritium*

A general description of the radioactive hydrogen isotope ^3H, tritium (Table 7.1), applied to hydrological studies is given by Fontes (1980, 1985). Tritium has been the most commonly used radioactive isotope for water dating. Since its half-life is relatively short (12.35 years, Table 7.1) it is well suited for dating of groundwater in catchments. It behaves similarly to deuterium in its fractionation, in that it follows the water phase, like the other 'water' isotopes described above, and may therefore give a direct age of the water. Tritium contents are expressed in tritium units (TU), where 1 TU = 1 tritium atom per 10^{18} atoms of ^1H (see Gat 1980).

The input of tritium via precipitation has been quite variable from before 1952 until around 1980. Tritium is generated naturally in the atmosphere by cosmic radiation, giving a natural tritium content in precipitation of less than 20 TU (Fontes 1980). Thermonuclear atmospheric experiments, which took place from 1952 to 1963 released large amounts of tritium, which very quickly became part of water molecules. The tritium concentration maximum was 1000-10,000 TU in precipitation around 1963, which is 100-10,000 times the natural level (Fontes 1980). Some of it still remains in the atmosphere. The tritium in the atmosphere today consists of three components:
1. Natural tritium;
2. Remaining bomb tritium which takes part in the hydrological cycle; and
3. Tritium leakage to the atmosphere from existing nuclear power plants.

The level of tritium in the atmosphere today is still much higher than the natural values prior to 1952.

Looking back over the last forty-five years, we have the following tritium events or 'markers': the very low tritium content prior to 1952, the marked jump around 1952, the maximum around 1963, and the relatively stable but low content after 1980.

Tritium activity in precipitation increases towards middle and high latitudes (Gat 1980). At any given location, the maximum activity of precipitation is observed in the springtime (spring peak) and corresponds to about three times the annual weighted mean (Gat 1980). The winter season is characterised by a minimum value of about one half this average. The length of transit time of wet air masses over continents or oceans leads to variations in the tritium content due to mixing with tritiated vapour and molecular exchange with the water surface, respectively. Put simply, continental precipitation is enriched in tritium with respect to marine rain. Since the tritium content of precipitation varies both in time and space, good historical background data for tritium in the local precipitation is necessary for residence time calculations of groundwater. However, even when such data exist, it can be complicated to use the tritium data directly for dating. It may be difficult to distinguish between

1963-water, in which only minor parts of the original tritium is left, and medium aged water from 1970 in which the initial tritium content was lower but which has undergone less radioactive decay.

Within the near future, the radioactive decay of bomb tritium will make it impossible to use the 1963 peak as a marker zone. It will make it difficult to date 'old' water and tritium will no longer be as useful for residence time calculations.

7.2.3.1 *Application*

Sampling groundwater at different depths in deep aquifers may detect the 1963 tritium maximum, giving a 'marker' horizon for groundwater older than ca. 30 years. One can also sample deeper until there is no traceable tritium, i.e. the water which was recharged before 1952. The depth of the mid-1960 peak may also be used for recharge rate calculations, because it gives information about the addition of water since that time. Between 1975 and 1990, many published groundwater age calculations and groundwater recharge rate estimations were based on tritium (e.g. Allison & Hughes 1974; Andersen & Sevel 1974; IAEA 1974; Foster & Smith-Carington 1980; Campana & Mahin 1985; Knott & Olimpio 1986; Siegel & Jenkins 1987; Poreda et al. 1990). The basic assumption for the estimations is that the flow is dominantly horizontal.

The more recent tritium values in precipitation and groundwater may form the basis for the age determination of water from the 1980's and onwards. Meinardi (1994) used the measured and calculated tritium values in precipitation and groundwater to estimate average residence times and recharge rates of shallow and deep groundwater in a number of Dutch aquifers. These calculations require good precipitation input data, since the background level of tritium is quite low, and may be of interest in many catchment studies. The interpretations will be greatly complicated unless it can be assumed that the flow in the unsaturated zone is primarily vertical.

Locally, higher tritium values can be found close to nuclear waste storage areas. Poreda et al. (1990) used tritium together with ^{3}He to identify groundwater components with transit time of 0.1 to 6 years in the Tennessee area.

7.3 CARBON ISOTOPES

In addition to ^{12}C, other carbon isotopes found in nature are ^{13}C and ^{14}C. ^{13}C is a stable isotope with a natural abundance of 1.11% of the total C and is useful for understanding the carbon cycle. ^{14}C is a radioactive isotope present in trace amounts that is produced primarily in the atmosphere. It has been found to be extremely useful for dating carbon-containing materials. ^{13}C is also used for correcting ^{14}C dates where isotopic fractionation has occurred.

7.3.1 *The ^{14}C isotope*

Dating by ^{14}C was developed in the 1950's (Libby 1955) for organic samples. The dominant source of ^{14}C is the atmosphere through the interaction of cosmic ray neutrons with the stable isotopes of oxygen, carbon, and nitrogen. The dominant reaction is with nitrogen (Faure 1986):

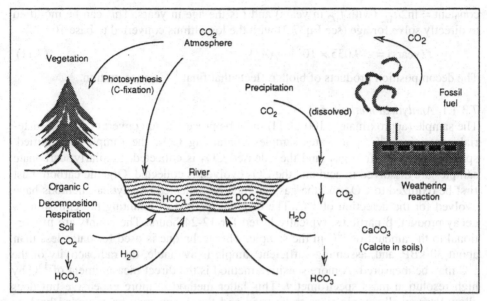

Figure 7.5. The carbon cycle, relevant for the origin of $\delta^{13}C$ and ^{14}C in catchment studies (after Mook 1980).

$$\ _0^1 n + \ _7^{14}N \rightarrow \ _6^{14}C + \ _1^1H \tag{7.8}$$

^{14}C from the atmosphere may then be incorporated into CO_2 and assimilated by biota, both terrestrial and marine. This balance in the production and loss of ^{14}C results in a steady state equilibrium of ^{14}C in the atmosphere that has been relatively constant for the past 7000 years. The atmospheric concentration of ^{14}C is about 1 atom of ^{14}C per 10^{12} atoms. Some natural variation of ^{14}C has occurred over time, but the most significant deviation in the recent past has been the input of ^{14}C as a result of atmospheric thermonuclear explosions. This resulted in a peak value of atmospheric ^{14}C in 1963. The concentration of ^{14}C may be expressed as percent modern carbon (pmc) where the ^{14}C of a sample is compared to the relative abundance of ^{14}C in nature.

Plants will also attain a steady state concentration of ^{14}C in their tissue, and animals feeding on the plants will attain a steady state level of ^{14}C. Upon the death of the plant or animal, input of ^{14}C ceases and begins to decay with a half-life of 5730 ± 40 years according to the reaction:

$$\ _6^{14}C \rightarrow \ _7^{14}N + \beta^- + Q \tag{7.9}$$

where β^- is a beta-particle and Q is the end point energy of the emitted β^- (156 keV). The loss of ^{14}C may be described by the equation:

$$A = A_0 e^{-\lambda t} \tag{7.10}$$

where A is the measured activity of ^{14}C, A_0 is the initial activity of ^{14}C, λ is the decay constant = $\ln 2/t_{1/2}$ (with $t_{1/2}$ in years) and t is the age in years. This can be modified to directly solve for age (see Eq. 7.2) with the logarithms converted to base 10:

$$t \text{ (years)} = -19.035 \times 10^3 \log (A/A_0) \tag{7.11}$$

The decomposition products of biota reflects that final ^{14}C concentration.

7.3.1.1 *Analytical methods for ^{14}C*

The sample (approximately 100-200 l) must be processed to convert or release material containing ^{14}C. For water samples containing CO_2, the sample is acidified, sparged with CO_2-free gas, and the released CO_2 is collected. Similarly, carbonate samples are dissolved in acid and the CO_2 evolved is collected. Organic carbon must first be oxidized to CO_2 and the gas can then be collected. Analytical methods have evolved for the detection of ^{14}C. The original method of counting the emission its decay product, β-particles, typically lasted for 12-24 hours. The counts are proportional to the amount of ^{14}C in the sample. This technique is used for dates less than about 30 kBP, and, assuming sufficient sample is available, the radioactivity of the ^{14}C may be measured. A more sensitive method is the direct measurement of ^{14}C by high resolution mass spectrometry. This latter method is more expensive but does allow very small sample sizes to be used and the dating may be extended back to about 100 kBP.

7.3.1.2 *^{14}C in groundwater*

Dating of groundwater by ^{14}C originating from dissolved carbonate species (CO_2, HCO_3^-, and CO_3^{2-}) has been used (Calf 1978; Mook 1980; Fontes 1985) and reliable ages in the range of 1000 to over 30,000 years have been obtained. This technique is generally not useful for shallow surface waters, but may be useful for groundwater, and surface waters which are fed by deep groundwater, with long residence times.

The source of carbon in groundwater is primarily atmospheric CO_2, due to decomposition of organic matter, dissolution of carbonate rocks, and volcanic activity. Reardon & Fritz (1978), Wigley (1978) and Mook (1980), have published reviews of processes that may alter ^{14}C activity in the unsaturated zone (Fig. 7.5). The dominant sources of ^{14}C in groundwater are discussed in the following.

1. CO_2 is produced primarily from the degradation of organic matter in soil, root respiration, and dissolution of limestone. The dissolved CO_2 may react with carbonate rocks[1] and release more dissolved carbonate products according to:

$$CaCO_3 + CO_2 \text{ (dissolved carbon dioxide)} + H_2O \rightarrow Ca^{2+} + 2HCO_3^- \tag{7.12}$$

In the equation above, if we assume that the ^{14}C is from decomposition of modern organic matter, it has a concentration of 100 pmc. Primary carbonate rocks have old carbon and the concentration of ^{14}C is 0 pmc. The dissolution products in the above reaction will have a value of 50 pmc;

2. The contribution of CO_2 from precipitation ($^{14}C = 100$ pmc) to water in a catchment is normally low compared to CO_2 produced in soils from decomposition and

---

[1] Secondary carbonate minerals, such as those formed in soils, may be younger than primary minerals and may have non-zero ^{14}C values.

respiration. An exception to this might be in alpine areas with sparse vegetation, where the atmospheric CO_2 may dominate;

3. CO_2 may also react with silicate minerals in weathering reactions, for example:

$$CaAl_2(SiO_4)_2 + 2 CO_2 + H_2O \rightarrow Ca^{2+} + Al_2O_3 + SiO_2 + 2HCO_3^- \qquad (7.13)$$

The ^{14}C content of the bicarbonate produced in this reaction will not be altered from the initial CO_2 since there is no loss or production of carbon;

4. Protons may also be released from organic matter in cation exchange reactions, and this in turn may dissolve carbonate minerals:

$$2CaCO_3 + 2H\text{-humic material} \rightarrow Ca\text{-humic material} +$$
$$Ca^{2+} + 2HCO_3^- \qquad (7.14)$$

Since the bicarbonate released is from the carbonate minerals, the ^{14}C activity is zero.

The amount of remaining ^{14}C is then used to determine the age of the organic matter. In addition to using ^{14}C for dating, ^{13}C is often measured in the same sample in order to correct for isotopic fractionation. Since ^{13}C is stable this may be used to estimate the original fractionation of ^{14}C in the sample; ^{14}C fractionation is expected to be double that of ^{13}C due to its greater mass.

7.3.2 *The ^{13}C isotope*

The stable isotope ^{13}C finds its greatest use in understanding the carbon cycle. The ^{13}C of HCO_3^- in the weathered zone is an important tool for understanding the carbonate geochemistry within a catchment. A very useful summary of this subject is given by Salomons & Mook (1986).

The sources of ^{13}C are analogous to those of ^{14}C: 1) biogenic carbon, 2) atmospheric carbon dioxide, and 3) carbonate rocks and minerals (Fig. 7.5).

^{13}C values are measured by mass spectrometry and it is much simpler to accurately determine isotopic ratios then absolute values. The $^{13}C/^{12}C$ ratio is therefore more commonly measured. The concentration is given as the $\delta^{13}C$ which compares the ratio of $^{13}C/^{12}C$ in the sample to a known standard multiplied by 1000, as defined in Equation (7.8). This gives the enrichment or depletion in parts per thousand (‰) compared to a standard: One of the most commonly used reference materials for ^{13}C is a limestone known as PDB (Peedee Belemnite).

$$\delta^{13}C = \frac{^{13}C/^{12}C_{(sample)} - ^{13}C/^{12}C_{(PDB)}}{^{13}C/^{12}C_{(PDB)}} \cdot 1000 \qquad (7.15)$$

7.3.2.1 *^{13}C in CO_2*

Currently, the average value of $\delta^{13}C$ in atmospheric CO_2 is approximately −8‰. The concentration has fallen recently from a value of −6.5‰ primarily due to the burning of fossil fuels, which are depleted in ^{13}C as discussed below in more detail. The concentration of CO_2 varies globally depending on the activity of the local vegetation. For example in forested ecosystems, the CO_2 might be elevated due to the release of CO_2.

7.3.2.2 ^{13}C in C3, C4 and CAM Plants

The original fractionation of ^{13}C in plants depends on the metabolic pathway of the plant species. Plants are classified into three classes, C3, C4, and CAM. Plants fractionate carbon isotopes by selective uptake of the 'lighter' CO_2, thereby resulting in organic matter that is depleted in ^{13}C relative to the atmospheric value of CO_2. C3 plants have a $\delta^{13}C$ of –25‰ to –32‰; these include trees, most shrubs and herbs, and grasses of cool regions (Cerling & Quade 1993; Ehleringer & Cerling 1995). C4 plants have a $\delta^{13}C$ of –10‰ to –14‰; these include many grasses of the prairie and savanna. CAM plants have an intermediate $\delta^{13}C$ and include cacti and other succulent plants.

7.3.2.3 Soil carbonates

Secondary carbonate minerals precipitated in soils reflect the $\delta^{13}C$ of the CO_2 in the soil atmosphere of the mineral; this is believed to be the case even when the alkalinity is due to dissolution of carbonate rocks. The reason for this is that the kinetics of carbonate precipitation are several orders of magnitude slower than the kinetics of CO_2 flux in typical soils (Cerling & Quade 1993). The carbonate minerals therefore provide an average index of the isotopic ratios in CO_2 during their formation. Isotopic fractionation must be accounted for as discussed below.

The fractionation factor, α, is defined as the increase or decrease of a given isotope for a given transition. The value at 0°C for the fractionation of ^{13}C in $CaCO_3$ in equilibrium with CO_2 is:

$$\alpha(^{13}C) = \frac{1 + \delta^{13}C_{CaCO_3}/1000}{1 + \delta^{13}C_{CO_2}/1000} = 1.01074 \qquad (7.16)$$

Since plants have lower $\delta^{13}C$ than the atmosphere, and, if the soil solutions have elevated CO_2 concentrations due to biological activity (and thus reduced $\delta^{13}C$ compared to atmospheric CO_2), the $\delta^{13}C$ of the carbonates would be expected to be lower than carbonates in isotopic equilibrium with atmospheric CO_2. One method to evaluate the stable isotope values is to consider what the values should be for $CaCO_3$ in equilibrium with atmospheric CO_2. The current global average value of the $\delta^{13}C$ of atmospheric CO_2 is approximately –8‰. Based on this, and using the α values given above, the expected value for $\delta^{13}C$ is 3.0‰. Marine carbonate rocks have $\delta^{13}C$ values between –4‰ and +4‰, with an average of +1‰. These values may change during metamorphism.

7.4 NITROGEN ISOTOPES

There are two stable isotopes of nitrogen in the environment, ^{14}N and ^{15}N (Table 7.1). Essentially, atmospheric nitrogen (N_2 forms ca. 80% of air) can be considered the principal reservoir for the nitrogen cycle. This external source may be regarded as in an overall steady state (Hübner 1986). The $\delta^{15}N$ is related to $(^{15}N/^{14}N)_{air}$ as the standard .

A summary paper on nitrogen isotopes in the natural environment was given by Létolle (1980) and $\delta^{15}N$ in the soil and biosphere is discussed in detail by Hübner

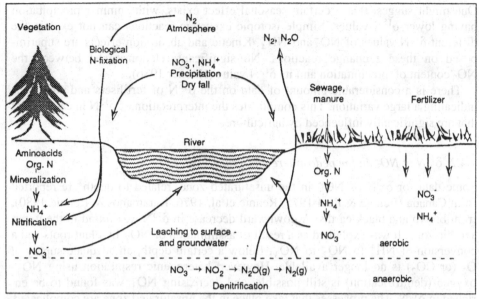

Figure 7.6. Schematic representation of the geochemical cycle of nitrogen in the upper lithosphere (after Létolle 1980).

(1986). Most studies of stable nitrogen isotopes in soils have been concerned with the distinction between fertiliser nitrogen and natural nitrogen.

The abundance and isotopic composition of nitrogen is essentially controlled by biologic rather than inorganic chemical processes. Very little is known about the $\delta^{15}N$ of sedimentary, igneous and metamorphic rocks.

In natural waters, nitrogen usually occurs as NO_3^-, NO_2^-, NH_4^+, and in amino acids (Fig. 7.6). The same compounds are found in soil, where most nitrogen is bound to organic matter. Each step in the chain of biochemical reactions may cause an isotopic fractionation, which means that the products of each steps may also have different isotopic compositions (Létolle 1980, Hübner 1986). However, the $\delta^{15}N$-values can not be used directly for the identification of specific processes. Only in very favourable situations is it possible to trace the history of a compound through its isotopic composition.

Two factors will control the $\delta^{15}N$ value of a chemical compound in a soil horizon:

1. Variations in (or constancy of) the $\delta^{15}N$ input values of the chemical compound. The input variables include precipitation, dry deposition and fertilisers.

2. Transformation processes of the particular chemical compound and related isotopic fractionation. Among these processes denitrification is very important.

All these variables have $\delta^{15}N$-values within a wide range.

7.4.1 *Nitrogen isotope variations in precipitation*

Freyer & Aly (1976) found that NH_4^+ in rains over western Germany had $\delta^{15}N$-values between –10 and –5‰, with the associated NO_3^- ranging from –5 to +5‰.

One might suggest that a certain seasonal effect exists, with summer precipitation having lower δ^{15}N-values. Simple isotopic exchange reactions can not explain the different δ^{15}N values of NO_3^- and NH_4^+. Kinetic and absorption effects are superimposed on these exchange reactions. No significant relations exist between the NO_3^- content of precipitation and its δ^{15}N value (Hübner 1986).

There is a considerable amount of data on the δ^{15}N of fertilisers and chemicals indicating a large variation. This complicates the interpretation of δ^{15}N in catchments that are significantly influenced by agriculture.

7.4.2 $\delta^{15}N$ of NO_3^- in the pedosphere and groundwater

Some data for δ^{15}N of NO_3^- in the unsaturated zone, related to depth are reported from Canada (Rennie & Paul 1975; Rennie et al. 1976, Karamanos & Rennie 1980), from brown and black earths. A downward decrease in δ^{15}N was found in the upper aerobic zone. It was explained as a result of assimilation of NO_3^- by plant roots and a conversion of NH_4^+ to NO_2^- to NO_3^-. Below a certain depth, an adequate supply of O_2 (or CO_2) is no longer available. However, heterogenic respiration using NO_3^--oxygen (denitrification) is still possible, and decreasing NO_3^- was found to be enriched in δ^{15}N. The processes that take place in the unsaturated zone are complicated. Since many of them involve isotopic fractionation as discussed above, it is not an easy task to interpret variations in δ^{15}N.

The δ^{15}N values of groundwater will depend on the nitrogen input to the groundwater and the reactions within the groundwater zone that involve nitrate. The δ^{15}N values of NO_3^- in groundwater reported by Aly (1975) show a decreasing δ^{15}N value with increasing NO_3^- content. The variable NO_3^- content results from denitrification, which causes an isotopic effect. The volatilisation of ammonia will give nitrate enriched in ^{15}N.

7.4.3 *Pollution studies in catchments and groundwater aquifers*

The relation between the δ^{15}N in streams and contamination by fertiliser nitrate was studied by Kohl et al. (1971) in Illinois. They related the nitrate content of drainage water to a mixing of nitrate produced by natural biochemical activity and unmodified fertiliser nitrate.

Kreitler et al. (1978) and Kreitler & Browning (1983) studied two superposed aquifers on Long Island, New York, and found a higher δ^{15}N value in the upper unconfined aquifer than in the lower confined aquifer. This was interpreted as a reflection of a higher influence from fertilisers in the upper aquifer.

Mariotti & Létolle (1977) studied a small agricultural area near Paris, which is drained by a small creek. Comparison of water from an unfertilised area, unpolluted water from a confined aquifer underlying the catchment, and nitrate from the fertilised agricultural area gave input for a δ^{15}N budget of the nitrate. The δ^{15}N values of the nitrate in the creek were quite uniform except during major precipitation events. During the summer peaks, high δ^{15}N values occurred in the run-off. This was interpreted as being due to the discharge of nitrate produced during oxidation of ammonia after fractional evaporation.

Aravena et al. (1993) used a combined nitrate oxygen-nitrogen isotope study to identify the specific nitrate component from septic systems in an aquifer where the nitrate content was generally high.

Kendall et al. (1995) also used a combined $\delta^{18}O$-$\delta^{15}N$ study of nitrate for the interpretation of episodic acidification in small catchments during early spring melt due to pulses of nitrate and hydrogen flushed into the stream. The combined oxygen-nitrogen isotope study was used to distinguish atmospherically derived nitrate from nitrate produced in the organic horizon during the winter.

7.5 SULPHUR ISOTOPES

Most studies of sulphur isotopes are based on ^{34}S and ^{32}S. Early work by Thode et al. (1949) established general trends for the isotopic composition of sulphur in the environment. Informative reviews are given by Krouse (1980, 1988) and Pearson and Rightmire (1980). The $\delta^{34}S$ value is related to the $^{34}S/^{32}S$ of Cañon Diablo Troilite (CDT), which has a $\delta^{34}S$ value around the average for sulphur in the environment. It should also be mentioned that the radioactive tracer, ^{35}S has been used as an environmental tracer. It has a half-life of 87 days and is suitable for the determination of short residence times (see Michel 1995).

During recent years, there has been a focus on sulphur in the environment in connection with the increased emissions of anthropogenic sulphur. The transport of anthropogenic sulphur can be followed, provided its isotopic composition is sufficiently different from its surrounding environment.

There is a great natural variation of sulphur isotope composition in the environment. Values of $\delta^{34}S$ between $-60‰$ and $+40‰$ are found in sulphur sources such as rocks, organic material, water, soil and air. Sulphur exists in many forms, and numerous terrestrial processes alter its isotopic abundance. Modern oceanic sulphate and evaporite minerals have $\delta^{34}S$ values of $+20‰$, while the $\delta^{34}S$ values of other evaporites vary with their age (Pearson & Rightmire 1980). Of particular interest is the isotopic composition of fossil fuel, because of the sulphate leakage to the atmosphere during which it becomes part of the hydrologic cycle. The $\delta^{34}S$ values of petroleum (oil and gas) may vary within a range of $-10‰$ to $+33‰$, coal within a range of $-30‰$ to $+30‰$ (Krouse 1980).

The $\delta^{34}S$ value is, among other processes, related to redox reactions. Sulphur has oxidation states from $-II$ to VI, where SO_4^{2-} generally dominates under oxidizing and H_2S/HS^- during reducing conditions. Isotopic fractionation occurs in many processes of the sulphur cycle (Fig. 7.7). Microbial transformation of sulphur compounds is to a large extent responsible for the wide variation in isotopic compositions that occur in nature. Kinetic isotope effects during reduction processes normally favour the lighter isotope (Tudge & Thode 1950), resulting in a depletion of ^{34}S in natural biogenic H_2S. Thus the remaining sulphate becomes enriched in ^{34}S.

The sulphur in air and water show large variations in the $\delta^{34}S$ isotopic composition. This is not surprising given that the atmosphere and hydrosphere receive sulphur compounds from many sources in which different processes of isotopic fractionation occur. As the isotopic 'backgrounds' vary spatially, the $\delta^{34}S$ values for the sources of input must be determined for the particular system studied (Krouse 1980). However, $\delta^{34}S$ measurements of streamwater sulphate, groundwater sulphate and

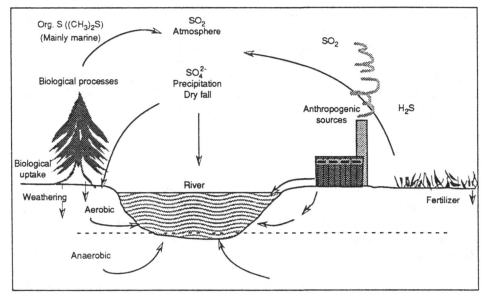

Figure 7.7. Simplified sulphur cycle (after Krouse 1980).

precipitation sulphate may nevertheless give information on the sulphur sources, geochemical processes and environments affecting them (Pearson & Rightmire 1980), especially when coupled with other chemical data (Jackson & Gough 1988).

7.5.1 Some examples of sulphur isotopes studies

Combined studies of sulphur and oxygen isotopes in forest ecosystems have shown that the dynamics of sulphur through the unsaturated zone are controlled by abiotic as well as biological processes in the upper part of the soil profile (Mitchell et al. 1988; Schoenay & Bettany 1989; Krouse et al. 1991; Mayer et al. 1992).

Mayer et al. (1995) studied the isotopic composition of sulphate in precipitation, seepage water and soil sulphate in some acid forest soils in Germany. The $\delta^{34}S$ values of inorganic sulphate in soil solution and solid phases were found to be close to those of precipitation, while there was a depletion of $\delta^{18}O$ values of sulphate in the upper 30 cm of the soil. This was explained as the result of mineralisation of carbon-bonded sulphur to SO_4^{2-}.

Also in agricultural areas, the ^{34}S studies have proved to be of interest. Knief et al. (1995) showed that the long-term use of sulphur-containing fertilisers influenced the quality and isotopic composition of sulphur fractions of the upper soil horizon and the rate of sulphate transport to deeper parts of the unsaturated soil.

Studies of the isotope ratios in run-off water have shown that most of the SO_4^{2-} leaving the catchment is 'old' SO_4^{2-} and that the SO_4^{2-} now entering the system is being exchanged with the SO_4^{2-} stores in the soil (Mörth & Torssander 1995).

Increased sulphur loadings in remote lakes low in sulphur, have been shown to stimulate S-reduction in the lake sediments resulting in an increased amount of sulphur in the sediments which are depleted in ^{34}S (Fry 1988).

Morgan (1992) points out that dissimilar S-reductions in a dystrophic system that is artificially treated with acid precipitation can store and neutralise part of the acidic sulphate added to the system.

7.6 CONCLUSIONS

In general, environmental tracers can be divided in two main groups (see Kendall 1990):

1. Hydrological isotopes, which are tracers of the water itself (O, D, T);
2. Biochemical isotopes, which are tracers of solutes or reactions (C, N, S).

These groups overlap, e.g. oxygen may form a part of sulphate studies in addition to hydrological budget studies, and carbon isotopes may be related solely to groundwater geology and not necessarily to biochemical processes.

Different methods are applied for the analyses of isotopes. A mass spectrometer is normally used for the most stable isotopes, but may also be used for the study of radioactive isotopes. Radioactive isotopes may also be analysed in reactors. The radioactive isotopes of interest have such a long half-life that the samples can be brought to the laboratory before analysis and reduce the need for expensive field equipment. The costs of analysis varies from quite unexpensive to very expensive. However, the cost will depend upon the isotope concentration, and become more expensive if the concentrations are low and an enrichment procedure is needed. The relatively low costs of analysis for oxygen, deuterium, nitrogen and sulphur isotopes has contributed to their wide usage in catchment studies.

Before carrying out a hydrological and/or geochemical study in a catchment, one should define the specific problem to be solved. One can then decide whether or not the study of any environmental isotope may help solve these problems. It is very important to choose the right isotopes for the specific study. The most obvious choice is an isotope of the same element as the element under consideration (e.g. ^{18}O, D or H in hydrologic studies or ^{15}N when pollution by nitrate is studied), or an isotope that behaves in a similar manner in soils or water as the component of interest. When the isotope is chosen, it is important to decide on the correct sampling procedure. No general rules can be given here; decisions will be based on budget, availability, etc.

Environmental isotopes are already today important parameters in hydrological, geochemical and biochemical catchment studies, as is clearly demonstrated by the reference list of this chapter. Only a few isotopes are straightforward to use, and in most cases the isotopic fractionation is so complicated that interpretation of the data depends on a very good understanding of chemical and hydrological processes. However, although there are many problems, environmental isotopes will probably become even more frequently used in the future, as they help understand processes that are not easily studied with other parameters. Some of the common isotopes, such as ^{18}O, may become as important as more conventional parameters, e.g. temperature, electrical conductivity and major chemical composition, are today. In particular, the combination of isotopes from the two different groups listed above, e.g. O in combination with C, N or S isotopes looks promising for future studies. New methods will give better quantitative results, and make it possible to measure smaller concentrations of the different isotopes. Obviously, other isotopes than those discussed in this chapter will also gain increased interest in the future.

ACKNOWLEDGEMENTS

Keith Beven and Allan Rodhe have commented upon the oxygen and hydrogen isotope text and have made valuable suggestions for alterations. Aslaug Borgan drafted the figures. David Segar kindly improved the English text. Financial support for the oxygen isotope studies in Norway were provided by the Norwegian Research Council. To all these persons and institutions we offer our sincere thanks.

REFERENCES

Allison, G.B. & Hughes, M.W. 1974. The use of environmental tritium to estimate recharge to a South-Australian aquifer. *J. Hydr.* 26: 245-254.

Aly, A.L.M. 1975. N-15 Untersuchungen zur anthropogenen Störung des natürlichen Stickstoffzyklus. Dissertation TH Aachen.

Andersen, L.J. & Sevel, T. 1974. Six years' environmental tritium profiles in the unsaturated and saturated zones, Grønhøj, Denmark. *Isotope Techniques in Groundwater Hydrology, Vol. I. IAEA-SM-182*: 3-20.

Anderson, M.G. & Burt, T.P. (eds) 1990. *Process Studies in Hillslope Hydrology.* New York: John Wiley & Sons Ltd.

Aravena, R., Evans, M.L. & Cherry, J.A. 1993. Stable isotopes of oxygen and nitrogen in source identification of nitrate from septic systems. *Ground Water* 31: 180-186.

Bazemore, D.E., Eshleman, K.N. & Hollenbeck, K.J. 1994. The role of soil water in stormflow generation in a forestred catchment: synthesis of natural tracer and hydrometric evidence. *J. Hydr.* 162: 47-76.

Calf, G.E. 1978. The isotope hydrology of the Mereenie sandstone aquifer, Alice Springs, Northern Territory, Australia. *J. Hydr.* 38: 343-355.

Campana, M.E. & Mahin, D.A. 1985. Model-derived estimates of groundwater mean ages, recharge rates, effective porosities and storage in a limestone aquifer. *J. of Hydr.* 76: 247-264.

Cerling, T.E. & Quade, J. 1993. Stable carbon and oxygen isotopes in soil carbonates. *Climate Change in Continental Isotopic Records:* Geophysical Monographs 78. American Geophysical Union.

Chapelle, F.H. & Knobel, Le R.L. 1985. Stable carbon isotopes of HCO_3^- in the Aquaia aquifer, Maryland: Evidence for an isotopically heavy source of CO_2. *Ground Water* 23: 592-599.

Craig, H. 1961. Standard for reporting concentrations of deuterium and oxygen-18 in natural waters. *Sci.* 133: 1833-1834.

Crouzet, E., Hubert, P., Olive, Ph. & Siwertz, E. 1970. Le tritium dans les mesures d'hydrologie de surface. Détermination expérimentale dy coefficient de ruissellement. *J. Hydrol.* 11: 217-229.

Dansgaard, W. 1964. Stable isotopes in precipitation. *Tellus* 16: 436-468.

DeWalle, D.R., Switstock, B.R. & Sharpe, W.E. 1988. Three-component tracer model for stormflow on a small Appalachian forestred catchment. *J. Hydr.* 104: 301-310.

Dincer, T. & Davis, G.H. 1984. Application of environmental isotope tracers to modeling in hydrogeology. *J. Hydr.* 68: 95-113.

Dincer, T., Payne, B.R., Florkowski, T. Martinec, J. & Tongiorgi, E. 1970. Snowmelt runoff from measurements of tritium and oxygen-18. *Water Resour. Res.* 6: 110-124.

Ehleringer, J.R. & Cerling, T.E. 1995. Atmospheric CO_2 and ratio of intercellular to ambient CO_2 concentrations in plants. *Tree Physiology* 15: 105.

Faure, G. 1986. *Principle of Isotope Geology.* 2nd edition. New York: John Wiley & Sons.

Fontes, J.Ch. 1980. Environmental isotopes in groundwater hydrology. In Fritz, P. & Fontes, J.Ch. (eds), *Handbook of Environmental Isotope Geochemistry, Vol. 1*: 75-140. Amsterdam: Elsevier.

Fontes, J.Ch. 1985. Some considerations on ground water dating using environmental isotopes. Hydrogeology in the Service of Man. *Mem. 18th Congr. of IAH*, Cambridge.

Fontes, J.Ch., Létolle, R., Olive, Ph. & Blavoux, B. 1967. Oxygène-18 et tritium dans le bassin d'Evian. In *Isotopes in Hydrology*, Intern. Atomic Energy Agency, Vienna: 401-415.

Foster, S.S.D. & Smith-Carington, A. 1980. The interpretation of tritium in the chalk unsaturated zone. *J. Hydr.* 46: 343-364.

Freidman, I. & O'Neil, J.R. 1977. Compilation of stable isotope fractionation factors of Geochemical interest. In Fleischer, M. (techn. ed.), *Data of Geochemistry*. USGS Prof. Papers 440-KK.

Freyer, H.D. & Aly, A.L.M. 1976. Seasonal trends of NH_4^+ and NO_3^- nitrogen isotope composition in rain collected in rural air. *ECOG IV, Amsterdam*, April 1976.

Fritz, P. & Fontes, J.Ch. (eds) 1980. *Handbook of Environmental Isotope Geochemistry, Vol. 1.* Amsterdam: Elsevier.

Fritz, P. & Fontes, J.Ch. (eds) 1986. *Handbook of Environmental Isotope Geochemistry, Vol. 2.* Amsterdam: Elsevier.

Fritz, P., Cherry, J.A., Weyer, K.U., & Sklash, M. 1976. Storm runoff analyses using environmental isotopes and major ions. In *Interpretation of Environmental Isotope and Hydrochemical Data in Groundwater Hydrology, Vol I*: 111-130. Vienna: Intern. Atomic Energy Agency.

Fry, B. 1988. Sulfate fertilization and changes in stable sulfur isotopic composition of lake sediments. In Rundel, P.W., Ehleringer, J.R. & Nagy K.A. (eds), *Stable isotopes in ecological research*, Ecological Studies 68: 445-453. Springer Verlag.

Gat, J.R. 1980. The isotopes of hydrogen and oxygen in precipitation. In Fritz, P. & Fontes, J. Ch. (eds), *Handbook of Environmental Isotope Geochemistry, Vol. 1*: 22-47. Amsterdam: Elsevier.

Gat, J.R. & Gonfiantini, R. 1981. Stable isotope hydrology. Deuterium and Oxygen-18 in the water cycle. *IAEA, techn. rep. ser. 210.*

Grip, H. & Rodhe, A. 1994. *Vattnets väg från regn til bäck.* Uppsala: Hallgren & Fallgren Studieförlag AG.

Gunyakti, A. & Altinbilek, H.D. 1993. Isotopic evaluation of hydrologic components in the Guvenc basin of Ankara. In Peters, N.E., Hoehn, E., Leibundgut, Ch., Tase, N. & Walling, D.E. (eds), Tracers in hydrology. Proceedings of the Yokohama symposium, *IAHS publ.* 215: 241-247.

Haldorsen, S. 1994: Oksygenisotoper og grunnvann. *Department of Soil and Water Sci. Agricultural Univ. of Norway, Report* 13/94 (33).

Haldorsen, S., Englund, J.-O., Jørgensen, P., Kirkhusmo, L.A. & Hongve, D. 1992. Groundwater contribution to a mountain stream channel, Hedmark, Norway. *Nor. geol. unders., Bull.* 422: 3-14.

Haldorsen, S., Englund, J.-O. & Kirkhusmo, L.A. 1993. Groundwater springs in the Hedmarksvidda mountains, related to the deglaciation history. *Nor. geol. tidsskr.* 73: 234-242.

Harum, T. & Frank, J. 1992. Hydrograph separation by means of natural tracers. In Hötzl, H. & Werner, A. (eds), *Tracer hydrology*: 143-146. Rotterdam: Balkema.

Hoefs, J. 1987. *Stable isotope geochemistry.* Second edition. Berlin: Springer Verlag.

Hooper, R.P. & Shoemaker, C.A. 1986. A comparison of chemical and isotopic hydrograph separation. *Water Resour. Res.* 22: 1444-1454.

Hübner, H. 1986. Isotope effects of nitrogen in the soil and biosphere. In Fritz, P. & Fontes, J.Ch., (eds), *Handbook of Environmental Isotope Geochemistry, Vol. 2*: 361-425. Amsterdam: Elsevier.

IAEA 1974. *Isotope techniques in groundwater hydrology 1974.* Proceedings of a symposium, Vol. I and Vol. II. Vienna: IAEA.

IAEA, 1983. Guidebook on nuclear technique in hydrology. *IAEA technical report series* 91. Vienna: IAEA.

Jackson, L.L. & Gough, L.P. 1988. The use of stable sulfur isotope ratios in air pollution studies: An ecosystem approach in South Florida. In Rundel, P.W., Ehleringer, J.R. & Nagy K.A. (eds), *Stable isotopes in ecological research*, Ecological Studies 68: 471-490. Springer Verlag.

Karamanos, R.E. & Rennie, D.A. 1980. Changes in natural N-15 abundance associated with pedogenic processes in soil, I. Changes associated with saline seeps; II. Changes on different slope positions. *Can. J. Soil Sci.* 60: 337-344, 365-372.

Kendall, C. 1990. Isotope hydrology. *Water Resources Division Training Memorandum No. 90.29.* Denver: U.S. Dept. of the Interior Geol. Surv.

Kendall, C. & McDonnell, J.J. 1993. Effect of intrastorm isotopic heterogeneities of rainfall, soil water and groundwater on runoff modeling. In Peters, N.E., Hoehn, E., Leibundgut, Ch., Tase, N. & Walling, D.E. (eds), Tracers in hydrology. Proceedings of the Yokohama symposium. *IAHS publication* 215: 41-48.

Kendall, C., Silva, S.R., Chang, C.C.Y., Burns, D.A., Campbell, D.H. & Shanley, J.B. 1995. Use of the $\delta^{18}O$ and $\delta^{15}N$ of nitrate to determine sources of nitrate in early spring runoff in forested catchments. In: *International symposium on isotopes in water resources management. IAEA-SM-336*: 56-57.

Kennedy, V.C., Kendall, C., Zellweger, G.W., Wyerman, T.A. & Avanzino, R.J. 1986. Determination of the components of stormflow using water chemistry and environmental isotopes, Mattole River Basin, California. *J. Hydr.* 84: 107-140.

Kimball, B.A. 1984. Ground water age determinations, Piceance Creek Basin. In Hitchon, B. & Wallick, E. I. (eds), *First Canadian/American Conference on Hydrogeology. National Water Well Association*, 267-283.

Knief, K., Fritz, P., Graf, W & Rackwitz, R. 1995. Sulphur biochemistry in soils managed agriculturally using sulphur-34. In International symposium on isotopes in water resources management. *IAEA-SM-336*: 58-59.

Knott, J.F. & Olimpio, J.C. 1986. Estimation of recharge rates to the sand and gravel aquifer using environmental tritium, Nantucket Island, Massachusetts. *U.S. Geol. Surv., Water-Supply Papers* 2297.

Knutsson, G. & Morfeldt, C.-O. 1993. *Grundvatten teori & tillämpning*. Solna: AB Svensk Byggtjänst.

Kohl, D.A., Shearer, G.B. & Commoner, B. 1971. Fertilizer nitrogen: contribution to nitrate in surface water in a Corn Belt watershed. *Sci.* 174, 1331-1334.

Kovac, L. & Drost, W. 1992. Statistical analysis of stable isotope and single well point dilution data for the determination of groundwater flow direction in the Reuss Valley (Switzerland). In Hötzl, H. & Werner, A. (eds), *Tracer Hydrology*: 193-196. Rotterdam: Balkema, Rotterdam.

Kreitler, C.W. & Browning, L.A. 1983. Nitrogen-isotope analysis of groundwater nitrate in carbonate aquifers: natural sources versus human pollution. *J. Hydr.* 61: 285-301.

Kreitler, C.W., Ragone, S.E. & Katz, B.G. 1978. $^{15}N/^{14}N$ ratios of ground-water nitrate, Long Island, N.Y. *Ground Water* 16: 404-409.

Krouse, H.R. 1980. Sulphur isotopes in our environment. In Fritz, P. & Fontes, J.Ch. (eds), *Handbook of Environmental Isotope Geochemistry, Vol. 1*: 435-471. Amsterdam: Elsevier.

Krouse, H.R. 1988. Sulfur isotope studies of the pedosphere and biosphere. In Rundel, P.W., Ehleringer, J.R. & Nagy K.A. (eds), *Stable Isotopes in Ecological Research*, Ecological Studies 68: 424-444. Springer Verlag.

Krouse, H.R., Stewart, J.B.W. & Grinenko, V.A. 1991. Pedosphere and biosphere. In Krouse, H.R. & Grinenko, V.A. (eds), Stable isotopes: Natural and anthropogenic sulphur in the enviroment. *Scope* 43: 267-306.

Lajtha, K. & Michener, R.H. (eds) 1994. *Stable isotopes in ecology and environmental science.* New York: Blackwell Sci. Publ.

Létolle, R. 1980. Nitrogen-15 in the natural environment. In Fritz, P. & Fontes, J.Ch. (eds), *Handbook of Environmental Isotope Geochemistry, Vol. 1*: 407-433. Amsterdam: Elsevier.

Libby, W. F. 1955. *Radiocarbon Dating, 2nd ed.* Chicago: University of Chicago Press.

Lindström, G. & Rodhe, A. 1986. Modelling water exchange and transit-times in till basins using oxygen-18. *Nord. Hydr.* 17: 325-334.

McDonnell, J.J. & Kendall, C. 1996. *Isotope Tracers in Catchment Hydrology.* New York: J. Wiley & Sons.

Mariotti, A. & Létolle, R. 1977. Application de l'étude isotopique de l'azote en hydrologie et hydrogeologie. Analyse des résultats obtenus sur un exemple précis: le bassin de Melarchez (Seine et Marne, France). *J. Hydr.* 33: 157-172.

Martinec, J., Siegenthaler, U., Oeschlager, H. & Tongiorgi, E. 1974. New insights into the run-off mechanism by environmental isotopes. In *Isotope Techniques in Groundwater Hydrology, Vol I*: 129-143. Vienna: IAEA, Vienna.

Mayer, B., Fritz, P., Knief, K. & Li, G. 1992. Evaluating pathways of sulphate between atmosphere and hydrosphere using stable sulphur and oxygen isotope data. Proc. Intern. Conf. *IAEA-SM-319/31.* Vienna: IAEA.

Mayer, B., Fritz, P., Prietzel, J. & Krouse, H.R. 1995. The use of stable sulfur and oxygen isotope ratios for interpreting the mobility of sulfate in aerobic forest soils. *Appl. Geochem.* 10: 161-173.

McDonnell, J.J. 1990. A rationale for old water discharge through macropores in a steep, humid catchment. *Water Resour. Res.* 26: 2821-2832

McDonnell, J.J., Bonell, M., Stewart, M.K. & Pearce, A.J. 1990. Deuterium variations in storm rainfall: implications for stream hydrograph separation. *Water Resour. Res.* 26: 455-458.

Meinardi, C.R. 1994. Groundwater recharge and travel times in the sandy regions of the Netherlands. *RIVM report 715501004.* Bilthoven: RIVM.

Michel, R.L. 1995. Use of sulphur-35 to study sulphur migration in the flat Tops Wilderness area, Colorado. In International symposium on isotopes in water resources management, *IAEA-SM-336*: 69-70.

Mitchell, M.J., David, M.B. & Harrison, R.B. 1988. Sulphur dynamics of forest ecosystems. In Howarth, R.W., Stewart, J.W.B. & Ivanov, M.V. (eds), Sulphur cycling on the continents: Wetlands, terrestrial ecosystems and associated water bodies. *Scope* 48: 215-259.

Mörth, C.M. & Torssander, P. 1995. Sulpur and oxygen isotope ratios in sulfate during an acidification reversal study at Lake Gårdsjön, western Sweden. *Water, Air and Soil Pollution* 79: 261-278.

Mook, W.G. 1980. Carbon-14 in hydrogeological studies. In Fritz, P. & Fontes, J.Ch. (eds), *Handbook of Environmental Isotope Geochemistry, Vol. 1*: 49-74. Amsterdam: Elsevier.

Mook, W.G., Groeneveld, D.J., Brouwn, A.E. & van Ganswijk, A.J. 1974. Analysis of a run-off hydrograph by means of natural ^{18}O. In *Isotope Techniques in Groundwater Hydrology, Vol I, IAEA-SM-182*: 145-155. Vienna: IAEA.

Morgan, M.D. 1992. Sulfur pool sizes and stable isotope ratios in HUMEX peat before and immediately after the onset of acidification. *Environ. Internat.* 18: 545-553.

Mulholland, P.J. 1993. Hydrometric and stream chemistry evidence of three storm flowpaths in Walker Branch watershed. *J. Hydrol.* 151: 291-316.

Newbury, R.W., Cherry, J.A. & Cox, R.A. 1969. Groundwater-streamflow systems in Wilson Creek experimental watershed, Manitoba. *Can. J. Earth Sci.* 6: 613-623.

Nyberg, L. 1995. Soil- and groundwater distribution, flowpaths and transit times in a small till catchment. *Comprehensive summaries of Uppsala dissertations from the Faculty of Science and Technology* 97.

Payne, B.R. & Yurtsever, Y. 1974. Environmental isotopes as a hydrogeological tool in Nicaragua. In *Isotope technique in Groundwater Hydrology, Vol I. IAEA-SM-182*: 193-202. Vienna: IAEA.

Pearson, F.J. (Jr.) & Rightmire, C.T. 1980. Sulphur and oxygen isotopes in aqueous sulphur compounds. In Fritz, P. & Fontes, J.Ch. (eds), *Handbook of Environmental Isotope Geochemistry, Vol. 1*: 227-258. Amsterdam: Elsevier.

Peters, N.E., Hoehn, E., Leibundgut, Ch., Tase, N. & Walling, D.E. (eds) 1993. Tracers in hydrology. Proceedings of the Yokohama symposium. *IAHS publication* 215.

Pinder, G.F. & Jones, J.F. 1969. Determination of the groundwater component of peak discharge from the chemistry of total runoff. *Water Resour. Res.* 5: 438-445.

Poreda, R.J., Cerling, T.E. & Solomon, D.K. 1990. Use of tritium and helium isotopes as hydrologic tracers in a shallow unconfined aquifer. In Kendall, C. (ed.), Isotope hydrology. *Water Resources Division Training Memorandum No. 90.29.* Denver: United States Dept. of the Interior. Geol. Surv.

Reardon, E.J. & Fritz, P. 1978. Computer modelling of ground water ^{13}C and ^{14}C isotopic compositions. *J. Hydr.* 36, 201-224.

Rennie, D.A. & Paul, E.P. 1975. Nitrogen isotope ratios in surface and sub-surface soil horizons. In *Isotope Ratio as Pollutant Source and Behaviour Indicators*: 441-453. Vienna: IAEA, Vienna.

Rennie, D.A., Paul, E.P. & Johns, L.E. 1976. Natural N-15 abundance of soil and plant samples. *Can. J. Soil Sci.* 56: 43-50

Rodhe, A. 1987. The origin of streamwater traced by oxygen-18. *Uppsala Universitet, Naturgeografiska institutionen, avd. för Hydrologi, Report Series A 41.*

Salomons, W. & Mook, W.G. 1986. Isotope geochemistry of carbonates in the weathering zone. In Fritz, P. & Fontes, J.Ch. (eds), *Handbook of Environmental Isotope Geochemistry, Vol. 2*: 239-270. Amsterdam: Elsevier.

Schlesinger, W.H. 1985. The formation of caliche in soils of the Mojave Desert, California. *Geochim. et Cosmochim. Acta* 49: 57-66.

Schoenay, J.J. & Bettany, J.R. 1989. ^{34}S natural abundance variations in prairie and boreal forest soils. *J. Soil Sci.* 40: 397-413.

Siegel, D.I. & Jenkins, D.T. 1987. Isotopic analysis of groundwater flow systems in a wet alluvial fan, southern Nepal. *IAEA-SM-299/118*: 475-482.

Sklash, M.G. 1990. Environmental isotope studies of storm and snowmelt runoff generation. In Anderson, M.G. & Burt, T.P. (eds), *Process Studies in Hillslope Hydrology*: 401-435. New York: John Wiley & Sons Ltd.

Sklash, M.K. & Farvolden, R.N. 1979. The role of groundwater in storm runoff. *J. Hydrol.* 43: 45-65.

Sklash, M.K., Farvolden, R.N. & Fritz, P. 1976. A conceptual model of watershed response to rainfall developed through the use of oxygen-18 as a natural tracer. *Canadian J. Earth Sci.* 13: 271-283.

Sklash, M.K., Beven, K.J., Gilman, K. & Darling, W.G. (in print). Isotope studies of pipeflow at Plynlimon, Wales, UK. *Hydrological Processes.*

Sletten, R.S. 1988. The formation of pedogenic carbonates on Svalbard: the influence of cold temperatures and freezing. *V Intern. Conf. on Permafrost Proceed, Trondheim:* 467-472.

Swistock, B.R., DeWalle, D.R. & Sharpe, W.E. 1989. Sources of acidic storm flow in Appalachian headwater stream. *Water Resourc. Res.* 25: 2139-2147.

Thode, G.H., Macnamara, J. & Collins, C.B. 1949. Natural variations in the isotopic content of sulphur and their significance. *Can. J. Res.* 27: 361-373.

Tudge, A.P. & Thode, H.G. 1950. Thermodynamic properties of isotopic compounds of sulphur. *Can. J. Res., Sect. B.* 28: 567-578.

Wels, C., Cornett, R.J. & LaZerte, B.D. 1990. Groundwater and wetland contributions to stream acidification: An isotopic analysis. *Water Resour. Res.* 26: 2993-3003.

Wigley, T.M.L., Plummer, L.N. & Pearson, F.J.Jr. 1978. Mass transfer and carbon isotope evolution in natural water systems. *Geochim. Cosmochim. Acta* 42: 1117-1139.

Yasuhara, M., Marui, A. & Kazahaya, K. 1993. An isotopic study of groundwater flow in a volcano under humid climatic conditions. In Peters, N.E., Hoehn, E., Leibundgut, Ch., Tase, N. & Walling, D.E. (eds), Tracers in hydrology. *IAHS Publication* 215: 179-186.

CHAPTER 8

Field instrumentation

ÅNUND KILLINGTVEIT
Department of Hydraulic and Environmental Engineering, Norwegian University of Science and Technology, Trondheim, Norway

KNUT SAND
SINTEF Norwegian Hydrotechnical Laboratory, Trondheim, Norway

NILS ROAR SÆLTHUN
Norwegian Institute for Water Research, Oslo, Norway

8.1 STREAMFLOW MEASUREMENTS

8.1.1 *Introduction*

Measurement of water stage and discharge is a central part of operational hydrology. This chapter covers the measurement of stage, discharge and the determination of rating curves, and methods for direct measurement of stage. Only the main principles are covered, and for details the reader is referred to specialized literature such as the WMO manuals (WMO 1994; Tilrem 1986a & 1986b, based on experience from Norway and developing countries), and the authoritative works by Herschy on streamflow measurements (Herschy 1978, 1985).

8.1.2 *Measurement of stage*

Stage is usually either used as an index for discharge estimation, or to give levels in reservoirs. Although the fundamental measurement techniques are the same in both cases, the practical considerations are usually different, and the demands on the installations and operation procedures differ. The measurement range is usually larger in reservoirs than in rivers (up to more than a hundred meters larger) while the rate of change is lower. This section will focus on stage used as a discharge index.

8.1.2.1 *Non-recording stations*

The traditional instrument for stage measurement is the staff gauge, as shown in Figure 8.1. The gauge will usually have marks for every centimetre, often with alternating black/white one centimeter wide marks. Such alternating markings make the gauge easier to read, but may lead to individual preferences of choosing even or odd values. Most investigations also show over-representation of stages with moduli 5 or 10. Waves and turbulence can often make it difficult to read stage with 1 centimetre accuracy. An easy way to overcome this difficulty is to attach a transparent tube with a narrow opening at the bottom end along the gauge, illustrated in Figure 8.1. The water level in this tube will be strongly damped compared to the fluctuations of the

Figure 8.1. Staff gauge.

free surface, but still representative. The staff gauge must have a hard and durable surface and markings, be resistant to corrosion, rot, mechanical damage (for instance by floating ice), UV decay and algae growth.

8.1.2.2 *Crest gauges*
As non-recording stations will usually only be read once or twice a day, flood crests will usually not be logged. To register crest level (i.e. the highest level of water), stations can be equipped with a simple crest gauge, consisting of a vertical tube with a small opening at the bottom end and a removable central measuring stick and floating granular cork inside. Some cork will adhere to the measurement stick, showing the highest water level since the last measurement.

8.1.2.3 *Recording stations*
Non-recording gauges are labour intensive, unsuitable in uninhabited areas, and give inadequate time resolution in streams with rapid variation in stage. The solution to this is to use recording gauges. The traditional recording station has a stilling well with a chart recording instrument measuring time with a mechanical clock and stage with a counter-weighted float. The stilling well is connected to the stream or reservoir via a tube, narrow enough to damp out rapid water level fluctuations, but large enough to avoid being easily clogged (Fig. 8.2). The main problems with mechanical stage recorders arise from negligence and improper maintenance. The most common failures are:
   – Poor trace due to faulty pen or pencil;
   – Missing data due to low paper capacity or an unreliable clock.
Traditionally, recorder charts have been read manually. This is time consuming,

Figure 8.2. Recording gauge.

even if only daily values are read. It is even more time consuming if timing errors have to be corrected. Today, charts will usually be converted to a digital format by systems based on digitization boards.

High precision mechanical instruments such as stage recorders are becoming increasingly expensive, while prices on electronic equipment are falling. This, together with the vulnerability of mechanical recorders to poor maintenance, has led to increased interest in electronic stage recorders. The transitional generation of loggers was the punching instrument; mechanical instruments like the graphical recorders, but using punched paper tape as a recording medium. The next generation was electrical equipment, recording on audio tape. Today the most common types are fully electronic instruments, recording on solid state memory, either in the form of exchangeable cassettes or cards, or in an internal memory. The storage units can be tapped locally or by telemetry, usually over standard serial interfaces. The most advanced stations are telemetering stations, transmitting either in real time or at fixed intervals. See Section 8.2 for more information on telemetering systems.

Electronic recording systems are presently quite reliable, and offer several advantages over mechanical equipment. The most important are probably:

– More precise timing;
– Exact synchronization of several variables if necessary;
– Less vulnerable to poor maintenance;
– Can combine a high time resolution with a long observation period; and
– Easy post-processing, as data are stored in digital form.

One drawback is the need for electrical power, but the power consumption of electronic devices and memory is decreasing, while battery capacities are increasing, so this is not a significant problem today. For many systems it was often difficult for observers to detect malfunctioning of the instrument, but now most systems have displays that can show measured stage, battery state, system time etc.

On an electronic recording instrument, stage from a float water-level sensor must be converted to a signal that can be electronically read and stored. The most common translating systems convert the angle of the float-wire pulley to resistance by means of a potentiometer, and then to digital form through an AD converter. An alternative is to convert the pulley position directly to digital form using a mechanical or photo-electric decoding device.

The use of electronic recording instruments enables the use of sensor types for water level recording other than float types. The most commonly used sensor type is the pressure transducer, which registers water pressure on a membrane. To avoid the influence of varying atmospheric pressure (the normal variations of atmospheric pressure correspond to one meter water head), either the other side of the membrane must be exposed to air at atmospheric pressure, or water pressure measurements must be combined with air pressure measurements. In reservoirs with large water heads, the thermal expansion of water with varying temperature will significantly influence the water pressure/water stage relation. The pressure transducer and air tube are usually protected by an anchored steel or PE (polyethylene)-pipe.

The main advantage of the pressure transducer compared to a float type stage recorder is that stilling wells are no longer needed, reducing the cost of installation dramatically. The absence of the stilling well also solves the problem with ice in the well. There have, however, also been serious operational problems with pressure transducers, such as drift away from the calibration curve, condensing water in the air tube etc.

8.1.3 *Discharge measurement methods*

The rating curve must be established and controlled by discharge measurements. In principle there are a number of methods available for measuring discharge, but the most commonly used are dilution gauging and the use of current meters.

8.1.3.1 *Volumetric measurement*

This is the most straightforward method of measuring discharge, but this method is unfortunately seldom applicable, as it requires:

1. Low discharge (less than 100 l/s); and
2. A site where it is possible to divert all water to a measuring vessel.

The principle of volumetric measurement is simply to measure the time it takes to fill a measuring vessel of known capacity. As the uncertainty of the timing and of the capacity of the measuring vessel should be easy to determine, it is easy to make an uncertainty analysis of the method. As this is swift, it is easy to repeat measurements.

8.1.3.2 *Dilution gauging*

The principle of dilution gauging is based on the fact that if the amount of a conservative tracer (e.g. NaCl) in the river is known, and the concentration can be measured, the discharge can be calculated. Two different techniques are used, constant injection and bulk injection. In the case of constant injection, the method for calculating the discharge is simply:

$$c = \frac{i}{q+i} \qquad (8.1)$$

where q is discharge and i is the injection rate (in the same measurement units), and c is the measured concentration. Solving for q yields:

$$q = i\left(\frac{1}{c} - 1\right) \qquad (8.2)$$

The bulk injection method is based upon injecting a known quantity of tracer into the river. Provided all of the tracer passes the measuring point, the following relationship holds:

$$\int_0^T i(t) \, dt c = I \qquad (8.3)$$

where I is the volume of tracer, T is the time necessary for all of the tracer to pass the dependent measuring point, and $i(t)$ is the time dependent tracer 'discharge' at the measuring point. If i is negligible compared to q, it is easily found that:

$$q = \frac{I}{\int_0^T c(t) \, dt} \qquad (8.4)$$

where $c(t)$ is the measured concentration.

The method can be used where the turbulence of the river provides full mixing of the tracer in the water (i.e. homogeneous concentration in a river cross-section) within a practical distance from the injection points. This makes the method suitable for highly turbulent streams, where the conditions may be difficult for current meter measurements. Thus, the two methods (constant and bulk injection) can be considered complementary.

The tracer must be easily detectable, chemically inert, and environmentally harmless. Any natural content of the tracer must be low compared to the injected amount, and stable. Fluorescent tracers such as rhodamine are quite common, but are difficult to used in turbid water.

In Norway, sodium chloride (NaCl) is the most commonly used tracer. Concentration is measured by reading changes in electrical conductivity of the river water using a standard conductivity meter, preferably logging, with adjustable range. The tracer is cheap and easily available, but the method requires care when dissolving the salt, and the calibration procedure is somewhat time-consuming. The use of bulk injection of sodium chloride was pioneered by Norwegian hydrologists Aastad and Søgnen (Aastad & Søgnen 1954). Sodium chloride has been used for discharges up to 3-400 m³/s.

Radioactive tracers have been used, and are accurate in small doses, but restrictions on the handling and application of such tracers limit their use.

8.1.3.3 *Use of current meters*
The use of current meters is the most widespread method for discharge measurements (see Section 8.1.5.10). The basic principle here is to measure the flux of water through a river cross-section by determining the velocity of water of a large number

of points. The most commonly used current meters are propeller type water velocity meters.

The cross-section is defined by a wire with distance markings stretched across the water or by a permanent or removable cableway. The current meter is operated from a cableway, bridge, boat or by wading, depending on the size of the river.

The discharge is calculated from the mean velocities in vertical cross-sections, usually at fifteen to thirty evenly spaced points. In each vertical, the estimate of the mean velocity is based on one or several measurements. In a 'standard' turbulent vertical velocity profile, the velocity at 0.4 times the water depth above the bottom corresponds to the mean profile velocity, so this point would be the obvious choice if only one measurement is to be taken. If two measurements are taken, they will normally be 0.2 times the water depth and 0.8 times the water depth, as the average of these velocities is close to the vertical mean (Fig. 8.3).

For the main current region of the section more points are usually measured – one close to the surface and one close to the bottom, and two to five in between. The mean velocity is obtained by graphical or numerical integration. The recommended practice in Norway is to measure the flow velocity at five points; near bottom, 0.2, 0.4, 0.8 times water depth, and near surface. When the current meter is suspended from a wire, it will drift somewhat if the current is strong, and this must be corrected for when the depth of the measuring point is calculated. The water velocity can display quite large fluctuations due to moving eddies, and it is important to allow a sufficiently long averaging time to even out such fluctuations. On the other hand, too long an averaging time will increase the total time used for discharge measurements unnecessarily. An averaging time, of at least 45 seconds is recommended.

Measurements from a safe ice cover is done through holes drilled through the ice, one hole at each vertical. Frazil ice disturbs the measurements, and should if possible be avoided.

When the mean velocities in the verticals have been determined, the total discharge is calculated by weighting each vertical velocity by a representative area.

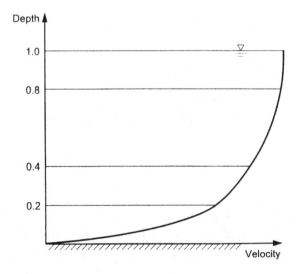

Figure 8.3. Vertical velocity profile.

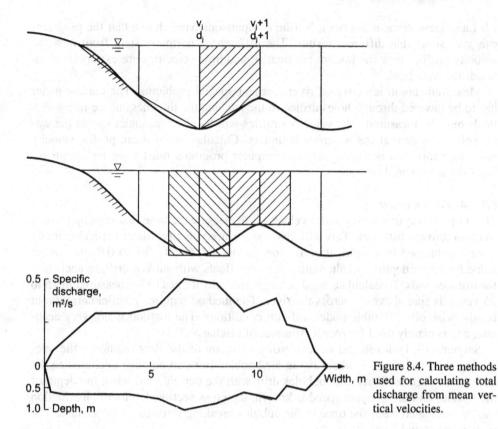

Figure 8.4. Three methods used for calculating total discharge from mean vertical velocities.

Three methods (Fig. 8.4) are commonly used:
- The mean section method;
- The mid-section area method; and
- The specific discharge method (Harlacher's method).

In the mean section method discharge is calculated for the trapezoidal segments between the verticals, using the mean of the velocities in the confining verticals as a representative velocity. The mid-section area method calculates the discharge in a rectangular segment embracing a vertical, using the velocity in that vertical as representative velocity. Harlacher's method calculates the discharge by integrating the specific discharge (velocity × depth) along the cross section. Harlacher's method is the standard method used in Norway.

The calculations can be made manually, graphically or using programmed calculators. For field-use programmable calculators or notebook type battery powered personal computers are convenient. Two examples of commercially available or free computer programs are FLYGEL[1] (NVE = Norges Vassdrags & Elektrisitetsvesen (Norwegian Water Resources and Energy Administration)) and CUMEC (The Dan-

---

[1] Names of computer programmes, instruments, manufacturers and companies are quoted for information only.

ish Land Development Service). Nordic comparisons have shown that the programs can give somewhat different results. The differences normally stem from how the velocity profiles near the bottom are treated, and how velocities are extrapolated towards the river bank.

Measurements in ice covered rivers present special problems. The current meter has to be lowered through holes drilled in the ice, and the thickness of ice in the verticals must be measured. The velocity profiles will be different under ice, as the water velocity is zero at the water/ice boundary. Calculations of mean profile velocity must take this into account, and the computer programs must have an option for handling ice covered sections.

8.1.3.4 *Float method*

The simplest way to measure water velocity is to measure the time floating objects take to cover known distances. This will give the surface velocity. Mean vertical velocity must be obtained by empirical correction factors – typically 0.7 to 0.8 (the higher value for larger depth). An alternative is to use floats with sunken drifting anchors – for instance rods. To calculate the discharge, the velocity has to be measured in 15 to 25 verticals spread evenly across the river. The method requires parallel current, but is otherwise often suitable under difficult conditions. The method is not very accurate, and is mainly used for rough estimates of discharge.

Sargent (1981) developed an interesting variation of the float method – the integrating float technique – where rising air bubbles from an air tube anchored to the bottom are used as floats. The bubbles drift with the current, and when the depth of the nozzles and the buoyant speed is known, the cross-sectional velocity distribution can be calculated from the trace of air bubbles breaking through to the surface. The trace is registered by photography.

8.1.3.5 *Moving boat*

The moving boat technique is a fast method to make discharge measurements in large rivers. A fast reacting current meter with a wane is attached to the boat with a rod, and measures the instant velocity one meter below the surface. The boat traverses the stream at 90° angles to the current, with a known velocity. The relative water speed and the direction are typically measured at twenty to thirty points across the river, while the depth is measured with an echo sounder. The true water velocity is calculated by resolving the apparent velocity vector in the transverse direction (boat speed) and the current direction (water speed). As the water speed is measured at one fixed level, a correction factor must be applied to get the mean velocity in the verticals (about 0.9 is common).

Usually several traverses are made, to reduce the uncertainty, and an accuracy of approximately 5% is claimed at the 95% confidence level.

8.1.3.6 *Ultrasonic and electromagnetic measurements*

While the discharge measurement methods mentioned so far are methods of discrete sampling, the ultrasonic (acoustic) method is capable of measuring mean velocity continuously, and the electromagnetic methods provide continuous measurement of discharge.

If a sound signal is transmitted through water having a velocity component parallel to the sound path, the travel time of the signal will depend on the water velocity. In the ultrasonic gauges, this principle is used to measure the mean water velocity. The velocity will be averaged only along the path defined by the line between the transmitter and the receiver, and depth variations are not considered. To cover the depth variations of the velocity, several gauges must be used, or the instruments must travel up and down. Ultrasonic gauges are usually permanent installations, and are mainly used where it is difficult to establish rating curves. The method is best suited for approximately rectangular sections. The signal path is usually about 45° to the current direction.

While the ultrasonic method measures the mean water velocity along a path, the electromagnetic method measures the mean velocity and total discharge by determining changes in the electromotive force induced in the water by the moving water in the earth magnetic field. The method requires large installations with probes buried under the river bed. It is particularly suited for rivers with unstable profiles or vegetation growth.

8.1.3.7 *Hydraulic structures*

Under given conditions, hydraulic structures have known stage/discharge relationships. They also provide a stable hydraulic control. Their use is limited by the installation cost, which increases rapidly with maximum discharge capacity. There is a wealth of designs of hydraulic control structures, and the criteria for choice of design are:
- Rating curve characteristics;
- Capacity/cost relationship;
- Available head difference, possibility for raising the upstream water level;
- Sediment load; and
- Fish migration through structure.

The two main categories are raised structures with full hydraulic jump, and submerged structures with partial jump (subcritical to critical/supercritical flow, and the reverse). In Norway the standard sharp edged 90° or 120° V-notch weir is the most commonly used structure of the first category, while the Crump weir is used where submerged structures are needed. In most Norwegian streams, the range of discharge is very large. To obtain a satisfactory resolution on low flow, while maintaining high total capacity, composite weir profiles are sometimes used. To avoid ice disturbing the rating curve during winter, V-notch weirs can be covered with removable lids to prevent icing in the profile. For a comprehensive description of the different types of control structures, the reader is referred to the specialized literature, for example Bos (1978) and Ackers et al. (1978).

8.1.4 *Stage-discharge relation*

When stage is used as an index on discharge, it is necessary to establish the stage versus discharge relationship, also called the rating curve. In the general case, the discharge (q) is dependent on stage and head slope, as illustrated for instance by the Manning formula:

$$q = M \cdot A \cdot R^{2/3} \cdot S^{1/2} \qquad (8.5)$$

where the head slope S can vary independently of stage, while the friction parameter M, the wetted area A and the hydraulic radius R are dependent on stage h; $M(h)$, $A(h)$, $R(h)$. If the slope can be considered constant, the discharge is a function of stage only (as long as the friction-stage and geometry-stage relations are unchanged).

The stage-discharge relationship is unique upstream a hydraulic control, as long as the distance is not so long that the slope is affected. Otherwise it will be unique if the stage gauging station is not affected by backwater effects, and the discharge does not vary so quickly that the floodwave significantly alters the slope. The last condition may arise in the case of artificial releases, or flash floods producing flood waves.

Even if the discharge is a function of stage only, this relationship may change with time. As illustrated by the Manning formula (Eq. 8.11a below), this may be caused by changes in the friction-stage relationship, $M(h)$, or the geometry-stage relationship; $A(h)$, $R(h)$. Changes in friction are usually caused by vegetation changes or ice cover, while changes in geometry are typically caused by ice cover formation, channel bed erosion or sedimentation, or hydraulic constructions or works. Ice and vegetation changes are seasonal, while most other changes are permanent. Changes may arise abruptly, usually caused by floods or construction works, but will often evolve over time.

8.1.4.1 *Rating curves*

For natural channels, the rating curve has to be established by discharge measurements over as wide a range as possible. The curve must also be checked regularly by follow up measurements, more frequently for unstable channels and hydraulic controls.

The most commonly used mathematical formulation of the rating curve is

$$q = k \, (h + c)^n \tag{8.6}$$

where k, c and n are constants. Most weirs will conform to this formula, but most composite weirs and natural controls will need two or more segments, each of the type above, to have an acceptable fit over all stage ranges.

$$q = k_j \, (h + c_j)^{nj} \quad \text{for} \quad h_{j-1} \le h < h_j \tag{8.7}$$

For the lowest segment, c_1 will be equal to $-h_0$, the lowest point in the weir or saddle point in the control. A rectangular weir has an n value close to 1.5, and a triangular weir 2.5. Most natural controls will have an n value between these two values.

Usually the power formula is fitted by logarithmic regression, in the form

$$\log q = k + n \log (h + c) \tag{8.8}$$

This requires c to be preset, as in case of the lowest segment, or to be optimized to produce the best fit. Standard least square logarithmic regression assumes errors to be proportional to the variable, and this is not an unlikely assumption – discharge measurement errors are approximately proportional to discharge. Fitting the curve may be done numerically or graphically. Both procedures are somewhat cumbersome on multisegment curves, as c_j usually has to be determined by trial and error, and h_{j-1} has to be determined in a separate step – by finding the intersection between segments. The segments do not necessarily intersect.

Although there is some physical rationale behind the power curve, fitting it to a natural control is primarily an exercise in curve fitting. The curve has many parameters as the number of segments increase (four extra parameters for each new segment), and the resulting curve is not smooth. Other mathematical formulations have been proposed, usually polynomials, but the problems with these are mainly that they cannot be used for extrapolating the rating curve; the extrapolation has to be done prior to the curve fitting. This is, in any case, a more sensible procedure.

8.1.4.2 *Extrapolation of the rating curve*
A rating curve cannot be calibrated to the possible extremes at the low or the high end, and the curve has to be extrapolated to cover the range of stages. If the stage of zero discharge is known, extrapolation at the lower end will not pose any problems. Extrapolation at the high end can easily introduce large errors, as the control section will often exhibit large changes and can even change position during flood stages. The value of flood discharge measurements cannot be overemphasised.

Extrapolation of the upper power curve segment is a common practice, but is not advisable for large extrapolations. Methods based on the description of the channel or control geometry are better. Most methods require at least that the relationship between the cross-sectional area and stage is known. The simplest method is based on the assumption that it is easier to extrapolate the mean velocity than the discharge. Discharge q is then calculated by the relationship

$$q = A(h) \cdot V(h) \tag{8.9}$$

where $A(h)$ is known and $V(h)$ is extrapolated.

Steven's method and similar methods are based on the Manning or Chezy formulas, which state that :

$$q \text{ versus } A(h) \cdot R(h)^{2/3} \quad \text{(Manning formula)} \tag{8.10a}$$

$$q \text{ versus } A(h) \cdot R(h)^{1/2} \quad \text{(Chezy formula)} \tag{8.10b}$$

should be close to linear relationships, as

$$\frac{q}{A(h) \cdot R(h)^{2/3}} = MS^{1/2} \quad \text{(Manning formula)} \tag{8.11a}$$

$$\frac{q}{A(h) \cdot R(h)^{1/2}} = CS^{1/2} \quad \text{(Chezy formula)} \tag{8.11b}$$

and C, M and S are usually relatively independent of stage. Steven's method is illustrated in Figure 8.5.

Where few discharge measurements are available, physical hydraulic scale models, or mathematical models for stationary flow such as HEC-2 (US Army Corps of Engineers, Hydrologic Engineering Center) may be used for rating curve extrapolation.

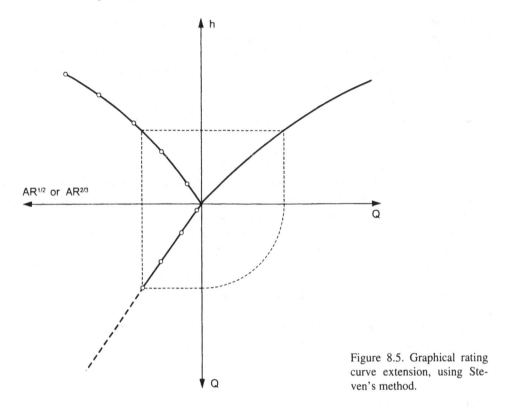

Figure 8.5. Graphical rating curve extension, using Steven's method.

8.1.4.3 *Discharge-stage-slope relationships*

As mentioned previously, backwater effects may cause the slope of the water head to vary for a given stage. As indicated by the Manning formula, the result is that discharge is not a unique function of stage, but of stage and slope. To determine the slope, it is necessary to use two-stage gauges. The distance between the gauges must be large enough to determine the slope with sufficient certainty. The rating curve is replaced by a surface or two-dimensional table, describing discharge as function of stage at one gauge and slope, or – more practical – stage at both gauges.

Since calibrating a two-dimensional surface requires more data than calibrating a one-dimensional rating curve, applying the Manning formula or hydraulic models (physical or mathematical) may be the most preferable method of establishing the relationship.

Backwater effects may be caused by tidal influence, downstream reservoirs, sluices etc. In principle such conditions affects the water level and slope a long distance upstream under subcritical flow conditions, the effect sinking asymptotically to zero, only to be broken by supercritical flow reaches.

8.1.4.4 *Effects of river ice*

River ice reduces the discharge at a given stage, compared to ice free conditions. This is caused by reduced active water conveyance area due to ice cover and at times bottom ice formation. In addition, the velocity profile is altered, due to increased

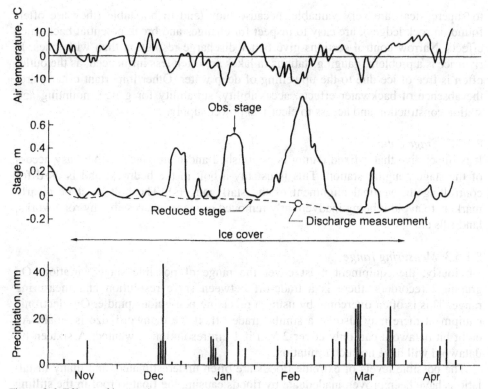

Figure 8.6. Effects of ice on river stage.

friction at the water/ice boundary. The problem cannot be solved by applying another rating curve, as the ice conditions and their effects on the stage changes with time. An example of ice influence is given in Figure 8.6.

Correction for the influence of ice must be based on discharge measurements during winter. This is time consuming and expensive, and the number of measurements is accordingly low. The interpolation between measurement points is a subjective procedure, based on knowledge about temperature and precipitation, and comparison with ice free gauging stations. An alternative is to interpolate between measurement points using a hydrological model.

Usually the ice correction is established by finding an 'equivalent' stage, allowing the standard rating curve to be used for winter conditions. The procedure is in principle the same as adjusting for vegetation growth (Stout's method), but the influence of ice is usually greater and more fluctuating through the season. Cold spells can give dramatic freeze-ups with stage increases that can resemble floods (Fig. 8.6).

8.1.5 *Practical considerations*

8.1.5.1 *Choosing a gauging station location*
A stable hydraulic control is the most important criterion for choosing a gauging stage location. Sites with hydraulic control where the flow changes from subcritical

to supercritical, are very valuable, because they tend to be stable (they are often found on rock ledges), are easy to inspect for changes and break potential backwater effects. Narrow control sections give better discharge resolution than wide ones. If river ice is a problem, gauging stations in lakes are often less influenced, as the outlet often is free of ice due to the upwelling of deep water. Other important criteria are: the absence of backwater effects, accessibility, suitability for gauge mounting and station construction and access to electrical power supply.

8.1.5.2 *Stage datum*

It is imperative that a fixed datum is established and maintained within easy access of the stage gauging station. This is usually a bolt in the bedrock, and is used for controlling the vertical alignment of the staff gauge(s). The bolt must be clearly marked, to avoid it being inadvertently removed or made inaccessible by road works, land fills etc.

8.1.5.3 *Measuring range*

Obviously, the equipment must cover the range of possible stage variation. On graphical recorders, there is a trade-off between stage resolution and measuring range. This is often overcome by using a reversing pen-guide spindle. On electronic equipment there may also be a similar trade off, if the dataword size is small. An eight bit dataword can only cover 2.5 m if 1 cm resolution is wanted. A sixteen bit dataword will have no such limitation.

The possible levels of extreme floods are often underestimated, and many installations have been proven inadequate by floods causing the float to roof in the stilling well, recording equipment to be drowned or the entire installation to be washed away. This is a lamentable situation, as very valuable data is lost in addition to the direct damages, to say nothing of how embarrassing such situations are to a responsible hydrological institution!

8.1.5.4 *Observation time*

Staff gauges, which are manually read, are usually read once a day at a fixed time, and this reading is often regarded as an unbiased estimate of the daily mean stage. This may not be true. Many rivers display a diurnal variation, which may bias the estimation of the daily mean stage. Such diurnal variations are caused by snow or glacier melt, convective afternoon showers, or peak releases from hydro-electric power plants. The remedy is either to have two or more observations a day, to find a representative observation time, or to switch to recording equipment. A special problem arises on manually operated reservoirs, where stage often is read in connection with gate operation. It will then be difficult to know for which time period the stage reading is representative.

8.1.5.5 *Controlling the stage*

The recorder must be checked against observed stage regularly. A recording station must therefore also be equipped with a staff gauge. If the stilling well is large enough, the staff gauge is often located inside the well. If the gauge is outside the well, it is important to be aware that the recording gauge will measure the actual head where the connecting tube emerges, while the staff gauge is usually located in quiet water. This will introduce a difference between the recorded level and the quiet water level equal to the velocity head

$$\Delta h_v = \frac{v^2}{2g} \tag{8.12}$$

where v is the mean water velocity in the vertical section over the connecting tube. The velocity head is not negligible, at 1 m/s velocity it amounts to 5 cm, at 2 m/s to 20 cm. As the rating curve will usually be established on the manually read stages, the velocity head water level reduction will lead to an underestimation of flood stages and discharge. The differences between registered and observed stage can also be misinterpreted as instrument drift when start and end stage (or other control stages) differ.

8.1.5.6 *Winter maintenance*
Measuring stage under winter conditions with ice covering the river introduces problems. For manual measurements an ice well has to be kept open at the gauge. In stilling wells the well has to be kept free of ice. If electric supply is available, this can be done by lowering a light bulb in the stilling well. Without electrical supply, the stilling well can be kept ice free by adding and maintaining a paraffin cover in the stilling well. This will give a higher level in the stilling well than in the river, due to the different densities of water and paraffin. Stilling wells dug in the ground will be less vulnerable to ice formation than pipe-type wells exposed to the air.

8.1.5.7 *Clogged pipe*
A common cause of operational errors is clogging of the connection pipe, giving constant or highly damped measured stage. It can take time before this is noticed, especially if the staff gauge used for control is inside the stilling well. A clogged pipe can often be opened by flushing.

8.1.5.8 *Regular inspection*
A stage station should be inspected at regular intervals by trained field technicians to check gauge alignment, maintenance etc. In addition, recording stations should be visited by local personnel in order to change recording charts, data cassettes etc., and to control stage. With the availability of electronic loggers with a large storage capacity, it may be tempting to increase the inspection interval. All operational experience indicates that this kind of equipment should also be inspected regularly, and that data cassettes should be changed at least every six months to avoid large data losses in the case of a malfunction.

8.1.5.9 *Sheltering of instruments*
Instruments, both mechanical and electronic, have differing requirements in terms of operational environment. It is important to consider these demands, especially on temperature and humidity, when instrument stands and shelters are designed. In addition it is also wise to take into account the risk of vandalism and theft, and secure the instruments accordingly. Information signs about the instruments and their purpose may also reduce the risk of damage.

8.1.5.10 *Choosing a section for current meter discharge measurements*
The main criteria for the selection of a section for current meter measurements used in Norway are:
– The difference in discharge between the stage measurement station and the discharge measurement must be considerably less than the measurement uncertainty;
– The flow must be parallel, and normal to the section;
– No significant backwater flow should be present;
– No boulders, giving irregular velocity profiles and abnormal flow components, should be present;
– Eddies of any kind should be avoided, if possible;
– The site should be easily accessible and safe, even during high floods;
– Water velocity should not be below 0.15 m/s, and not above 2.5 m/s.

At some stations, it may be practical to have separate measuring sections for high and low discharges.

8.2 AUTOMATIC DATA ACQUISITION SYSTEMS IN HYDROLOGY

8.2.1 *Introduction*

The use of hydrological models for forecasting runoff has increased rapidly during the last ten years. In Norway, forecasting inflow for hydro-electric power utilities have become especially important. To be able to use the models efficiently for real time flow forecasting in hydro-electric power systems, observations must be made available for model input and updating as quickly as possible. This is even more important for flood forecasting, where efficient collection, transmission and quality control is mandatory to be able to issue reliable forecasts. For small catchments it may be necessary to collect meteorological and hydrological data at hourly intervals or less. In most data collection systems built in Norway during the last years, hourly intervals have become the standard.

The increasing need for automatic data acquisition systems has led to considerable efforts in research and development to find optimal system design and improve reliability. This work has led to the development of sensors, electronics, software and complete systems which have been tested for extended periods both in experimental and operational networks. Even if most of these systems are tailored for use in inflow forecasting, the structure and components are similar to other automatic data acquisition systems used for hydrological data.

8.2.2 *Main structure and system components*

The main components of an automatic data collection system for hydrological and meteorological data are shown in Figure 8.7. The components are explained below:
Sensor. Only a few meterological and hydrological phenomena can be measured directly. In most cases one observes the phenomena indirectly by observing the effects in a suitable way. For measurements of wind for example one can observe how clouds, smoke or trees are affected. More practically, one can use a vane anemometer and observe how this is rotated by the wind. This rotation can be observed directly,

mechanically or electrically, and by a suitable transformation give the wind speed. Similarly one can observe the wind direction by a wind vane, temperature by changes in electrical resistance, precipitation as weight of rain and snow and so on. The combination of what is influenced and the equipment that records the effect is called a sensor.

Field station. This is the outermost link in the data collection system. It is usually equipped with one or more sensors and is designed to operate in localities with no mains power supply. Power is normally supplied by batteries or solar cells. Data transmission to the next link (the outstation) is usually done by simplex radio but can also be by line. Measurements and transmissions are triggered by commands from an internal clock at predetermined time steps, usually once an hour. The field station has to be able to withstand extreme climatic variations. To conserve power at the field stations, the power is switched on only for the period from the measurement sequence is started until the data has been transmitted to the outstation.

Outstation. An outstation is primarily a connecting link between one or more field stations and the master station. It can, however, also have its own sensors similar to the field station. Since an outstation will often serve several field stations, it is sometimes also called a concentration station. It will normally be located where mains power supply and an established communication network, such as a telephone or power line are avalailable. An outstation is normally required to receive and store data from the fieldstation(s) for at least one week without losing any data. Transmission of data to the master station is normally done by duplex communication via public telephone lines or through other communication channels. The data transmission from outstation to central station is initiated and controlled from the master station.

Master station. The master station (sometimes also called *Central station* or *Base station*) is the final destination for data collected at fieldstations and outstations. The master station can establish communication with several outstations, either at regular time intervals or upon user request. The master station will also normally have soft-

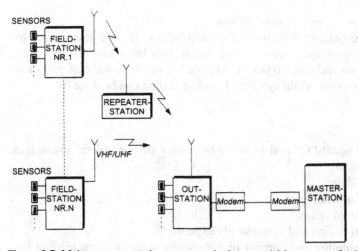

Figure 8.7. Main components in an automatic data acquisition system for hydrometric data.

ware for data quality control, error correction and a database for data storage and re-trieval. The master station can also send data on to other computers, for example for use in runoff forecasting models.

Repeater station. The repeater station is a station used to establish radio commu-nication between a field station and an outstation where a direct radio path is not possible. Data are received from the field station and forwarded unchanged to the outstation. The signal strength and frequency may be changed, however. The re-peater station is usually placed on hills or mountain tops. One repeater station may serve several field stations.

Modem. Modem is an abbreviation for modulate/demodulate. A modem is a de-vice which allows computers to communicate along telephone lines. The modulator converts digital signals into audio frequency tones, and the demodulator transforms the audio tones back into digital signals in the other end.

8.2.3 *Hydrological data for hydropower design and operation*

8.2.3.1 *Purpose of data collection*
The purpose of the hydrological data collection may vary, but at least three different purposes can be identified:
– Runoff data for the design of new plants;
– Measurements for the control and verification of concession rules given by the government;
– Data for the economic optimization of operation.

These three different uses of data have different requirements for time resolution in data and speed in data acquisition. Data for operational use must generally be made available much quicker than for the two other types of usage. Data, which are being used for load or inflow forecasting, must be collected in real-time or near real-time. This can only be accomplished by using automatic systems. For the other two uses, manual or off-line data collection using data-loggers or analog recorders may be cheaper and more reliable.

8.2.3.2 *Local recording or remote transmission*
Traditionally most hydrometric data has been recorded locally and transmitted by mail for later processing. Remote transmission has mainly been used for recording water levels in reservoirs and other types of data needed for the control of gates and valves in hydro-electric power plant systems. Local recording can be done by:
– Manual readings;
– Analog recording;
– Data loggers.

Even if data are not needed for real-time use there may still be several advantages using an automatic system:
– Quicker access to data;
– Easier processing of data;
– Early identification of errors;
– Observers are hard to find and increasingly expensive;
– Increased time resolution and number of parameters makes manual reading im-practical.

Due to these and other reasons the use of automatic systems is increasing, even for data types that are not used immediately. In many cases it can also be beneficial to extend an automatic system to record additional data types.

8.2.3.3 *Types of data needed*

A number of different hydrological and meteorological parameters can be of interest, depending on the planned use. The most important types of parameters for use in runoff forecasting are in italics below:

– Meteorological parameters: *Precipitation, air temperature,* wind speed, solar radiation, air humidity.

– Hydrological parameters: *Stage, streamflow,* water temperature, snow depth, *snow water equivalent, snowmelt, evaporation,* ground water level.

Precipitation, air temperature and wind speed (for correction of precipitation data) are commonly used as input to runoff forecasting models. Snow water equivalent and streamflow are used for model updating. Radiation and air humidity can be useful for the computation of evaporation and snowmelt.

For water quality monitoring and chemical mass balance studies a number of different water quality parameters can be of interest. Sensors for automatic measurements are available for parameters such as: pH, turbidity, conductivity, colour, light penetration, ammonium, nitrogen, phosphorus, sulphate, chloride.

8.2.4 *Data transmission*

Hydrometric data are usually collected at remote sites, far from the operational center where they will be used. Thus, transmission links have to be established to transmit data in an economical and reliable way. The user requirement specifications will determine the degree of reliability needed for a given system. This may, in turn, determine the most economical transmission type. A number of different transmission channels may be used for hydrometric data transmission:

– Hard wire links: Private land lines, private (rented) telephone wires, public telephone network, power lines;

– Radio: Radio line transmission, VHF/UHF-transmission;

– Satellite telemetry;

– Meteor-burst telemetry.

8.2.4.1 *Hard-wire links*

Private land lines require the user to lay his own cables and maintain these. Such lines are only feasible over short distances or for control systems where a very high degree of reliability can be economically justified. For hydro-electric power utilities, the use of power line communications may be an alternative, especially if hydrometric data can be sent on existing communication channels via power lines.

Private (or rented) telephone wires provide a high standard and high capacity, but they are also expensive. For hydrometric data such lines are only feasible where the user requirements for speed and reliability are very high.

Public telephone network lines usually meet the transmission needs for most hydrometric users. By using automatic dialling equipment at the master station, a large number of outstations may be called sequentially from one telephone. Since the volume of data transferred normally is low (typically 1-10 sensors with one measurement each hour for each station) the data transfer from each outstation is very fast. A modem is needed at each outstation and at the master station. Today modems with transmission speed of 2400 and up to 9600 baud are readily available at a low price, and even higher speeds are becoming possible using special modems.

8.2.4.2 *Radio transmission*

Radio communication provides a good alternative to hard wire links, provided that frequency allocations are available. Both VHF and UHF equipment is used for data transmission. All radio transmission equipment must meet government performance specifications for parameters such as transmitter power, frequency stability, spurious emissions and signal distortion. It may be difficult to obtain a frequency allocation, especially if alternative communication links such as telephone lines exist. In planning a radio-based system it is very important to undertake detailed studies of radio paths between the planned station sites. Much of this work can be done using topographic maps, but in marginal cases direct measurements of signal strength should be carried out. Radio equipment is vulnerable to lightning strikes, particularly where tall aerial masts are located on hill tops. Radio transmission can be of *simplex, semi-duplex* or *full duplex* type:

– *Simplex* is data transmission where signals can be passed one way only;

– *Semi-duplex* is a two-way transmission where data can be sent both ways but not simultaneously;

– *Full duplex* or two-way transmission of data. A full duplex link allows transmission and reception of information simultaneously in both directions.

Simplex transmission is the simplest type, and is commonly used for radio transmission between a field station and the outstation. Half or full duplex transmission is used between master station and outstations, where the master station requests data and the outstation responds by sending back observed data. Full duplex transmission is the most reliable, but also the most expensive. With duplex communication two stations can exchange control information to verify that a message has been correctly transmitted. If errors are detected, the receiving station can ask for retransmittal until the transmission is OK. For simplex transmission a good strategy is to transmit a number of identical messages so that the receiving station can compare them to each other. At least three messages should be used. Tests performed in mountainous terrain in Norway under difficult conditions has shown that this can improve communication reliability substantially.

8.2.4.3 *Satellite telemetry*

Hydrologists can use satellite telemetry to collect data from networks of remote hydrometric stations. Such systems are both reliable and cost effective, and there are almost no restrictions on the geograpic area that can be covered. Three operational satellite networks are now readily available for environmental data transmission, the ARGOS-system, the GOES-system and the METEOSAT-system.

The polar-orbiting ARGOS-system consists of three main components: The terrestrial radios that transmit environmental data, the radio receiver/transmitter on the satellite and the radio reception, processing and distribution system at receiving ground stations. The transmitting radios in the ARGOS-system are known as Platform Terminal Transmitters (PTT's). They are small, battery-operated radio transmitters which cost approximately $2000 each (in 1993). They transmit short data messages in less than one second at random time intervals (100-200 sec) at a frequency of 401.650 MHz. When a message is received by the satellite it is immediately retransmitted at a different frequency, and can then be picked up by a Direct Readout Ground Station (DRGS). The message is also stored and retransmitted later when one of the main receiving ground stations are within radio range. From here data can be sent by mail, telex or modem to the user. The ARGOS-system has proved to be both reliable and economical, and has been used for hydrometric data transmission from many remote areas.

The GOES-system (Geostationary Operational Environmental Satellite) is operated by the United States. The data transmission operates in the same way as described for the ARGOS-system. Radio transmitters operating in the GOES-system are known as Data Collection Platforms (DCP's) and they can be operated in three modes: self-timed (simplex), random adaptive (simplex) and interrogated (duplex). The GOES-system is most widely used in North- and South America.

The METEOSAT-system is operated by the European Space Agency (ESA). This system is functionally equivalent to the GOES-system, and offers the same type of services. By using plug-in modules, data from up to 120 sensors can be collected, processed, formatted and transmitted by METEOSAT. The data can be received directly by the user by installing a 1.5 m dish antenna, or from the main groundstation in Darmstadt, Germany.

8.2.4.4 *Meteor-burst communication*
This communication type utilizes ionized meteorite trails in the upper stratosphere (80-120 km) to establish a radio link. The ionized meteor trails reflect radio waves back to earth, making communications between stations up to 2000 km apart possible. This communication system is primarily used in remote areas where other communication types are not easily available. A large snow measurement program in USA (SNOTEL) uses meteor burst communication to collect data from over five hundred stations in the western part of the USA.

8.2.4.5 *Communication protocols*
A communication protocol is a set of computing rules for ordering, transmission and error control for data communication. Both hardware and software protocols can be defined. The lack of standard communication protocols has long been a serious problem for environmental data collection systems. Different suppliers have developed their own proprietary protocols which usually can only communicate with their own type of hardware. This can lead to difficulties where equipment from several suppliers is being used together, or when a user wants to purchase equipment from a new supplier. In Norway the hydro-electric industry has initiated the specification of two important software protocols for data communication to minimize such problems: The EDC-protocol and the ELCOM-protocol. Both are considered important for the standardization of data-acquisition systems for hydro-electric utilities in Norway.

8.2.5 Hydrometric data for hydropower systems operation: Two examples

The first test-systems used in Norway for automatic transfer of data to hydro-electric power stations were put into operation around 1980. These systems have since been replaced by more modern equipment and a large number of new systems are planned or are now in operation. At present at least twenty automatic hydrometric systems are being operated by Norwegian hydro-electric power companies, and the number is increasing steadily. Two different systems that both have been in operation for several years are presented below.

8.2.5.1 The Orkla hydro-electric power system (KVO)

One of the first and most extensive automatic hydrometric data collection systems was installed in the Orkla river in central Norway when the Orkla development with its five hydro-electric power stations was completed in 1980. The hydrometric data system was planned to supply data for a number of runoff forecasting models, integrated with short and long term planning software. All data are transmitted automatically to the main control centre at Berkåk, from where all the power stations in the river are controlled. The system consists of the following components (as of 1/1-1988), (Fig. 8.8):

– 1 Master station in the control centre at Berkåk;
– 7 Outstations, 3 with their own sensors connected;
– 8 Field stations;
– 1 Repeater station;
– 48 Sensors.

The first version of the electronic equipment was installed in 1981, and proved not to be sufficiently reliable for operational use. It has since been replaced by more modern equipment. The field stations and outstations in the present system are made by the Norwegian firm, A/S Scan–matic. This firm has also delivered similar equipment to a substantial number of other power companies.

Data transmission from field stations to outstations is by simplex VHF-radio. One repeater station was nessecary to obtain radio communication from the Øvre Dølvad station to the outstation at Nonshaugen. From outstations to master station public telephone lines are used, with automatic dialling equipment. Power supply is from solar cells and batteries at the field stations, and mains at the outstations and the master station. Most of the sensors are made by the Norwegian firm Aanderaa (of Bergen, Norway), except for the weighing type precipitation gauges which are made by Belfort (USA) and Geonor (Norway).

8.2.5.2 The Alta hydro-electric power system

The Alta hydro-electric power scheme was built by, and is operated by, Statkraft (since 1986). From 1992, a new automatic hydro optimization system was put into operation, partly replacing an older system. It includes a large and very complex data acquisition system.

The system today consists of the following main components (Fig. 8.9):

– The master station with a HYDMET database installed on a Sun workstation, an IBM-compatible PC and a WX200 process control computer from Siemens, all collecting data automatically from outstations and field stations;

Figure 8.8. Main components in the Orkla automatic data acquisition system.

– 7 Outstations (4 from EDAS, 3 from Scan–matic);
– 5 Field stations (all from Aanderaa);
– 36 Sensors.

The older EDAS-equipment was used even though data communication became more complex. The master station is located in the central operational centre for the Alta power stations. All data are sent to the HYDMET database on a Sun workstation. Data from the old EDAS-system is transferred to the HYDMET database by

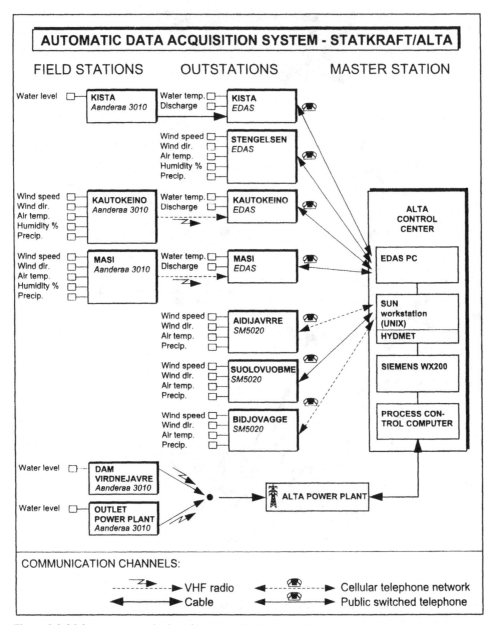

Figure 8.9. Main components in the Alta automatic data acquisition system.

two computer programs, one transmitting the data as a file and the other reading the data into the database. Data is transferred from the power station and reservoir by the Siemens process control system, and from the Siemens computer into HYDMET on the Sun workstation by ELCOM. The data transmission paths are complex and a number of different transmission channels are used to transmit data to the three receiving computers at the master station in the Alta control centre. The field stations in Kautokeino and Masi transmit data by VHF-radio to outstations located in a populated area. The field station at Kista (in the Alta river) transfers data to its outstation by line. The outstations at Kista, Stengelsen, Kautokeino and Masi all communicate with the old EDAS Master station by public telephone. The outstations at Bidijavrre and Bidjovagge use cellular telephones for data transmission, while the Suolovuobme outstation sends data by ordinary telephone. These three stations have no field stations and are all equipped with the same types of sensors: air temperature, precipitation, wind speed and wind direction. The field stations at the Virdnijavre dam and in the river below the tailrace outlet both are connected to the process control system in the Alta power station. Data are transmitted together with the process control signals to the process control computer. From here data are sent by ELCOM to the Sun workstation.

8.2.6 *Recommendations*

When an automatic data acquisition system is being planned, some of the most important things to consider are:
– Try to coordinate measurements for different purposes (historical and operational) within the same system;
– Coordinate meteorological and hydrological measurements;
– Work out a detailed user specification for the system *before* detailed planning is started;
– Make plans for system operation, maintenance and sensor recalibration *in advance*.

8.2.6.1 *User requirement specifications*
The preparation of a detailed user specification is an important first step in the design of an automatic data acquisition system. The users of the system must identify the essential objectives of the system, and decide on standards for operational reliability. A user specification should identify what the user wants from the system, not specific equipment details. Possible future needs should be taken into consideration, as for example the need for an increased number of sensors or increased time resolution. Some of the main steps in the user specification are commented below.

8.2.6.2 *Accuracy requirements*
In Table 8.1 a selected list of hydrological and meteorological parameters are presented, together with recommended units and accuracy requirements as defined for standard meteorological and hydrological observations in Norway. Generally it is recommended to adhere to these standard requirements, even if the accuracy is not needed at present for the planned use. The reason is that data series may in the next run be used in other types of applications, new model developments or in combination with other standard meteorological observations. In these applications the accuracy requirements may be higher than in the first applications.

8.2.6.3 *Time resolution*
For most purposes a time step of one hour is adequate. Only in very rapid rivers and

Table 8.1. Recommended units and standards for some hydrometeorological parameters (adapted from Kolderup Jensen et al. (1992) and WMO (1994)).

Parameter	Unit	Range	Accuracy	Resolution	Recalibration interval (Years)
Water level (stage)	cm	–	± 1 cm	1 cm	–
Stream discharge	m³/s	–	–	–	–
Snow depth	cm	–	± 1 cm	1 cm	–
Snow water eqivalent	mm	–	± 5 mm	1 mm	–
Evaporation	mm	0-100 mm/d	± 0.1 mm < 10 mm/d ± 2% > 10 mm/d	0.1 mm	–
Air temperature	°C	–60-+60°C	± 0.1°C	0.1°C	2 years
Precipitation	mm	0-300 mm	± 0.1 mm < 10 mm/d ± 2% > 10 mm/d	0.1 mm	0.5 year
Radiation	watt/cm²	300-3000 nm	–	–	1 year
Wind speed	m/s	0-75 m/s	± 0.5 m/s < 5 m/s	0.1 m/s	Group test
Wind direction	°	0-360°	± 5°	5°	Group test
Relative humidity	% Rh	10-100% Rh	± 5% (< 50%) ± 2% (> 50%)	1% Rh	2 years
Atmospheric pressure	hPa	Average ± 100 hPa	± 0.1 hPa	0.1 hPa	3 years

small catchments (e.g. urban areas) more frequent measurements might be needed. Increasing the time resolution will increase the total volume of data, and put more strain on database and quality control systems. If integrated or averaged values within the timestep are needed, the sensors may have to perform continuous measurements and process them to compute such values. Examples are global radiation, precipitation and snow melt where hourly sums are needed, or wind speed and air temperature where hourly averages are needed.

8.2.6.4 *Environmental considerations*
The operating environment for electronic equipment has to be considered carefully. Some of the most important environmental parameters to be considered are:
- – Temperature (low/high/cycling/shock);
- – Air humidity (condensation);
- – Shock/Vibrations (during transport, installation and operation);
- – Biological effects (insects, rodents, fungus);
- – Chemical effects (salt spray, acid precipitation,...);
- – Electromagnetic interference.

8.2.6.5 *Power supply*
Field stations are always required to operate from local power supply, using internal batteries combined with solar cells or wind generators. For field stations with low energy consumption the use of internal batteries can be the most economical solution. With higher energy consumption the use of solar cells has proved to be reliable and economical. Some types of field stations, e.g. from Aanderaa can be equipped with solar cells integrated in the station hardware, making installation and operation

very simple. When using solar cells at high latitudes the available solar radiation during the winter period has to be considered carefully. Snow and freezing rain may clog the solar cells and reduce power capacity. The solar cell output and/or battery voltage can be monitored and transmitted together with sensor data for control.

At outstations and at the master station a mains power supply is usually available but a backup power supply may still be needed. One of the most frequent failure is due to electrical mains failure. A battery backup with automatic switching and trickle recharging may be needed to ensure reliable operation.

8.2.6.6 *Maintainance and sensor calibration*

Both manual and automatic systems need regular control and maintainance to ensure high data quality. It is important to make such plans in advance, to make sure that manpower and money are available to perform regular inspections and preventive maintainance. Most types of sensors needs regular maintainance and recalibration, especially sensors with mechanical parts such as wind cup anemometers. The Norwegian Meteorological Institute (DNMI) has given recommended intervals between recalibration for most types of sensors mentioned here. These recommendations are listed in Table 8.1 together with other specifications for different sensor types.

8.2.6.7 *Documentation*

A good documentation of the system is vital for maintainance and repair of the system. Make sure that all details about sensors, cables, electronics, software and power supply are documented and kept available for later use. Equally important, update the documentation when the system has been changed!

REFERENCES

Aastad, J. & Søgnen, R. 1954. Discharge measurements by means of a salt solution, 'The relative dilution method'. *Proc. Association Internationale d'Hydrologie Scientifique*, Rome.

Ackers, P., White, W.R., Perkins, J.A. & Harrison, A.J.M. 1978. *Weirs and Flumes for Flow Measurement*, John Wiley & Sons, Chichester.

Bos, M.G. (ed.) 1978. *Discharge Measurement Structures* Delft Hydraulic Laboratory, publ. 161, Delft.

Herschy, R.W. (ed.) 1978. *Hydrometry, Principles and Practices*, John Wiley & Sons, Chichester.

Herschy, R.W. 1985. *Streamflow Measurement*, Elsevier, London.

Kolderup Jensen, C., Killingtveit, Å., Larsen, O., Nesse, L. & Sundøen, K. 1992. *Data collection equipment for meteorology and hydrology*, 4th ed., Water System Management Association, Asker.

Sargent, D.M. 1981. The development of a viable method of streamflow measurement using the integrated float technique. *Proc. Inst. Civil Eng.*, 71(2), pp. 1-15.

Tilrem, Ø.A. 1986a. Level and Discharge Measurements under difficult Conditions. *WMO, Operational Hydrology*, Rep. 24.

Tilrem, Ø.A. 1986b. Methods of Measurement and Estimation of Discharges at Hydraulic Structures. *WMO, Operational Hydrology*, Rep. 26.

WMO 1994. *Guide to Hydrological Practices*. World Meteorological Organization, WMO-168, Geneve, ISBN 92-63-15168-7. Ed. 5.

3. Integrated catchment studies

Geochemical processes, weathering and groundwater recharge in catchments
O.M. Saether & P. de Caritat (eds) © 1997 Balkema

CHAPTER 9

Catchment mass balance

JAMES I. DREVER
Department of Geology and Geophysics, University of Wyoming, Laramie, USA

9.1 INTRODUCTION

Mass balance simply means a budget. In the context of catchment studies, it is a budget that describes fluxes of solutes into and out of a catchment, and assigns the solutes to specific sources and sinks such as atmospheric deposition, biomass change, or bedrock weathering. Although the concept is simple, identifying and quantifying the various contributions to the individual fluxes is not easy, and the final budget often has ambiguities and uncertainties.

9.2 TERMS IN THE MASS BALANCE EQUATION

Budgets can be constructed at different levels of detail. As an example, a geochemist would generally view 'biomass uptake' as a single process, but would probably try to resolve 'chemical weathering' into a detailed set of mineral reactions. A biologist, on the other hand, would tend to regard 'chemical weathering' as a single process, but subdivide the biomass into different compartments.

A simple view of some processes occurring in a catchment is shown in Figure 9.1. For this system, we can write a mass balance equation:

$$\text{solutes in outflow} = \text{solutes from atmosphere} + \text{solutes from weathering} \pm \text{solutes from change in biomass} \pm \text{change in exchange pool} \qquad (9.1)$$

In writing this equation, we have already made the assumption that our time-scale is long enough that changes in water storage can be ignored. The question of time-scale is very important in mass-balance studies: it is rarely possible to construct a meaningful catchment budget for a time-scale of less than a year. On the shorter time-scale, biomass effects tend to be large, whereas they tend to average out over an annual cycle. Let us consider the individual terms in the above equation.

241

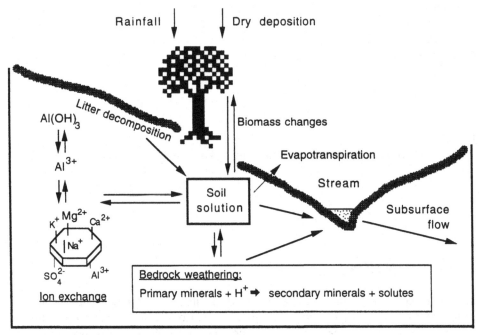

Figure 9.1. Some processes affecting solute outputs from a catchment.

9.2.1 *Solutes in outflow*

Questions of gauging and chemical analysis are discussed elsewhere in this book (Chapters 5, 6 and 8). The information that is needed is the volume of water leaving the catchment and its chemical composition. The main assumption that is made is that the gauged outflow represents all the water leaving the catchment and that discharge via groundwater is negligible. It is difficult to test this assumption. One approach is to test the input-output budget for chloride balance, assuming that there are no sources or sinks for chloride in the catchment (i.e. that chloride is a conservative tracer). The problems with this approach may include uncertainties as to whether precipitation gauges accurately measure the amount of incoming precipitation (blowing snow can be a problem), a highly variable chloride concentration in precipitation (making it difficult to quantify precisely the annual input), and unknown amounts of dry deposition of chloride (Probst et al. 1990). The assumption of no sources or sinks for chloride on a relatively short time-scale may also be questionable.

In cold environments, there may be additional practical problems with measuring outflow. The greatest water flux and solute flux occur at the time of snowmelt, a time when access may be difficult and instrumentation may be affected by intermittent freezing.

9.2.2 *Solutes from the atmosphere*

Atmospheric deposition occurs in the form of rain and snowfall, condensation from mist or fog, and dry deposition. Dry deposition in turn consists of solid particles and gases such as SO_2 and NO_x that are taken up directly by vegetation or moist foliage. The term *occult deposition* is sometimes used to cover dry deposition and deposition from mist or fog. Occult deposition is difficult to quantify; one approach is to measure *throughfall*, which is rain collected after it has passed through the forest canopy. The solutes in throughfall can be attributed to:

1. Solutes in the incoming precipitation;

2. Solutes from dry deposition and deposition from fog on the foliage of the trees; and

3. Solutes translocated from the soil by the trees and then leached from the foliage.

It is possible to distinguish *approximately* between occult deposition and translocation on the basis of tree physiology (Matzner 1986), but ambiguities remain, particularly for trace elements. As an example, Table 9.1 shows the composition of open-field rainfall and throughfall in an old spruce stand in the Vosges Mountains of eastern France. The corresponding fluxes in units of keq ha^{-1} y^{-1} are also shown. The conversion between concentration and flux is affected by the fact that about 26% of the rain that falls on the forest canopy evaporates without ever reaching the ground. The dry/occult deposition flux represents the difference between rainfall and throughfall, corrected for elements translocated through the trees.

It is clear from Table 9.1 that occult deposition may be a major input into a catchment. It is quite sensitive to the amount and type of vegetation present. Conifers are much more effective in trapping atmospheric solutes than deciduous trees, and trees are more effective than grassland. Occult deposition of sulfate can be a major factor in acidification of soils and surface water (see Chapter 12).

In areas close to the ocean, the atmospheric input of solutes is dominated by sea salts. The sea salt input decreases with increasing distance from the ocean. Pollution from human activities causes increased concentrations of sulfate (primarily from fossil fuel burning), nitrate (from fossil fuels and agriculture) and ammonium (primarily from agriculture).

Table 9.1. Mean concentrations of rainfall and throughfall, and fluxes of elements from the atmosphere into an old spruce stand, Vosges mountains, eastern France (data from Probst et al. 1990).

	NH_4	Na	K	Mg	Ca	H	Cl	NO_3	SO_4
Concentrations (µeq/L)									
Bulk precipitation	19.1	10.0	2.8	4.5	11.9	33.9	12.5	24.1	41.5
Throughfall	36.9	46.4	52.7	17.8	65.5	114.8	63.4	78.3	185.0
Fluxes (keq/ha/y)									
Bulk precipitation	0.270	0.142	0.039	0.063	0.168	0.480	0.177	0.340	0.58
Throughfall	0.385	0.484	0.550	0.185	0.683	1.197	0.661	0.817	1.932
Difference	0.115	0.342	0.511	0.122	0.515	0.717	0.484	0.477	1.352
Occult deposition*	0.115	0.342	0.102	0.061	0.412	1.282	0.484	0.477	1.352

*Difference corrected for elements translocated through the trees.

9.2.3 *Changes in the exchange pool*

Soils contain exchangeable cations and anions that are in equilibrium with the soil solution. As the composition of the soil solution changes, ions will be exchanged between the solid phase and solution. If the soil solution composition does not change with time, adsorbed ions will not change either and ion exchange will make no contribution to the solute budget. In the short term, changes in solution composition occur as a result of precipitation events, evaporation, and growth cycles of plants. It is often assumed that these changes average out over an annual cycle so that, on a time scale longer than a year they cause no net change to the exchange pool and hence no net contribution to solute flux. This assumes that year-to-year variations in solution chemistry (caused, for example by alternating wet and drought years) are insignificant.

If there is a permanent change in the chemistry of atmospheric input, this will cause a unidirectional change in the exchange pool. If, for example, the concentration of H^+ in precipitation increased, the H^+ (or Al^{3+} generated from dissolution of $Al(OH)_3$ in response to the input of H^+) would displace other cations (commonly mostly Ca^{2+}) from exchange sites, and these cations would appear in the outflow from the catchment (e.g. Reuss & Johnson 1986). If the composition of precipitation remained constant, the composition of the exchange pool would gradually adjust to the new input and a new steady state, in which there was no net change in the exchange pool over time, would be established. The time taken by the exchange pool to adjust to a change in atmospheric input is generally quite long – on a time-scale of decades – because the reservoir of exchangeable ions in the soil is generally large compared to the annual input from the atmosphere.

In catchments that are influenced by acid deposition, one of the major tasks of modeling is to distinguish between cations from weathering of minerals and cations displaced from exchange sites. This is very difficult to determine. If the catchment does not receive a large input of anthropogenic acidity from the atmosphere, it is generally assumed that the exchange pool is in steady state, and is not a net contributor of solutes.

9.2.4 *Changes in the biomass*

As plants grow, they extract inorganic nutrients from the soil solution and incorporate them into plant tissue. The stoichiometry is approximately (Schnoor & Stumm 1985), shown in Equation (9.2).

Plant growth affects the budgets of the major cations and protons and will affect both the exchange pool and the net output from a catchment. When plants die and decompose, the process is reversed and the elements are returned to the soil. This topic is discussed in Chapter 2.

If a forest is in steady state, that is to say the growth of new vegetation is exactly balanced by the death and decay of old vegetation, the biomass will be neither a source nor a sink in the mass balance equation. However, in forested catchments, the biomass is rarely in a steady state. Even without human intervention such as tree cutting or planting, forests typically go through cycles of gradual biomass increase interrupted by catastrophic events, such as fire or disease, that result in rapid biomass

$$
\left.\begin{array}{l}
800\,CO_2 \\
6\,NH_4^+ \\
4\,Ca^{2+} \\
1\,Mg^{2+} \\
2\,K^+ \\
1\,Al(OH)_2^+ \\
1\,Fe^{2+} \\
2\,NO_3^- \\
1\,HP_2O_4^- \\
1\,SO_4^{2-} \\
H_2O
\end{array}\right\} \xrightarrow{\text{photosynthesis}} \text{biomass} + 16\,H^+ + 804\,O_2
\qquad (9.2)
$$

loss. It is only on a very long time scale, a scale of centuries, that the biomass of a forest can be considered in steady state. Hubbard Brook, New Hampshire, in the northeastern United States, is an example of a catchment where the biomass term is large compared to the weathering term (Fig. 9.2). According to the data of Likens et al. (1977), the net uptake of Ca in the form of biomass increment and forest floor (organic) increment was 45% of the amount released by weathering. For potassium, the biomass increment was 86% of the amount released by weathering: the net outflow of K from the catchment (runoff − precipitation) was only 14% of the K calculated to be released by chemical weathering. More recent measurements have decreased the biomass uptake term relative to the weathering term, but the conclusion

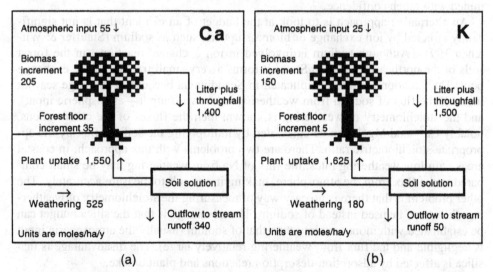

Figure 9.2. Calcium (a) and potassium (b) fluxes at Hubbard Brook Experimental Forest, New Hampshire, USA (data from Likens et al. 1977; after Drever 1988).

remains: in forested catchments with silicate bedrocks, the biomass uptake term is likely to be important.

The biomass term in Equation (9.1) can be measured directly by measuring the rate of growth of trees and the chemical composition of different tissues. This requires long-term measurements over several years, implying a significant investment of time and resources. Even if the long-term average biomass uptake rate is known, the rate may vary from year to year depending on weather conditions, particularly the availability of moisture.

In principle, the biomass uptake term can be determined by difference if the other terms in Equation (9.1) are known (Velbel 1986). This approach is rarely useful because the other terms can rarely be measured independently with sufficient accuracy.

9.2.5 *Chemical weathering*

There is no simple direct way of measuring the contribution of chemical weathering to the solutes measured at the output of a catchment. The most common approach is to assign values to (or simply ignore) the biomass change and ion exchange terms in Equation (9.1) and determine the weathering term by difference. The utility of this approach depends very much on the catchment. In catchments where the bedrock weathers rapidly, the weathering term is much larger than the other two, so there is no problem. Where atmospheric deposition has not been significantly affected by pollution, neglecting the ion exchange term should not introduce much error, and in catchments that are largely unvegetated, the biomass term should be negligible. Unfortunately, the catchments that are of greatest interest, particularly in the Nordic countries, are those that are characterized by low weathering rates, the presence of forests (often complicated by changes in land use in historic times), and significant anthropogenic inputs of acid (Wright 1988). In these situations, the weathering term may be small compared to the other terms in Equation (9.1) and it cannot be determined with much confidence.

An alternative approach is to look at the budget of an element that is not significantly affected by ion exchange or biomass uptake, such as sodium (Stauffer & Wittchen 1991). Although sodium is involved in ion exchange reactions, in the forest soils of the northeastern United States it forms a very small fraction of the exchange pool. The situation is more complicated in Scandinavia because of the large sea salt input. If the flux of sodium from weathering is known (output – atmospheric input), and the stoichiometry of weathering is known, then the fluxes of the other cations, notably Ca^{2+} and Mg^{2+}, can be calculated by multiplying the sodium flux by the appropriate stoichiometric ratio. There are two problems with this approach: in coastal areas with low weathering rates, the flux of Na from weathering may be small compared to the flux from the atmosphere, making it difficult to measure accurately. The other problem is that there is no easy way of measuring the stoichiometry of weathering. Silica can be used instead of sodium. The advantage is that the silica budget can be established with more certainty than that of sodium (usually the atmospheric input is negligible and the flux from weathering relatively large). The disadvantage is that silica is affected by adsorption-desorption reactions and plant uptake.

In some catchments, the isotopic composition of strontium on exchange sites is different from that of bedrock minerals (Miller et al. 1993). This can be used to determine the proportion of the Sr in catchment outflow that is derived from each of the two sources. If it is assumed that Ca^{2+} behaves exactly as Sr^{2+}, then the Ca^{2+} in the outflow can be assigned to the two sources in the same ratio as for Sr^{2+}. Since Ca^{2+} is the most important cation released by weathering, establishing a budget for Ca^{2+} is almost equivalent to establishing a budget for weathering as a whole.

9.3 MASS BALANCE AND MINERAL WEATHERING

9.3.1 *The Sierra Nevada, California, USA*

The classic example of the use of mass balance to characterize weathering reactions was presented by Garrels & Mackenzie (1967). They interpreted the mean compositions of perennial and ephemeral springs emerging from granite and granodiorite bedrock on the east side of the Sierra Nevada of California in terms of atmospheric input and bedrock weathering. Ion exchange and biomass uptake were ignored. The compositions of the springs is shown in Table 9.2 and Figure 9.3.

9.3.1.1 *Ephemeral springs*
The two sets of springs were considered separately. The calculation procedure for the ephemeral springs (i.e. flowering only parts of the year) was as follows (Table 9.3):

1. The atmospheric input (which is minor) was subtracted;

2. All the remaining Na and Ca were ascribed to the weathering of plagioclase feldspar to kaolinite. The Na/Ca ratio of the feldspar chosen was equal to the Na/Ca ratio in the water, which was reasonably consistent with the composition of the feldspar in the bedrock. The corresponding amounts of silica and bicarbonate were calculated;

3. All the Mg was ascribed to the weathering of biotite (actually end-member

Table 9.2. Mean values for compositions of springs of the Sierra Nevada (after Garrels & Mackenzie 1967).

	Ephemeral Springs ppm	Ephemeral Springs µmol/L	Perennial Springs ppm	Perennial Springs µmol/L	Difference µmol/L
Ca	3.11	78	10.4	260	182
Mg	0.70	29	1.70	71	42
Na	3.03	134	5.95	259	125
K	1.09	28	1.57	40	12
HCO₃	20.0	328	54.6	895	567
SO₄	1.00	10	2.38	25	15
Cl	0.50	14	1.06	30	16
SiO₂	16.4	273	24.6	410	137
Al	0.03		0.02		
Fe	0.03		0.03		
F	0.07		0.09		
NO₃	0.02		0.28		
TDS	36.0		75.0		
pH (median value)	6.2		6.8		

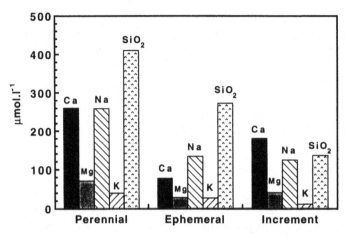

Figure 9.3. Major solutes in springs from granites and granodiorites, Sierra Nevada. Ephemeral springs flow for only part of the year; perennial springs flow year-round. The 'Increment' is the difference between the perennial and the ephemeral springs, and is inferred to represent the additional solutes acquired by deeper water circulation (data from Garrels & Mackenzie 1967).

Table 9.3. Mass balance calculation for the ephemeral springs of the Sierra Nevada (after Garrels & Mackenzie 1967). Units are $\mu moles.l^{-1}$.

	Na	Ca	Mg	K	HCO$_3$	SO$_4$	Cl	SiO$_2$
Concentrations in spring	134	78	29	28	328	10	14	273
Corr. for atmos. input	110	68	22	20	310	0	0	270

Alteration of plagioclase to kaolinite: $177\ Na_{0.62}Ca_{0.38}Al_{1.38}Si_{2.62}O_8 + 246\ CO_2 + 367\ H_2O =$

$123\ Al_2Si_2O_5(OH)_4 + 110\ Na^+ + 68\ Ca^{2+} + 246\ HCO_3^- + 220\ SiO_{2(aq)}$

Remainder	0	0	22	20	64	0	0	50

Alteration of biotite to kaolinite: $7.3\ KMg_3AlSi_3O_{10}(OH)_2 + 51\ CO_2 + 26\ H_2O =$

$3.7\ Al_2Si_2O_5(OH)_4 + 22\ Mg^{2+} + 7.3\ K^+ + 51\ HCO_3^- + 15\ SiO_2$

Remainder	0	0	0	13	13	0	0	35

Alteration of K-feldspar to kaolinite: $13\ KAlSi_3O_8 + 13\ CO_2 + 19.5\ H_2O =$

$6.5\ Al_2Si_2O_5(OH)_4 + 13\ K^+ + 26\ SiO_2 + 13\ HCO_3^-$

Remainder	0	0	0	0	0	0	0	9

Overall reaction: $177\ Plagioclase + 0.073\ Biotite + 0.13\ Kspar\ (+H_2O + CO_2) =$

$1.33\ Kaolinite + 1.10\ Na^+ + 0.68\ Ca^{2+} + 0.22\ Mg^{2+} + 0.20\ K^+ + 2.70\ SiO_2 + 3.1\ HCO_3^-$

phlogopite) to kaolinite. The corresponding amounts of K, silica and bicarbonate were calculated;

4. The remaining K (which was a small amount) was ascribed to the weathering of K-feldspar to kaolinite. The corresponding amounts of silica and bicarbonate were calculated.

These three reactions could 'explain' all the major solutes in the water (the minor amount of silica remaining was within the precision of the analyses). Bicarbonate balance is inevitable from charge balance, but there is no such constraint on silica. From the excellent balance obtained, several conclusions were drawn:

1. The reactions chosen were a reasonable description of what was occurring in nature;

2. Because the three reactions accounted for all the observed silica in solution, silica was derived from the breakdown of silicate minerals and not from dissolution of quartz;

3. About 80% of the rock-derived solutes came from the dissolution of plagioclase alone. Even though K-feldspar was abundant in the rocks, little dissolution occurred. This illustrates the selectivity of weathering discussed in Chapter 1: more reactive minerals contribute disproportionately to the solute flux.

9.3.1.2 *Perennial springs*

The perennial springs (flowing throughout the year) (Table 9.2 and Fig. 9.3) represent water that has circulated deeper through the bedrock and had more time to react with minerals. Garrels & Mackenzie (1967) made the assumption that when the water first infiltrated, it acquired the composition represented by the ephemeral springs, and then the deeper circulation provided additional solutes. The 'increment' in Figure 9.3, the difference between the compositions of the ephemeral and perennial springs, represents the additional solutes acquired by deeper circulation. Chemically, the increment is not the same as the ephemeral springs (Fig. 9.3), which means that different weathering reactions must determine its composition. The ratio of Ca to Na is higher, and the ratio of SiO_2 to Na lower. In calculating a budget, Garrels & Mackenzie (1967) assumed the same feldspar composition as that used to interpret the ephemeral springs. (An alternative approach might have been to assume that the deeper waters interacted with a more calcic plagioclase, but there was no geological reason to suggest this.) The lower ratio of SiO_2 to Na implied that more silica was retained in the solid phase than was the case for the ephemeral springs. This was interpreted to mean that a secondary phase with a higher Si/Al ratio than kaolinite was being formed as a secondary phase. The phase hypothesized was a beidellite (montmorillonite in the original terminology of Garrels & Mackenzie 1967), which was consistent with the idea (Chapter 1, Fig. 1.4) that more concentrated solutions would be in equilibrium with a smectite rather than kaolinite. A reaction was then written (Table 9.4) for plagioclase altering to a mixture of kaolinite and beidellite, and the relative proportion of the two phases was adjusted to give exact balance for silica. This left the problem of excess Ca, which was attributed to 'minor amounts of carbonate encountered en route.' The balance for the perennial springs is less satisfying than that for the ephemeral springs. Is a smectite really forming, and what is its chemical composition? Is calcite really present, or does the excess calcium come

Table 9.4. Mass balance calculation for the perennial springs of the Sierra Nevada (after Garrels & Mackenzie 1967). Units are μmoles.l^{-1}.

	Na	Ca	Mg	K	HCO$_3$	SO$_4$	Cl	SiO$_2$
Initial concentration (perennial-ephemeral)	125	182	42	12	539*	15	16	137
Corr. for atmos. input	109	167	42	12	539	0	0	137

Alteration of biotite to kaolinite: $14\ KMg_3AlSi_3O_{10}(OH)_2 + 98\ CO_2 + 49\ H_2O =$
$$7\ Al_2Si_2O_5(OH)_4 + 42\ Mg^{2+} + 14\ K^+ + 98\ HCO_3^- + 28\ SiO_2$$

	Na	Ca	Mg	K	HCO$_3$	SO$_4$	Cl	SiO$_2$
Remainder	109	167	0	–2	441	0	0	109

Alteration of plagioclase to kaolinite: $38\ Na_{0.62}Ca_{0.38}Al_{1.38}Si_{2.62}O_8 + 52\ CO_2 + 78\ H_2O = 26$
$$Al_2Si_2O_5(OH)_4 + 24\ Na^+ + 14\ Ca^{2+} + 52\ HCO_3^- + 47\ SiO_{2(aq)}$$

	Na	Ca	Mg	K	HCO$_3$	SO$_4$	Cl	SiO$_2$
Remainder	85	153	0	–2	389	0	0	62

Alteration of plagioclase to smectite: $137\ Na_{0.62}Ca_{0.38}Al_{1.38}Si_{2.62}O_8 + 162\ CO_2 + 162\ H_2O =$
$$81\ Ca_{0.17}Al_{2.33}Si_{3.67}O_{10}(OH)_2 + 85\ Na^+ + 38\ Ca^{2+} + 162\ HCO_3^- + 62\ SiO_{2(aq)}$$

	Na	Ca	Mg	K	HCO$_3$	SO$_4$	Cl	SiO$_2$
Remainder	0	115	0	–2	227	0	0	0

Dissolution of calcite: $115\ CaCO_3 + 115\ H_2O + 115\ CO_3{}^-{}_2 = 115\ Ca^{2+} + 230\ HCO_3^-$

	Na	Ca	Mg	K	HCO$_3$	SO$_4$	Cl	SiO$_2$
Remainder**	0	0	0	–2	–3	0	0	0

Overall reaction: $175\ Plag + 14\ Biotite + 115\ Calcite\ (+H_2O + CO_2) =$
$$81\ Smectite + 33\ Kaol + 109\ Na^+ + 167\ Ca^{2+} + 42\ Mg^{2+} + 12\ K^+ + 539\ HCO_3^- + 137\ SiO_2$$

*Bicarbonate number was adjusted slightly to make cations and anions balance. **These numbers are effectively zero.

from some other source? We shall return to these questions later in this chapter (Section 9.4).

The importance of the Garrels & Mackenzie (1967) paper here is not so much the conclusions as the concept of explaining quantitatively the composition of a water in terms of mineral weathering reactions. The same approach has been used subsequently in numerous quantitative studies. The type of calculation can be generalized and solved by matrix methods (Velbel 1986; Bowser & Jones 1993), or by computer codes such as BALANCE (Parkhurst et al. 1982) and NETPATH (Plummer et al. 1991). Mathematically speaking, if the number of solid phases available as reactants or products is equal to the number of independent solutes used in the mass balance calculation, a balanced weathering reaction can generally be written. The solution to calculations of this type is generally non-unique, and judgment is required in selecting the most plausible balance for a particular situation.

9.3.2 *Absaroka mountains, Wyoming, USA*

Miller & Drever (1977) applied the same approach as Garrels & Mackenzie (1967) to a river draining andesitic volcanic rocks in Wyoming. They also analyzed soils and bedrock to see if the behavior of the solid phases was consistent with that predicted from the runoff chemistry. The chemistry of runoff was quite similar to that of the Sierra Nevada springs (Fig. 9.4). The Ca concentration was lower, suggesting, per-haps, that calcite was absent, and the ratio of SiO_2 to Na was lower, which could be interpreted as formation of relatively more smectite and less kaolinite in the Absaro-kas.

According to the calculation procedure of Garrels & Mackenzie (1967), the Na, Ca, and SiO_2 concentrations in the water could be 'explained' by the equation:

$$Na_{0.6}Ca_{0.4}Al_{1.4}Si_{2.6}O_8 + 2.69\ H_2O + 1.25\ CO_2$$

$$\text{plagioclase}$$

$$= 0.18\ Al_2Si_2O_5(OH)_4 + 0.22\ Ca_{0.33}Al_{4.67}Si_{7.33}O_{20}(OH)_4 \qquad (9.3)$$

$$\text{kaolinite} \qquad \text{smectite (beidellite)}$$

$$+ 0.6\ Na^+ + 0.32\ Ca^{2+} + 1.25\ HCO_3^- + 0.63\ H_4SiO_4$$

The water chemistry thus predicts that smectite and kaolinite should be forming in approximately equal proportions. However, kaolinite was a rare mineral in the soils; the only common secondary mineral was smectite. Furthermore, the smectite present was a montmorillonite, which can be represented approximately by the formula:

$$Ca_{0.33}(Mg, Fe)_{0.67}Al_{3.33}Si_8O_{20}(OH)_4 \qquad (9.4)$$

and is quite different from the beidellite assumed by Garrels & Mackenzie (1967) (Fig. 9.5). A further major problem was that the composition of the clay fraction of the soil (Fig. 9.6) was such that a reasonable mass balance for the equation:

$$\text{Bedrock minerals} + \text{atmospheric input} =$$

$$\text{soil minerals} + \text{solutes in runoff} \qquad (9.5)$$

could not be written.

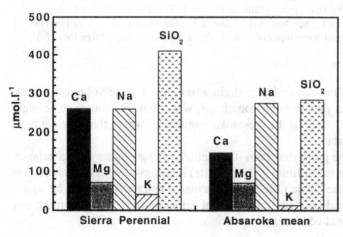

Figure 9.4. Mean compos-ition of runoff in the Absaroka Mountains, Wyoming (Miller & Drever 1977) compared to that of the Sierra Nevada perennial springs (Garrels & Mackenzie 1967).

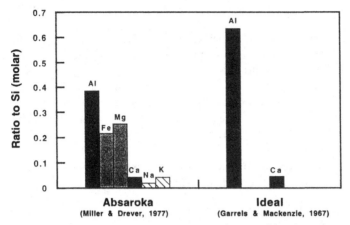

Figure 9.5. Comparison between the measured composition (as ratios to silica) of smectite from the soils of the Absaroka Mountains compared to the idealized composition assumed in the calculations of Garrels & Mackenzie (1967).

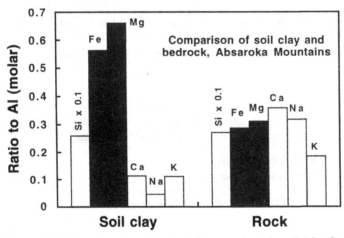

Figure 9.6. Ratios of major cations to Al (mean values) in soil (clay fraction) and bedrock from the Absaroka Mountains, Wyoming (data from Miller & Drever 1977). The fact that the ratios of Fe and Mg to Al in the soil clay are higher than in the bedrock implies that the clay cannot be derived directly from the bedrock without some input of Fe and Mg, or an unreasonably large loss of Al.

Miller & Drever (1977) concluded that slight alteration of large volumes of rock below the soil zone, rather than the soil zone itself, was the major source of solutes. This conclusion may apply only to the Absaroka mountains, where the andesite bedrock is porous and permeable.

This study illustrates the point that even if an apparently satisfactory mass balance (bedrock minerals altering to kaolinite and beidellite) is achieved with the calculation procedure of Garrels & Mackenzie (1967), the calculated reactions may not be an accurate representation of reality. Actual analyses of minerals from the catchment can provide important additional constraints.

9.3.3 *Adirondack mountains, New York, USA*

The Adirondack mountains have been the site of many studies related to acid deposition. The question of mineral weathering was addressed in detail by April et al. (1986). They focused particularly on two apparently similar catchments, Woods Lake and Panther Lake. Despite the close proximity and apparent similarity of the catchments, Woods Lake is 'acidic', with a typical pH in the outlet between 4.5 and 5, whereas Panther Lake is 'neutral', with a typical outlet pH near 7. It was suggested that the difference was related to thicker tills in the Panther Lake basin, which allowed for more chemical weathering. April et al. (1986) used three different mass-balance approaches to estimate weathering rates:

1. The present-day (1978-1980) flux of solutes leaving the basins, which reflects the present-day weathering rate ± ion exchange reactions ± the effect of biomass change.

2. Depletion of weatherable minerals in soil profiles in the basins. The abundance of weatherable minerals such as hornblende and plagioclase decreases towards the surface of the soil, presumably as a result of chemical weathering. If it is assumed that there has been no physical erosion or lateral transport, the decrease towards the surface reflects the integrated effect of weathering since the soil was first exposed at the surface, presumably at the end of the last glaciation 14,000 years ago. Changes in soil volume associated with weathering can be corrected for by normalization to a mineral that is not weathered significantly, such as ilmenite. By this procedure, a mean flux of cations over the last 14,000 years can be calculated. This will be a 'true' weathering rate in the sense that it is unaffected by plant uptake or cation exchange.

3. The change in the bulk chemistry of the soil with depth. This method is strictly analogous to the mineral depletion method above, but is based on bulk chemistry (normalized to an element such as titanium) rather than mineralogy.

A comparison of the cation fluxes calculated from methods 1 and 3 is shown in Table 9.5, and a comparison of mineral weathering rates for the Panther Lake catchment calculated from methods 1 and 2 is shown in Table 9.6. The results from the different methods clearly disagree. According to the soil chemistry profiles, the rates of weathering in the two catchments are indistinguishable, whereas the present-day cation fluxes differ by a factor of 8. The rate of hornblende weathering at Panther

Table 9.5. Cation fluxes (eq/ha/y) from Woods Lake and Panther Lake basins in the Adirondack mountains (after April et al. 1986).

| | Net present-day stream flux | | Calculated from bulk composition of soil | |
	Woods	Panther	Woods	Panther
Ca	167	1149	178	100
Mg	28	294	122	84
K	–21*	32	129	150
Na	24	204	189	108
Total	198	1679	618	449

*K input from the atmosphere was greater than the stream output.

Table 9.6. Mineral weathering rates (kg/ha/y) for Panther Lake basin calculated from net present-day stream flux and from mineral depletion curves in soil (after April et al. 1986).

	From present-day stream flux	From mineral depletion curves
Hornblende	312	8
Plagioclase	7	32
K-feldspar	−22	31

Lake calculated from the stream flux is about 40 times the rate calculated from the mineral depletion curves.

The differences between the present and long-term rates at Woods Lake could be explained by a gradual decrease over time in the amount of weatherable minerals exposed to incoming precipitation. As reactive minerals were dissolved, the rate of weathering would decrease. There is no easy or unique explanation for the discrepancies at Panther Lake. One possible explanation is that weathering is taking place deeper in the hydrologic system than the 1 m of soil analyzed by April et al. (1986). The discrepancy between the present-day and long-term apparent weathering rates for hornblende suggest that ion exchange may be a significant present-day source of calcium. An alternative explanation would be that the weathering rate at Panther Lake has increased in response to the recent increase in acid deposition. There is no obvious reason, however, why mineral weathering would increase in one catchment and not the other.

This study illustrates one way in which solid phases can be used as a constraint on a mass balance calculation. It also illustrates the difficulty in distinguishing between ion exchange and weathering as a source of cations.

9.3.4 *Sogndal, Norway*

Sogndal was the site of an experiment in which two small catchments on gabbroic gneiss were artificially acidified by application of acid (H_2SO_4) on one, and a mixture of H_2SO_4 and HNO_3 on the other over a four-year period (Frogner 1990). The experiment was part of the 8-year RAIN Project (Wright et al. 1986) directed towards understanding the effects of acid deposition on surface waters. Catchments adjacent to the acidified catchments served as controls. The area was characterized by treeless alpine vegetation and thin (about 30 cm) sandy soils. In addition to the field studies, laboratory studies were conducted on the dissolution of minerals from the soil. The discussion here is based on Frogner (1990).

Acidification caused an increase of about 45% in the annual fluxes of both base cations (dominantly Ca^{2+}) and silica, which suggests an increase in the weathering rate of plagioclase (plagioclase was the dominant source of solutes). Sodium, however, showed no significant increase, which suggests no increase in the plagioclase weathering rate. Immediate increases in Ca and Mg concentrations associated with applications of acid were clearly related to ion exchange. The question is whether the change in annual flux represents increased weathering or ion exchange. The change in the exchange pool, even over the four years of the experiment, would have been too small to measure directly. One could argue that the increased silica flux clearly indicates increased weathering, and the sodium data are inconclusive, or that the silica flux represents pH-dependent desorption and the sodium data are to be believed. The important point, in my opinion, is that even in a small, intensively studied catchment with several years of data, the question cannot be unambiguously resolved.

9.4 THE PROBLEM OF EXCESS CALCIUM

In many catchments underlain by silicate rocks, particularly high-elevation catchments where there has been little weathering and soil development since the Pleistocene glaciation, Ca makes up a larger proportion of the cations in runoff than can be explained by stoichiometric weathering of feldspar. Two examples are shown in Figures 9.7 and 9.8. The Sierra Nevada perennial springs of Garrels & Mackenzie (1967) discussed above are an example of the phenomenon. Garrels & Mackenzie (1967) attributed the excess to dissolution of calcite, without really discussing the issue.

9.4.1 *South Cascade Glacier, Washington, USA*

A fairly extreme example is the South Cascade Glacier area of Washington State in the northwest United States (Drever & Hurcomb 1986). The compositions of waters draining different lithologies in the area are shown in Figure 9.9. The relatively high abundance of Ca and K, and low abundances of Na and SiO_2 are striking. Plagioclase in the bedrock ranged in composition from about $An_{28}(Na_{0.72}Ca_{0.28}Al_{1.28}Si_{2.72}O_8)$ to $An_{55}(Na_{0.45}Ca_{0.55}Al_{1.55}Si_{2.45}O_8)$. Weathering of plagioclase should produce waters with Na/Ca ratios of about 1 or greater, so stoichiometric weathering of plagioclase

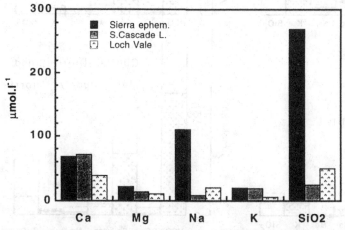

Figure 9.7. Compositions of South Cascade Lake (Drever & Hurcomb 1986) and mean runoff from Loch Vale (Mast et al. 1990) compared to the Sierra Nevada ephemeral springs (Garrels & Mackenzie 1967), all corrected for atmospheric input. Note that the ratios of all other solutes to calcium are lower in South Cascade Lake and Loch Vale.

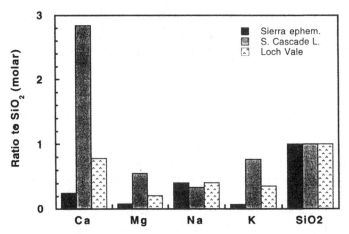

Figure 9.8. Compositions from Figure 9.7 plotted as ratios to SiO$_2$ to show 'anomalously' high concentrations of Ca compared to other solutes in South Cascade Lake and Loch Vale.

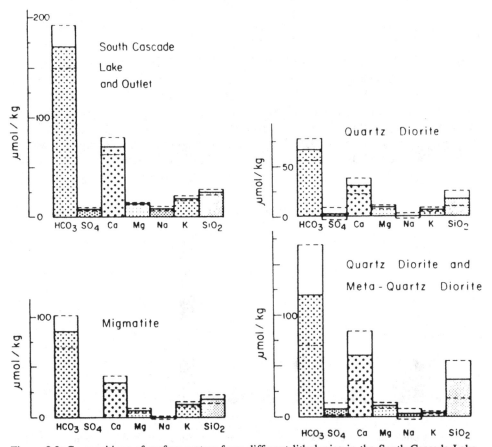

Figure 9.9. Compositions of surface waters from different lithologies in the South Cascade Lake basin, corrected for atmospheric input. Error bars represent one standard deviation of mean surface water composition (after Drever & Hurcomb 1986).

could not be the source of the calcium in the water. Drever & Hurcomb (1986) concluded that the only reasonable source of Ca, without large amounts of Mg or silica, was dissolution of calcite. They proposed a mass balance equation:

$$28.4 \, CaCO_3 + 5.9 \, KMg_{1.75}Fe_{1.25}AlSi_3O_{10}(OH)_2 + 1.5 \, Ca_2Fe_{2.5}Mg_{2.5}Si_8O_{22}(OH)_2$$

calcite   biotite   actinolite

$$+ \, 0.9 \, FeS_2 + 81.2 \, H_2O + 48.9 \, CO_2 + 2.9 \, O_2 =$$

$$4.5 \, Ca_{0.28}Mg_{1.5}Fe_{1.25}Al_{1.3}Si_{2.7}O_{10}(OH)_2 + 0.6 \, Fe_2O_3 + 30.2 \, Ca^{2+} \qquad (9.6)$$

vermiculite

$$+ \, 7.3 \, Mg^{2+} + 5.9 \, K^+ + 77.3 \, HCO_3^- + 1.8 \, SO_4^{2-} + 17.5 \, H_4SiO_4$$

In fact, calcite was present in the catchment as veins, as a deposit in joint planes, in occasional marble bands, and as a superficial subglacial deposit associated with regelation processes under glacial ice. The high (relatively) concentration of K was attributed to alteration of biotite to vermiculite, which is often an important reaction under cold conditions.

9.4.2 *Loch Vale, Colorado, USA*

The Loch Vale catchment in Colorado, USA, also shows an excess of calcium over what would be predicted from feldspar weathering (Mast et al. 1990; Figs 9.7 and 9.8). On the basis of the compositions of the minerals present in the catchment, Mast et al. (1990) came up with a mass-balance equation for weathering:

$$\left. \begin{array}{l} 125 \text{ plagioclase} \\ 29 \text{ biotite} \\ 8 \text{ chlorite} \\ 113 \text{ calcite} \end{array} \right\} + \text{atmos input} = \left. \begin{array}{l} 74 \text{ kaolinite} \\ 30 \text{ smectite-illite} \\ 12 \text{ Al(OH)}_3 \\ 52 \text{ FeO(OH)} \end{array} \right\} + \text{solutes} \qquad (9.7)$$

Mast et al. (1990) were unable to find any macroscopically visible calcite in the catchment, but demonstrated, using cathodoluminescence microscopy, that very fine-grained calcite was present in the bedrock in hydrothermally altered zones and in grain boundaries adjacent to plagioclase. They argued that deglaciation had occurred relatively recently, and rates of physical erosion were high, so that there was enough of this calcite in contact with meteoric waters to supply the observed excess of calcium. Over time, accessible calcite would be depleted and it would no longer serve as a significant source of calcium. One could use this argument, speculatively, to explain the excess calcium in the Sierra Nevada springs. The shallow, near-surface rocks (ephemeral springs) had been extensively flushed and no calcite remained. The deeper rocks (perennial springs) were less flushed and some calcite still remained.

M.A. Velbel (pers. comm.), on the other hand, explained the excess calcium in the runoff in Loch Vale on the basis of selective weathering of isolated regions of mafic amphibolite (amphibole and calcic plagioclase) in the bedrock. This alternative explanation cannot be ruled out on the basis of presently available information.

9.4.3 *Discussion*

The question of excess calcium is important because it is almost universal in high-elevation lakes in North America (Turk & Spahr 1990; Stauffer 1990) and also in waters of Scandinavia. If we are to model surface water chemistry successfully, we need to understand the underlying reactions. Several mechanisms could give rise to it:

1. Weathering of calcite in the bedrock. I would argue that, on the scale of a catchment, calcite will always be present initially in a granitic rock. It may be present in veins, in zones of hydrothermal alteration, or along grain boundaries as a result of deuteric alteration or simply closed-system alteration in the presence of groundwater. (Petrographic descriptions of rocks are often misleading; petrologists systematically avoid sampling altered rocks, which are the ones most likely to contain calcite.) Although calcite may be present initially in a granitic rock, it will be rapidly depleted as the rock is exposed to meteoric water. Whether enough calcite remains at the present time to influence water chemistry will vary from catchment to catchment. It is most likely to occur in catchments that have recently been glaciated and that are undergoing rapid physical erosion.

2. Selective weathering of more calcic plagioclase in the bedrock. Clayton (1988) showed that zoned plagioclases in the bedrock of the Idaho Batholith did not weather uniformly: the calcic cores dissolved while the more sodic rims remained intact. He used this observation to explain why the Ca/Na ratio in the water was higher than that of the average plagioclase in the rock. This is a plausible mechanism *provided the Ca/Na ratio in the water is not higher than the highest ratio present in the feldspar*. Some authors have claimed, on the basis of Clayton's (1988) results, that the albite and anorthite components of a feldspar could weather independently: the anorthite component could dissolve, leaving the albite component behind. I believe this is wrong: when a feldspar alters, Ca and Na are released in the same ratio as they are present in the feldspar (although a minor amount of Ca may be retained in a secondary smectite). But feldspars of different composition alter at different rates.

3. Reactive Ca-containing silicates such as pyroxenes, amphiboles, or epidote present in small concentrations weather rapidly and release amounts of Ca that are disproportionate compared to the amount of the mineral present. This is certainly the case in some catchments. Such reactions are often difficult to constrain by mass-balance equations because the secondary products are often smectites or vermiculites, whose compositions are poorly known. The overall plausibility of this mechanism must be assessed for each catchment individually.

4. Calcium comes from cation exchange sites. Calcium is greatly favored over sodium on the exchange sites of clays in dilute solutions. Thus the Ca/Na ratio of ions on exchange sites in a soil will be much higher than the Ca/Na ratio of plagioclase in the underlying bedrock. If exchangeable ions are displaced by, for example, an increased input of acidity from the atmosphere, the result will be an addition of Ca to runoff water. This is a plausible explanation for catchments affected by anthropogenic acidity (perhaps Scandinavia). It is not a plausible explanation for the Rocky Mountain region of North America, where the input of anthropogenic acidity is minor. Excess Ca could also be derived from a net decrease in biomass, but catchments showing a net decrease in biomass are relatively rare.

Documenting the source of calcium is an essential part of catchment mass balance studies, particularly for modeling future responses to acid deposition. The predictions of a mechanistic model will be quite different depending on which source is assumed for calcium.

9.5 CONCLUSIONS

The solutes leaving a catchment are ultimately derived from the atmosphere or mineral weathering in the bedrock, but they may be stored on exchange sites or as part of the biomass. Changes in storage on exchange sites or the biomass will affect concentrations in runoff. In many environments, mineral weathering is the dominant source of solutes; release from exchange sites and biomass change being relatively trivial. However, intensive studies of catchments have been established largely in areas where surface waters and soils are susceptible to acid deposition. Susceptible areas are those in which mineral weathering rates are slow, potentially too slow, to neutralize the input of acidity from the atmosphere. The bedrock is commonly made up of resistant silicate minerals. They also tend to be located in areas that are currently receiving significant deposition of anthropogenic sulfuric and nitric acid. In such catchments, the contributions from mineral weathering, cation exchange, and biomass change may be comparable in magnitude and it is very difficult to quantify each separately.

Once the contribution from mineral weathering has been isolated, it is relatively straightforward to calculate a set of mineral weathering reactions that are consistent with the measured solute fluxes. However, the set of reactions is non-unique and needs to be constrained as far as possible by detailed studies of primary and secondary minerals from the individual field site.

9.6 FUTURE DIRECTIONS IN RESEARCH

In this discussion I have stressed the difficulty of separating out the effects of mineral weathering, biomass change, and cation exchange. Resolving this question requires extensive data collection over an extended period of time. During the 1980s, several intensively monitored catchments were established, mostly in the context of understanding the effects of acid deposition. Several of these catchments have been 'manipulated' by artificial acidification or exclusion of acid input by a roof (e.g. Rasmussen et al. 1993). There have usually been intensive studies of soil at these sites. The amount of detailed information available from these studies over a 5 to 10 year period will allow well-constrained mass balance budgets to be established. This information is now slowly coming out in the literature; in the next few years we will have a much better understanding of the processes occurring, at least at these particular catchments.

Quantification of the various processes in a catchment requires extensive data collection over a long period of time. Unfortunately, research on acid deposition is, in North America at least, out of fashion. Funding for catchment studies has decreased and many of the long-term measurement programs are being terminated.

A separate line of research that will expand in the next few years is the use of isotopic tracers (see Chapter 7). The isotopic composition of strontium can be used to assign Sr in runoff to cation exchange (as mentioned above) or to specific mineral weathering reactions (Blum et al. 1994). Sulfur isotopes can be used to trace sources and pathways of sulfur (Thode 1991), and carbon isotopes for sources of carbon. Other isotopic systems will be applied in the future.

REFERENCES

April, R., Newton, R. & Coles, L.T. 1986. Chemical weathering in two Adirondack watersheds: Past and present-day rates. *Geol. Soc. Amer. Bull.* 97: 1232-1238.

Blum, J.D., Erel, Y. & Brown, K. 1994. $^{87}Sr/^{86}Sr$ of Sierra Nevada stream water: Implications for relative mineral weathering rates. *Geochim. Cosmochim. Acta* 57: 5019-5025.

Bowser, C.W. & Jones, B.F. 1993. Mass balances of natural waters: Silicate dissolution, clays, and the calcium problem. *Biogeomon* Symposium on Ecosystem Behaviour: Evaluation of Integrated Monitoring in Small Catchments, Prague, Czech Republic, Abstracts: 30-31.

Clayton, J.L. 1988. Some observations of the stoichiometry of feldspar hydrolysis in granitic soils. *J. Environmental Quality* 17: 153-157.

Drever, J.I. 1988. *The Geochemistry of Natural Waters*, 2nd Edition. Englewood Cliffs: Prentice-Hall.

Drever, J.I. & Hurcomb, D.R. 1986. Neutralization of atmospheric acidity by chemical weathering in an alpine drainage basin in the North Cascade Mountains. *Geology* 14: 221-224.

Frogner, T. 1990. The effect of acid deposition on cation fluxes in artificially acidified catchments in western Norway. *Geochim. Cosmochim. Acta* 54: 769-780.

Garrels, R.M. & Mackenzie, F.T. 1967. Origin of the chemical compositions of some springs and lakes. Equilibrium Concepts in Natural Water Systems. *Am. Chem. Soc. Adv. Chem. Ser.* 67: 222-242.

Likens, G.E., Bormann, F.H., Pierce, R.S., Eaton, J.S. & Johnson, N.M. 1977. *Biogeochemistry of a Forested Ecosystem.* New York: Springer-Verlag

Mast, M.A., Drever, J.I. & Baron, J. 1990. Chemical weathering in the Loch Vale watershed, Rocky Mountain National Park, Colorado. *Water Resources Research* 26: 2971-2978.

Matzner, E. 1986. Deposition/canopy interactions in two forest ecosystems of Northwest Germany. In H.W. Georgii (ed.), *Atmospheric Pollutants in Forested Area*: 247-462. Dordrecht: Reidel.

Miller, E.K., Blum, J.D. & Friedland, A.J. 1993. Determination of soil exchangeable-cation loss and weathering rates using Sr isotopes. *Nature* 362: 438-441.

Miller, W.R. & Drever, J.I. 1977. Chemical weathering and related controls on surface water chemistry in the Absaroka Mountains, Wyoming. *Geochim. Cosmochim. Acta* 41: 1693-1702.

Parkhurst, D.L., Plummer, L.N. & Thorstenson, D.C. 1982. BALANCE--a computer program for calculating mass transfer for geochemical reactions in ground water. US Geological Survey Water-Resources Investigations Report 82-14.

Plummer, L.N., Prestemon, E.C. & Parkhurst, D.L. 1991. An interactive code (NETPATH) for modeling *net* geochemical reactions along a flow *path*. US Geological Survey Water-Resources Investigations Report 91-4078.

Probst, A., Dambrine, E., Viville, D. & Fritz, B. 1990. Influence of acid atmospheric inputs on surface water chemistry and mineral fluxes in a declining spruce stand within a small granitic catchment (Vosges Massif, France). *J. Hydrology* 116: 101-124.

Rasmussen, L., Brydges, T. & Mathy, P. (eds) 1993. *Experimental Manipulations of Biota and Biogeochemical Cycling in Ecosystems-Approach, Methodologies, Findings.* Brussels: Commission of the European Communities.

Reuss, J.O. & Johnson, D.W. 1986. *Acid Deposition and the Acidification of Soils and Waters.* Ecological Studies 59, Springer-Verlag.

Schnoor, J.L & Stumm, W. 1985. Acidification of aquatic and terrestrial systems. In W. Stumm (ed.), *Chemical Processes in Lakes*: 311-338. New York: Wiley.

Stauffer, R.E. 1990. On granite weathering and the sensitivity of alpine lakes to acid deposition. *Limnology and Oceanography* 50: 1109-1130.

Stauffer, R.E. & Wittchen, B.D. 1991. Effect of silicate weathering on water chemistry in forested, upland, felsic terrane of the USA. *Geochim. Cosmochim. Acta* 55: 3253-3271.

Thode, H. G. 1991. Sulfur isotopes in nature and the environment: An overview. In: H.R. Krouse & V.A. Grinenko (eds), *Stable Isotopes: Natural and Anthropogenic Sulphur in the Environment*: 1-26. New York: Wiley.

Turk, J.T. & Spahr, N.E. 1990. Rocky Mountains: Controls on lake chemistry. In D.F. Charles (ed.), *Acid Deposition and Aquatic Ecosystems: Regional Case Studies*: New York: Springer-Verlag.

Velbel, M.A. 1986. The mathematical basis for determining rates of geochemical and geomorphic processes in small forested watersheds by mass balance: Examples and implications. In S.M. Colman & D.P. Dethier (eds), *Rates of Chemical Weathering of Rocks and Minerals*: 439-451. New York: Academic Press.

Wright, R.F. 1988. Influence of acid rain on weathering rates. In A. Lerman & M. Meybeck (eds), *Physical and Chemical Weathering in Geochemical Cycles*: 181-196. NATO ASI Series 251. Kluwer.

Wright, R.F., Gjessing, E., Christophersen, N., Lotse, E., Seip, H.M., Semb, A., Sletaune, B., Storhaug, R. & Wedum, K. 1986. Project RAIN: changing acid deposition to whole catchments. The first year of treatment. *Water, Air, Soil Pollution* 30: 47-63.

CHAPTER 10

Natural organic matter in catchments

JAMES F. RANVILLE & DONALD L. MACALADY
Department of Chemistry and Geochemistry, Colorado School of Mines, Golden, USA

10.1 INTRODUCTION

Natural organic matter (NOM), in its broadest sense, refers to the complex and chemically and physically diverse substances that result directly or indirectly from the photosynthetic activity of plants. In the geosciences, we typically limit the use of this term, however, to substances derived from the partial decay of detrital materials originating from terrestrial and aqueous plants. This definition includes everything along the diagenetic pathway from living plants to fossil fuels such as coal and petroleum.

For the purposes of this paper, a further limitation is required. In this discussion, nothing further along the diagenetic pathway than materials roughly resembling peat will be considered. In other words, the NOM discussed here has retained a considerable portion of its original oxygen and hydrogen. It may have picked up a bit of additional sulfur or metal along the diagenic pathway or in other geologic processes not accurately described as diagenic. It's the kind of stuff you're likely to find as you dig in your garden, walk through a bog or fen, or examine the colored substances found naturally in lakes, rivers and groundwaters.

This chapter is *not* primarily about the mass balances for the transport of organic materials in catchments. Rather, it concerns the origin and nature of NOM as defined above, and the role of NOM in the transport and transformations of inorganic and organic matter, natural and anthropogenic, in terrestrial and aquatic systems. Thus, the goal of this discussion is to provide the background and technical understanding necessary for the evaluation of the role of NOM in the many mass balances that are important to catchment studies. Hopefully, it will also stimulate the reader to additional study into the nature of the involvement of NOM in geochemical processes in aquatic systems.

The text of the paper is followed by a general bibliography that should facilitate further study. Specific references and notes from the text are listed separately.

10.2 THE NATURE AND ORIGIN OF NATURAL ORGANIC MATTER

Natural organic matter has been classified according to a wide variety of schemes, one of which involves the origin of the plants that serve as a starting material for the NOM (Fig. 10.1). *Pedogenic* refers to NOM from terrestrial plants and microorganisms, including material leached from soils into an aquatic system; *aquogenic* refers to NOM originating from the decomposition of aquatic organisms (macrophytes, algae, bacteria) within the water body. For most catchments that do not include a large lake, river or reservoir, and for all soils, pedogenic materials can be expected to dominate. The NOM in these systems primarily results from the transport of NOM from the surrounding watershed via surface runoff and/or groundwater input. Primary productivity for small streams is generally low and therefore the amount of *aquogenic* NOM is small. For large lakes, and, of course, the open oceans, aquogenic materials constitute the overwhelming majority. In this case the NOM is formed in-situ by the degradation of biomass formed by primary productivity. Another pair of terms often used to refer to a similar distinction among types of organic materials is *autochtonous* and *allochtonous*. The former describes material formed in-situ and the latter, material transported into the water body. Some authors are careful to distinguish between these pairs of terms (Buffle 1984), particularly with reference to dissolved or colloidal organic materials.

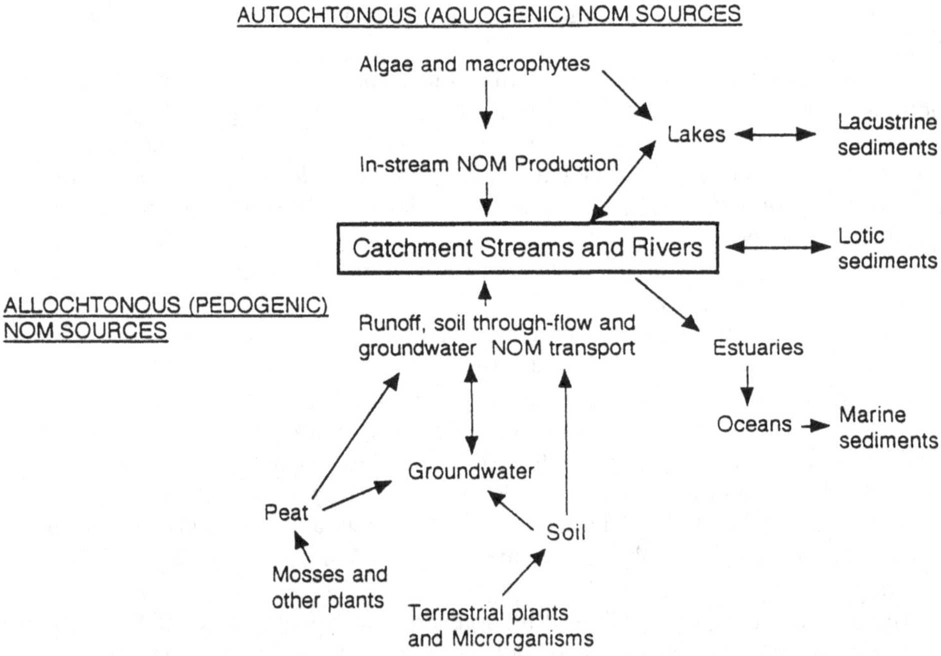

Figure 10.1. Some of the many possible flowpaths for NOM (adapted from Aiken et al. 1985).

10.2.1 *Nature of natural organic matter*

10.2.1.1 *Humus – Humic substances – Humic and fulvic acids*

The incomplete turnover of plant litter in most of the world's ecosystems results in the accumulation of soil organic matter. Accumulation rates for carbon vary from about 0.2 g $C/m^2/yr$ in polar desert environments up to 15 g $C/m^2/yr$ in cool, wet locations such as boreal forests (Ugolini 1968; Bockheim 1979). Much of this accumulated material is soil humus, which is extremely resistant to further degradation. Radiocarbon dating techniques indicate ages of several thousand years for soil organic matter in lower soil profiles (Schlesinger 1977; O'Brien & Stout 1978).

Humus is a complex mixture of both aliphatic and aromatic molecules with a wide variety of chemical compositions and molecular weights. The classification of humus is therefore based on an operational chemical characterization which is often referenced to a solubility scheme originally developed by soil scientists (Fig. 10.2, see also Stevenson 1986). Soil or sediment is first extracted with an alkaline solution (e.g. NaOH). The insoluble organic matter remaining after alkaline extraction is termed *humin*. The portion of the soil solubilized by the alkali is subsequently treated with acid to low pH (< 2). The material which precipitates in this step is termed *humic acid*, while the organic material remaining in solution is called *fulvic acid*. The precipitated humic acid is sometimes further differentiated into *hymatomelanic acid* (alcohol soluble portion of humic acid), *gray humic acid* (precipitated from an alkaline humic acid solution with added electrolyte), and *brown humic acid* (soluble in alkaline solution with added electrolyte), but these latter categories are not widely referred to in geochemistry.

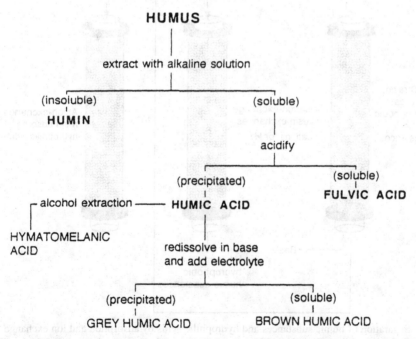

Figure 10.2. Scheme for the extraction of humic substances (adapted from Stevenson 1982).

Over the past two decades, dissolved organic carbon (DOC) in aquatic systems has been redefined in terms of a separation scheme based on retention characteristics on macro-reticular non-ionic resins (Fig. 10.3, Thurman & Malcolm 1981). In the most commonly applied method, the hydrophobic characteristics of different humic fractions are exploited to enable separation after adsorption (at pH = 2) on a column of a non-ionic acrylic resin. Application of a pH gradient to the material adsorbed onto the columns provides a separation of the aquatic NOM into fractions identifiable as fulvic and humic acids (MacCarthy et al. 1979), which typically account for about 50% of the total DOC (Fig. 10.4).

Regardless of the isolation/definition scheme, humic and fulvic acids constitute a major fraction of any NOM sample. Soil NOM typically has a larger humic/fulvic ratio, and smaller contents of identifiable non-humic materials. Humic and fulvic acids account for widely varying proportions of aquatic NOM, from around 25% for sea water to almost 90% for some wetlands and highly colored rivers. The remaining NOM in aquatic samples consists of more hydrophilic organic acids and a variety of identifiable non-humic compounds including amino acids, some simple sugars and several small organic acids (Thurman 1985).

Much of the effort to chemically and physically characterize NOM has focused on the humic and fulvic acid fractions. In general, aquatic NOM can be described as a polymolecular array of organic molecules which contain about 50-60% carbon, 4-6% hydrogen, 23-40% oxygen, 0.3-6% nitrogen (by weight), with traces of phos-

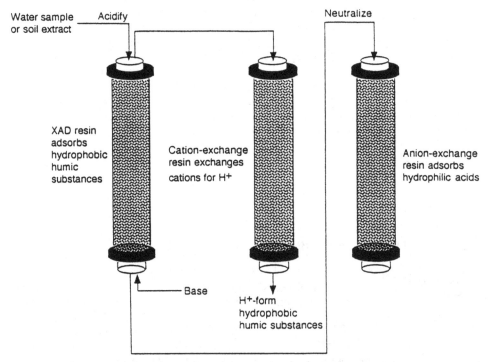

Figure 10.3. Separation of humic substances and hydrophilic acids by adsorption and ion exchange chromatography (adapted from Thurman 1985).

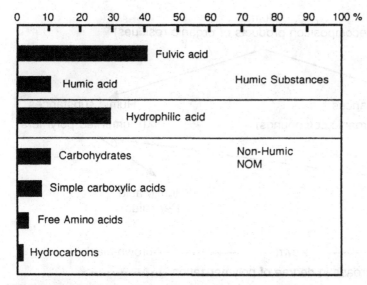

Figure 10.4. Dissolved organic carbon histogram for an average river water with a DOC of 5 mg/L (adapted from Thurman 1985).

phorus, sulfur and 'ash' (residue after combustion). Soil NOM is similar in elemental composition, with naturally larger proportions of highly insoluble and/or hydrophobic constituents such as lipids, proteins and polysaccharides. Isolates of soil NOM generally have considerably higher ash contents (Fig. 10.5).

In addition to elemental composition, NOM fractions have been characterized according to molecular weight distributions, the presence of certain functional groups, acid-base and redox characteristics, aromatic/aliphatic character, chelating/complexation abilities, and trace element composition. An excellent summary of many characterizations of aquatic NOM fractions is given by Thurman (1985). A similarly useful source for soil NOM is given by Stevenson (1982).

The utility of such analytical efforts, for the purposes of this paper, is limited by the fact that the humus fraction of NOM has been further fractionated into fulvic acid, humic acid, etc. prior to analysis. It is a given fact that, for the purposes of studies into processes which occur in nature, one is interested in the behavior of NOM as it exists in situ, not in the behavior of the humic or fulvic acid fraction of that NOM. Thus, the large amount of scientific endeavor which has been expended to isolate and characterize NOM fractions, though extremely useful in providing a basis for the understanding of the physical and chemical characteristic of NOM, is not directly applicable to the considerations of this paper. The brief look at some of these characteristics which follows is intended to provide a general picture of the types of interactions we might expect between NOM and the surrounding geochemical matrices.

The fractions of humus from soil or aquatic environments called fulvic and humic acids differ primarily in their hydrophobicity and solubility characteristics, with concomitant differences in average molecular size and mass. In general, fulvic acids are more soluble, smaller in average molecular weight, less aromatic, and more highly

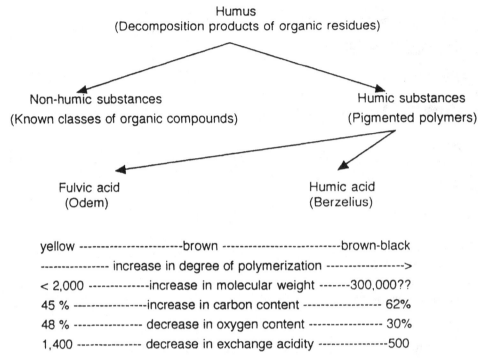

Figure 10.5. Classification and chemical properties of humic substances (adapted from Stevenson 1982).

charged than humic acids. Fulvic acids also typically have higher oxygen content, with higher carboxylic acid (COOH) and lower aromatic hydroxyl (ArOH) content than humic acids (Hayes et al. 1989).

Fulvic acids are therefore probably more representative of aquatic NOM, especially the 'dissolved' fraction (Malcolm 1985). A typical 'molecule' of fulvic acid might contain one COOH per 6 carbon atoms, with about 65% aliphatic (cf. aromatic) carbons (Thurman 1985). Fulvic acid and NOM in general are known from Electron-Spin Resonance (ESR) evidence to produce free radicals in aqueous solutions (Senesi & Steelink 1989). The most likely groups to contribute this free radical character are quinones, and quinone functionalities have recently been shown by ^{13}C Nuclear Magnetic Resonance (NMR) evidence to be present (Thorn et al. 1992). The emerging picture of aquatic NOM is one of polyfunctional organic acids containing significant metal-complexing and oxidation-reduction capabilities. Further, various portions of individual fulvic and humic acid 'molecules' are expected to exhibit significant hydrophobic character, more so for humic acids and particulate NOM (PNOM) than for the 'dissolved' fraction of aquatic NOM (DNOM). Finally, the dominance of carboxylic acid and phenolic functional groups in NOM means that the net 'surface' charge of aqueous NOM and the overwhelming majority of soil or sediment particles, which contain NOM-dominated surfaces, can be expected to be negative at essentially all environmentally relevant pH values (Tipping & Cooke 1982).

10.2.2 *Origin of natural organic matter*

10.2.2.1 *Terrestrial organic matter*

The major pools in the global carbon cycle are represented in Figure 10.6. For the purposes of this paper we are primarily concerned only with the production of vegetation and the subsequent decay and transport of plant-derived organic matter. Terrestrial detritus is a general term which refers to dead plant parts which are in various stages of decomposition. The production of plant litterfall has well-understood geographical patterns which roughly follow global patterns in net primary production. In a global sense, production of terrestrial detritus varies inversely with latitude from tropical to boreal forests (Lonsdale 1988).

Most terrestrial detritus results from either the above-ground portions or the root systems of plants, and is thus located in the upper layers of the soil and hence properly termed pedogenic. Here detritus is subject to a variety of decomposition processes, mostly through heterotrophic metabolism involving bacteria, fungi and other microfauna (Swift et al. 1979). One result of such processes is the release of CO_2, H_2O and nutrients such as nitrogen and phosphorus compounds (i.e. decay, which is the reverse of photosynthesis). Another decay process, more important for the purposes of this paper, is the production of a group of substances which are highly resistant to further aerobic or anaerobic decomposition, collectively known as *humus*. These early diagenetic processes can be viewed in terms of the rapid turnover of the majority of the detrital material near the soil surface and the much slower production, accumulation and turnover of humus in the deeper soil layers (Schlesinger 1977).

A kinetically simple, first-order model for the rate of decomposition of forest litterfall provides a basis for comparing a variety of ecosystems, allowing the estima-

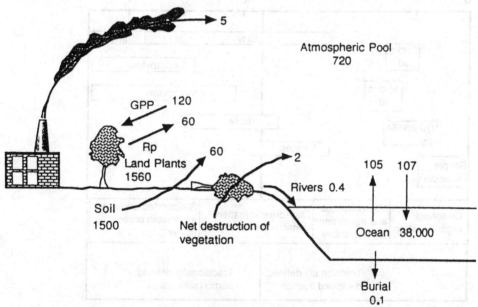

Figure 10.6. The present-day global carbon cycle. All pools are expressed in units of 10^{15} gC and all fluxes in units of 10^{15} gC/yr (adapted from Schlesinger 1991).

tion of a mean residence time for plant debris as *1/k*, where *k* is the first order decay rate constant (yr^{-1}). Values for many tropical rain forest ecosystems, for example are greater than 1.0, indicating little surface accumulation and rapid turnover of organic matter (Cuevas & Medina 1988). In contrast, some peat lands have *k* values of ca. 0.001 (Olsen 1963). Esser et al. (1982) estimated a global mean value of *k* = 0.33 for soil surface litter. In various geographical and ecological regions, the rate of decomposition may be limited by a variety of environmental factors, including temperature, moisture and the origin of the litter material. The role of litter material in soil genesis and soil profiles is discussed in some detail in Sections 10.3.1 and 10.3.2.

10.2.2.2 *Aquatic organic matter*

Dissolved and colloidal forms

When discussing natural organic matter in aquatic systems we need to distinguish between organic carbon, the generally measured parameter, and organic matter, used as a general concept. In general, the carbon content of most organic matter is roughly 50% by weight. So it is relatively routine to assume the organic matter concentration in a system is roughly twice the organic carbon concentration.

Organic matter present in raw natural waters may be considered to occur physically as an essentially continuous series which we will refer to as the dissolved/colloidal/particulate organic matter continuum (Fig. 10.7). While the pure end-member modes are easily conceptualized as: a) Dissolved (solvated) organic molecules; and b) Macroscopic solid particles which may occur with or without

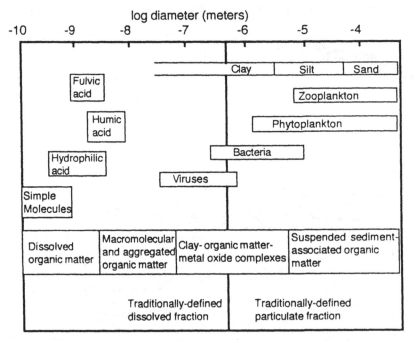

Figure 10.7. Continuum of particulate and dissolved organic carbon in natural waters (adapted from Thurman 1985).

sorbed ions or molecules, the intermediate colloidal mode requires further explanation. The word 'colloid' as used here collectively identifies the intermediate continuum of modes. Organic matter species which may be referred to as 'colloidal' include:

a) Discrete chemical species with sufficient size or mass to behave as colloids (macromolecules);

b) Aggregates of smaller organic molecules; and

c) Organic matter associated with (usually by sorption) compositionally distinct, colloid-size particles, such as clays or oxides.

Traditionally, filtration through a 0.45 μm pore-size membrane, although this pore-size is quite arbitrary, has been widely used as a means of separating particulate from 'dissolved' species. Such an operational definition ignores the continuum of suspended particle sizes and the issue of colloids entirely. For environmental biogeochemical investigations this separation may be insufficient to fully describe the processes involved in both organic matter and contaminant transport. Unfortunately, studies in which one is attempting to rigorously differentiate between dissolved, colloidal, and particulate organic matter in natural waters have only recently been carried out (Ranville et al. 1991).

Dissolved natural organic matter
Dissolved NOM, in its most rigorous sense, refers to those compounds which are fully solvated by water. This consists of a wide range of compounds which generally share the properties of relatively low molecular weight (< 1000 daltons) and significant numbers of polar and/or ionizable functional groups. Most significant in this group are the complex mixtures of polyelectrolytes called humic substances which are described above. Humic and fulvic acid are the major classes of humic substances and are defined by their pH-dependent solubility and their adsorption onto non-ionic acrylic resins. Another class of complex mixtures is known as hydrophilic acids. Hydrophilic acids are similar to fulvic acid but are more soluble, higher in oxygen content, and less readily isolated on acrylic resins. Other major components of dissolved NOM are simple organic compounds, the major classes being carbohydrates, carboxylic acids, amino acids, and hydrocarbons (Thurman 1985).

Experimentally, dissolved NOM is often determined by measurements of dissolved organic carbon (DOC). In general DOC can be converted to dissolved NOM by multiplying by a factor of two. Most modern DOC determinations rely on either chemical or thermal oxidation of the organic matter to CO_2 followed by detection using spectrophotometric or coulometric techniques (see, for example, Lee & Macalady 1989). DOC measurements are complicated by the low values often encountered, differing oxidation efficiencies for components of DOC, and the presence of interfering species such as chloride (Hedges & Lee 1992). Distinction between particulate organic carbon (POC) and DOC is made by analyzing raw and filtered samples. As previously pointed out, the result of this two-phase classification scheme is that most DOC and POC values will contain variable amounts of colloidal organic carbon (COC). The amount of COC will vary depending on the source of NOM and the aqueous environment.

Colloidal natural organic matter

The major components classified as colloidal organic matter are: high-molecular weight compounds, aggregates of smaller molecules, organic coatings on other colloidal-sized particles (i.e. clays and oxides), and viral and bacterial cells (viable or senescent). These colloidal materials can be classified as either hydrophilic or hydrophobic colloids. Macromolecular organics such as humic and fulvic acids, polysaccharides, proteins, peptides, and amino acids tend to be polar and maintain their colloidal stability via interactions between charged functional groups and water molecule dipoles. They are best classified as hydrophilic colloids. For hydrophobic solids, permanent dispersion or suspension is maintained by the random thermal activity of water molecules (Brownian movement). In order for this mechanism to be effective, the particle must be sufficiently small to allow spatially uneven bombardment by water molecules and must have at least some electrostatic charge. If not for electrostatic repulsions, the hydrophobic nature of these particles would tend to force them together during particle collisions and promote aggregation which would lead to larger and larger particle sizes and eventual destruction of the colloidal system. These electrostatic repulsions promote coulombic repulsion between particles and also allow interactions with water molecules, thus making the colloids effectively more hydrophilic. The importance of organic coatings to establish a substantial negative charge on mineral colloids and thereby impart significant stability has been widely demonstrated (Tiller & O'Melia 1993).

The 'molecular weight' ranges of aquatic fulvic and humic acids are generally considered to be approximately 500-1000 and 1000-5000 daltons respectively (Thurman 1985). These values, along with the significant number of ionizable functional groups, suggest that fulvic acid may best be considered a dissolved species, while humic acid molecules probably span the dissolved-colloidal interface. It should be noted that these molecular weights are generally determined on de-salted, isolated fractions of NOM. Fulvic and humic acids may occur in natural waters as aggregates with much higher apparent molecular weights and exist in colloidal forms. Aggregation is enhanced by the presence of multi-valent cations which reduce the electrical double layer. Ultrafiltration studies of NOM have suggested that a significant portion of NOM exists as colloids with 'molecular weights' exceeding 10,000 daltons. In general it is not known whether this colloidal NOM exists as discrete, high-molecular weight molecules or represents some of the other forms of colloidal NOM mentioned above such as aggregates or NOM coatings. Ultrafiltration results must be interpreted carefully as artifacts can occur due to processes such as charge repulsion by the ultrafilter membrane, clogging of the membrane, and aggregation at the membrane surface (Buffle et al. 1992). Some of the experimental difficulties of using ultrafiltration, particularly membrane fowling, have been obviated by the introduction of tangential-flow filtration methods (Figs 10.8 and 10.9, Ranville et al. 1991).

Some of the colloidal NOM is present as adsorbed coatings on mineral particles. Particles with very similar negative surface charge, regardless of underlying mineralogy, dominate in most aquatic environments, a reflection of the universal presence of organic coatings (Hunter 1980). These NOM coatings significantly modify the ability of mineral colloids to partition metal and organic contaminants in natural waters. The negative charge imparted by these coatings also facilitates the transport of colloids and associated contaminants in surface and ground waters by preventing ag-

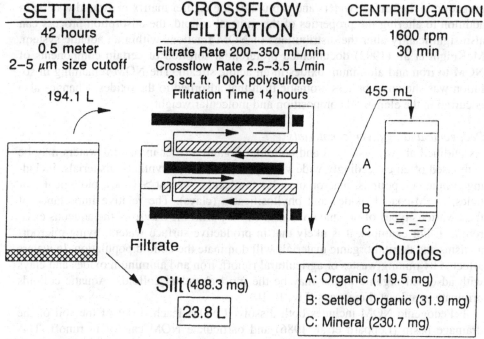

SETTLING	CROSSFLOW FILTRATION	CENTRIFUGATION
42 hours	Filtrate Rate 200–350 mL/min	1600 rpm
0.5 meter	Crossflow Rate 2.5–3.5 L/min	30 min
2–5 μm size cutoff	3 sq. ft. 100K polysulfone	
	Filtration Time 14 hours	

194.1 L

455 mL

Filtrate

Silt (488.3 mg)

23.8 L

Colloids

A: Organic (119.5 mg)
B: Settled Organic (31.9 mg)
C: Mineral (230.7 mg)

Figure 10.8. Diagram showing the operation of cross-flow ultrafiltration for collection of a colloid sample (from Ranville et al. 1991).

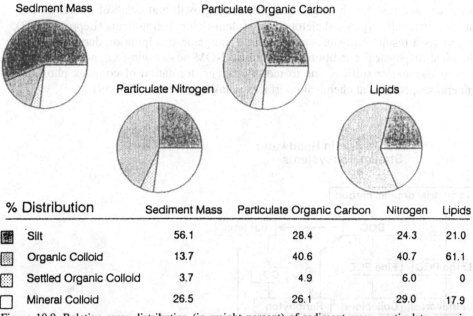

Sediment Mass

Particulate Organic Carbon

Particulate Nitrogen

Lipids

% Distribution	Sediment Mass	Particulate Organic Carbon	Nitrogen	Lipids
Silt	56.1	28.4	24.3	21.0
Organic Colloid	13.7	40.6	40.7	61.1
Settled Organic Colloid	3.7	4.9	6.0	0
Mineral Colloid	26.5	26.1	29.0	17.9

Figure 10.9. Relative mass distribution (in weight percent) of sediment mass, particulate organic carbon, nitrogen, and lipids in the silt, organic colloid, settled organic colloid, and mineral fraction from Pueblo Reservoir (from Ranville et al. 1991).

gregation in the surface waters and capture by the aquifer matrix in ground waters. In addition to altering the properties of the mineral colloid, the adsorption process can also significantly alter the distributions of NOM fractions within an aquatic system. McKnight et al. (1992) documented preferential sorption of certain components of NOM to iron and aluminum oxides in an acidic stream. The NOM remaining in solution was shown to be less aromatic than that adsorbed to the oxides. Changes also occurred in the elemental composition and molecular weight.

Pedogenic and aquogenic materials

As outlined above, dissolved and colloidal organic matter in natural waters may be composed of an exceedingly wide variety of natural and synthetic materials, including simple compounds, macromolecules (molecular weight > 1000), biological particles, NOM-coated oxides and phyllosilicates (clays). The relative importance of these various types of organic colloids will vary with the nature of the aqueous environment. For example, it is likely that in productive surface waters, living microorganisms and detrital organic materials will dominate the colloid population. In waters affected by mining wastes or agricultural runoff, iron and aluminum oxides and clays with adsorbed organic matter may be the most abundant colloids. Aquatic colloids originate in at least three ways (Fig. 10.10):

1. Pedogenic NOM includes both dissolved NOM leached out of the soil of the drainage basin (Leppard et al. 1986) and particulate NOM carried in runoff. This latter form, at least in streams and rivers, is comprised principally of plant fragments and woody debris, and may be chemically degraded during transport to the aquatic system (Ranville et al. 1991).

2. Aquogenic NOM, on the other hand, forms from two distinct types of processes: a) As a result of biological activity (e.g. growth and death of microbes, production of fibrils, organic skeletons and protein-rich cell fragments (Leppard 1992), and b) As a result of in-situ aqueous phase inorganic precipitation due to changing chemical or physical conditions followed by NOM adsorption (e.g. precipitation of iron oxides and/or sulfides due to redox changes, formation of complex phosphate mineral suspensions at chemical interfaces in lakes (Buffle et al. 1989).

Organic Carbon in Headwater Stream Ecosystems

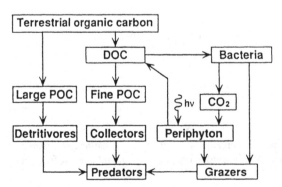

Figure 10.10. Schematic diagram representing carbon cycling in a headwater stream (from Ranville et al. 1991).

10.3 GEOCHEMICAL REACTIONS OF NATURAL ORGANIC MATTER

The geochemistry of NOM is as complex and varied as the diverse origins and nature of the material suggest. In this section, some of the more important geochemical interactions of NOM with unaltered geological systems are briefly reviewed. For a more complete discussion of any of the topics in this section, please refer to the bibliography.

10.3.1 *Weathering and natural organic matter in catchments*

The occurrence of specific interactions involving NOM in the chemical weathering of soils and minerals is difficult to determine because such interactions are universally accompanied by a complex array of biological and biochemical reactions as well as simple hydrolysis (Fig. 10.11). The role of microbial processes in weathering has been discussed in detail by Robert & Berthelin (1986). It is the opinion of these and other (e.g. Loughnan 1969) authors that the direct involvement of NOM in chemical weathering is extremely limited; carbonic and other mineral acids produced by biological activity being the primary agents in chemical weathering.

Nevertheless, both low-molecular weight biochemical compounds and humic and fulvic acids have been implicated in the degradation of mineral matter in nature. The ability of many microorganisms, particularly lichen fungi, to bring about the weathering of rocks and minerals due to the synthesis of biochemical chelating agents is well-known (Stevenson 1982). Specific weathering effects of low-molecular weight organic acids of natural origin are well-documented (Bennett & Siegel 1987; Mast & Drever 1987; Schalascha et al. 1967), with both the acidic and chelating characteristics of the acids implicated in weathering processes. Antweiler & Drever (1983) correlated the weathering of volcanic ash with the DOC of percolating waters. The DOC

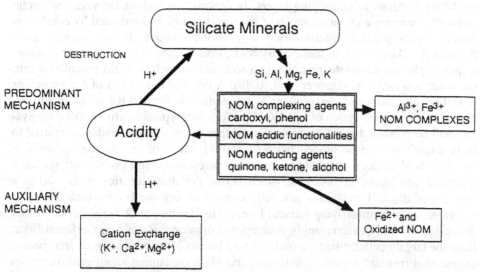

Figure 10.11. Mechanisms in the alteration of silicates by organic acids (adapted from Robert & Berthelin 1986).

consisted predominately of humic materials but contained traces of oxalate, acetate and formate with pH values between 4.3 and 5.2.

Data obtained by Baker (1973) show that humic acids exhibit an activity of the same order as that of several simple organic acid chelating agents in the weathering of a number of minerals and metals. The ability of NOM, more specifically humic substances, to decompose common soil minerals such as biotite, muscovite, illite and kaolinite has been demonstrated by a number of authors, including Huang & Keller (1971), Schnitzer & Kodama (1976) and Tan (1980). Because of their low molecular weights and abundance in soil solutions, fulvic acids may be particularly effective in dissolving silicate minerals (Stevenson 1985). Zunino & Martin (1977) have advanced a detailed concept of the involvement of humic substances in weathering and subsequent translocation of trace elements to biological systems.

An additional role ascribed to humic substances in weathering processes depends upon their redox properties. The reduction of iron minerals, causing a release of ferrous iron into solution, is an obvious possibility. However, little evidence has been reported for direct involvement of NOM in redox transformations of minerals. The role of NOM in redox transformations of metals, however, is well-established (see below).

10.3.2 *Development of natural organic matter profiles in catchments*

The general pattern of NOM content in a soil profile includes a gradation from the highly organic layers containing litterfall from the surface vegetation to the lower highly mineral layers containing relatively little organic content. The precise pattern for a particular catchment depends upon a variety of geographical, ecological and mineralogical factors, as well as the extent of agricultural and other anthropogenic disturbances (Fig. 10.12). For example, in typical tropical forest soils, the organic matter is largely recycled in the surface soil zones with little or no transport of soluble NOM fractions to underlying layers. In contrast, in regions between the Arctic and cooler temperate climates, much of the forested area is dominated by coniferous forests, which produce litterfall rich in phenolic compounds and organic acids (Cronan & Aiken 1985). In these ecosystems, decomposition is also slower and correspondingly less complete, resulting in a soil solution rich in NOM percolating into the lower soil horizons (Dethier et al. 1988). A detailed discussion of the variations in soil development and concomitant NOM profiles is beyond the scope of this review, but a brief description of soil NOM profiles in a typical Northern forest ecosystem will be presented as an example. For further discussion the reader is referred to Schlesinger's treatise on biogeochemistry (1988). Much of the discussion below is based on Schlesinger's treatment of temperate forest soils. Soil profiles are typically separated into layers, or horizons, which define the characteristics of the soil as a function of depth. Forest soils generally contain an organic layer which is clearly separated from the underlying mineral layers. The organic, or O, layer can be further divided into zones of increasingly decomposed organic materials on the forest floor, from the largely cellular surface material to a humus layer consisting of amorphous, degradation-resistant organic substances, and often containing significant quantities of minerals. Differentiation of sublayers within this surface zone is often difficult, with large regional and seasonal variations. In temperate and boreal coniferous for-

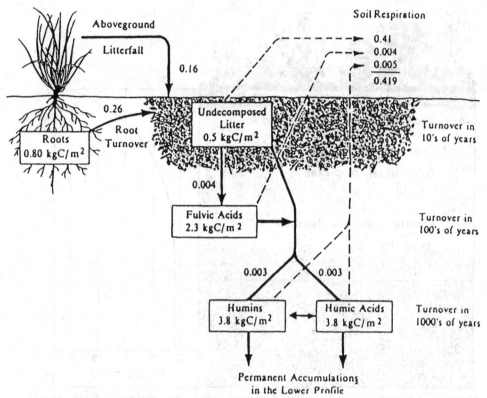

Figure 10.12. Turnover of litter and soil organic fractions in a grassland soil. Flux estimates are in kg C/m^2 yr (from Schlesinger 1991).

ests, this organic layer often accumulates into a thick organic mat, known as a moor, that is sharply differentiated from the underlying soil (Romell 1935). Many Arctic soils are waterlogged and contain primarily organic material in the entire rooting zone (peatland soils, or Histosols) (see Chapter 2).

Beneath this organic layer lies the upper mineral soil, designated the A horizon, which has a significant NOM fraction, and varies from several cm to 1 m in thickness. This is a zone of eluvial processes, characterized by significant weathering by organic acids contained in the water percolating through the forest floor. Iron and aluminum are typically removed by chelation with NOM. Such downward movement of Fe and Al in conjunction with NOM is known as *podzolization* (Antweiler & Drever 1983; Chesworth & Macias-Vasquez 1985). Known throughout the world, podzolization is particularly intense in temperate to boreal coniferous forest ecosystems (Fig. 10.13). In such soils, the pH of the soil solution is often as low as 4.0 (Dethier et al. 1988). When podzolization is particularly pronounced, a whitish layer of nearly pure quartz, the E horizon, can form at the base of the A horizon (Pedro et al. 1978).

Substances leached from the A and E horizons are deposited during soil development in the underlying, or B horizon, defined as a zone of deposition or illuvial proc-

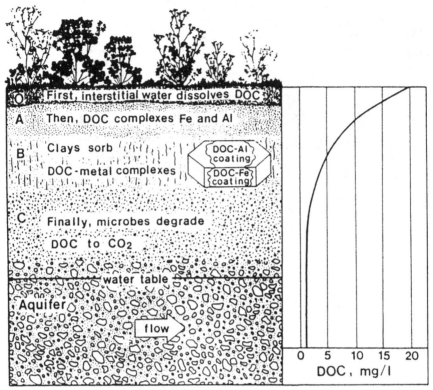

Figure 10.13. Podzolization and decrease of organic carbon in interstitial water of soils (from Thurman 1985).

esses, where secondary clay minerals accumulate. These clay minerals retard the downward movement of soluble NOM fractions that are carrying Fe and Al (Greenland 1971). Soils of varying degrees of podzolization are characterized by B horizons which contain varying amounts of clay and organic matter. Spodosols are highly podzolized soils which have an upper B horizon (B_s) which is dark and rich in Fe and NOM. Less highly podzolized soils may have B horizons which vary from nearly pure clays to zones which are orange-red to yellow from varying iron contents. In forests in New England (USA), the accumulation of NOM in the B_s horizon appears to control the loss of solution-phase NOM to streams (McDowell & Wood 1984). Beneath the B horizon, the C horizon is characterized by soil material with little NOM content. Depending upon whether or not the soil has developed from local materials, the C horizon may or may not resemble the underlying bedrock. In any case, carbonation weathering is generally the dominant process in the C horizon (Ugolini et al. 1977). In many temperate regions, anthropogenic activities such as agriculture and subsequent erosion substantially alter these forest profiles. For example, in the Piedmont region of the southeastern US, the forest floor often resides directly atop the B horizon due to such erosion. Acid rain can also substantially alter this picture through the introduction of strong acids such as sulfuric acid into the soil solution, particularly with respect to the weathering of aluminum.

The NOM content of aqueous systems associated with a given catchment is largely controlled by the soil processes outlined above. The export of NOM from a catchment is largely through dissolved NOM and colloidal NOM in surface and ground waters. The NOM content of ground water is generally a direct result of the nature and extent of the soil processes, though some NOM is imported into a given catchment through rainfall, which typically contains about 1 mg C/l above the forest canopy and 2-3 mg C/l below the canopy. (Rainfall that drips from leaf and plant matter may contain 5-25 mg C/l (Thurman 1985). Because of variations in soil development processes and hydrological considerations, the measured NOM content of ground water and streams draining various catchments shows dramatic regional variations. For example, streams and ponds in very sandy regions, which show little retention of NOM in soils, may have up to several hundred mg/l of organic carbon in the dissolved NOM and colloidal NOM forms. Shallow ground waters in peat lands or bogs may have similarly high NOM contents (see below).

However, NOM concentrations in both surface and ground waters are generally much lower than these extremes (Fig. 10.14). In a broad study of ground water NOM concentrations, Leenheer et al. (1974) analyzed 100 groundwater samples from a variety of aquifer types in the United States, finding a median DOC concentration of 0.7 mg C/l, with a range of 0.2 to 15 mg C/l. Their study and others show that the majority of groundwaters have a DOC concentration of 2 mg C/l or less, provided areas dominated by kerogen or petroleum and oil-field brines are excluded. Surface waters in rivers and streams show generally higher NOM levels, with significant fractions often present as dissolved NOM, colloidal NOM and larger particulate forms. Thurman (1985) lists dissolved NOM concentrations, expressed as DOC, ranging from 2 mg C/l in Arctic and alpine streams to above 25 mg C/l in streams draining swamps or wetlands. Inclusion of particulate NOM forms can increase these

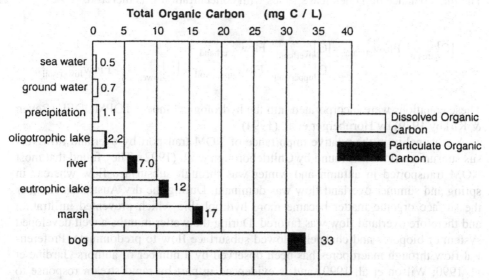

Figure 10.14. Approximate concentrations of dissolved and particulate organic carbon in natural waters (adapted from Thurman 1985).

organic stream loadings considerably. Particulate organic carbon loadings have been estimated to show a range of 1 to 30 mg C/l for 99% of all rivers (Meybeck 1981, 1982).

Thus, the export of NOM by ground and surface water flow is typically a relatively minor process in terms of the overall effect on catchment mass balance. The fraction of pedogenic NOM generated in catchments which is exported via ground and surface water flow is difficult to estimate in a general sense, but these fractions can be estimated for a given catchment given adequate data on NOM production, ground and surface water compositions and hydrology. The problems associated with such estimates are discussed briefly in the following section of this review. Concomitant roles of such exported NOM on the mass balances of other natural and anthropogenic materials may, as discussed in subsequent sections of this review, be a more significant consideration than the fractional export of NOM may indicate.

10.3.3 *Hydrological controls on the transport of natural organic matter*

Transport of pedogenic NOM to streams is controlled by the amount and pathways of water movement through the catchment. Both these variables show seasonal variations which are translated to changes in the NOM concentration and possibly composition of streams. Flow entering a stream is the sum of the overland and subsurface inputs (see also Chapter 3):

$$\text{Flow}_{\text{stream}} = \text{Flow}_{\text{overland}} + \text{Flow}_{\text{subsurface}} \qquad (10.1)$$

The subsurface component may be more accurately described as two components (Hornberger et al. 1994)

$$\text{Flow}_{\text{subsurface}} = \text{Flow}_{\text{upper soil}} + \text{Flow}_{\text{lower soil}} \qquad (10.2)$$

The mass balance of a chemical species with concentration C is therefore

$$[C]_{\text{stream}} \cdot \text{Flow}_{\text{stream}} = [C]_{\text{overhand}} \cdot \text{Flow}_{\text{overhand}} + \\ [C]_{\text{upper soil}} \cdot \text{Flow}_{\text{upper soil}} + [C]_{\text{lower soil}} \cdot \text{Flow}_{\text{lower soil}} \qquad (10.3)$$

These equations were incorporated into the hydrological model TOPMODEL (Beven & Kirkby 1979) by Hornberger et al. (1994).

A seasonality in the relative importance of NOM transport by overland flow versus subsurface flow was found by Chittleborough et al. (1992). They found that most NOM transported in autumn and winter was through subsurface flow whereas in spring and summer overland flow was dominant. During the dry Australian summer the surface organic matter became more hydrophobic which prevented infiltration and therefore overland flow was favored. During the wetter months a well developed system of biopores and channels allowed subsurface flow to predominate. Preferential flow through macropores has been observed by a number of authors (Jardine et al. 1990; Wilson et al. 1990) and is evidenced in part by rapid stream response to rainfall input. These processes have a number of potential effects on the amount and nature of NOM. Overland flow is likely to be enriched in particulate NOM and possibly colloidal NOM. Flow through macropores reduces the amount of contact between the NOM and the soil matrix and may limit, but not eliminate, the interaction and immobilization of NOM by soil surfaces (Jardine et al. 1989). Preferential flow may also allow enhanced particulate NOM and colloidal NOM transport and has been shown to allow transport of significant quantities of soil minerals (Chittleborough, in prep) which may carry adsorbed NOM.

Because of the variability of NOM in various soil horizons, it is clear that the relative amounts of flow in the two subsurface compartments defined in the TOP-MODEL will have major influence on NOM transport (Hornberger et al. 1994). During low baseflow conditions (Fig. 10.15), water travels primarily through the lower soil compartment where NOM concentrations are lower. This NOM is likely composed of less reactive components as interaction with soil minerals will have occurred in the upper horizons. During high flow conditions (Fig. 10.16), a water table

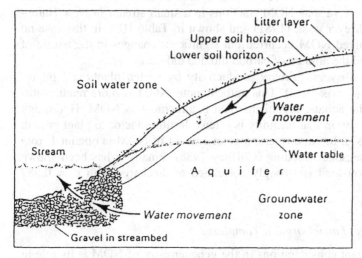

Figure 10.15. Schematic diagram of soil-water flushing under low baseflow conditions (from Hornberger et al. 1994).

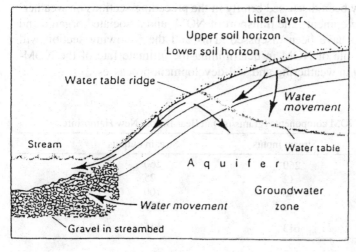

Figure 10.16. Schematic diagram of soil-water flushing under high-flow conditions (from Hornberger et al. 1994).

ridge may form at the stream-aquifer boundary thus allowing water movement through the upper soil compartment. Not only will the NOM concentration likely be higher, but this NOM may represent more reactive compounds formed from leaf litter leachates, which have had minimal interaction with the soil matrix. This 'flushing' of high NOM concentrations from upper soil horizons is particularly noticeable in systems where the stream hydrology is dominated by snowmelt. NOM builds up from microbial activity in the soil during the winter and is rapidly released during snowmelt. NOM concentrations are highest during the rising limb of the spring hydrograph (Hornberger et al. 1994).

Aquogenic NOM may also be influenced by hydrology but is in general more directly related to biogeochemical processes occurring within the stream. Major processes occurring are the utilization of coarse particulate NOM by heterotrophic organisms to produce finer particulate NOM and dissolved NOM and the mineralization of NOM to carbon dioxide. In small streams, net primary production is often small so that most of the NOM available for respiration is allochthonous (Fisher & Likens 1973). A mass balance of various NOM fractions in a small stream in New Hampshire was obtained by Meyer et al. (1981) and shown in Table 10.1. In this case no net production of dissolved NOM occurred but significant changes in the nature of the particulate NOM and the amount of carbon dioxide did occur.

In larger streams and rivers, primary productivity by rooted plants and phytoplankton is significant (Lewis 1988). This autochtonous NOM is more readily utilized and recycled by the aquatic organisms than is allochtonous NOM. Hydrology plays a role in productivity in that turbidity is often a limiting factor to plant growth (Edwards & Meyer 1987). In large rivers, a great amount of NOM is obtained from the floodplain during seasonal flooding (Cuffney 1988). Finally it has been shown that total organic carbon load is strongly correlated to discharge with $r^2 = 0.987$ (Schlesinger 1991).

10.3.4 *Redox chemistry of metal-organic complexes*

One of the most important considerations in the geochemistry of NOM is its role in the transport and transformations of metals. The interactions of NOM with base and soil minerals has already been discussed briefly in the preceding sections on weathering and soil profile development. The transport of NOM, and associated organic and inorganic materials also have been discussed. This and the following section will more fully explore the role of NOM in determining the ultimate fate of the NOM-chelated metals released in weathering and soil-development processes.

Table 10.1. Yearly flux of NOM components (grams/m^2) in Bear Brook, New Hampshire.

	Stream inputs	Stream outputs
Dissolved NOM	260	260
Fine particulate NOM	12	25
Coarse particulate NOM	340	100
Gases	1	230
Total	613	615

NOM plays an important role in the cycling of iron in surface waters. Reduction of Fe(III) to Fe(II) in oxygenated waters by NOM can occur by both photolytic and non-photolytic reactions. Non-photolytic reactions involving either NOM or specific organic compounds such as oxalate can promote the dissolution of iron oxides by forming soluble iron complexes. These may in turn be reduced in solution by species such as sulfide. Catalysed reductive dissolution of Fe(III) oxides by Fe(II) oxalate complexes is another important mechanism. Organic compounds such as ascorbate can form inner-sphere surface complexes with Fe(III) oxides and directly transfer electrons. Photolytic reduction of Fe(III) oxides is greatly enhanced by formation of surface complexes with electron-donating ligands such as oxalate. These and other processes involved in the cycling of iron in surface waters are illustrated in Figure 10.17, taken from Stumm (1992).

The role of NOM in the redox chemistry of iron in groundwaters has been demonstrated by the work of Liang et al. (1993). In laboratory column and in situ groundwater studies using waters with low concentrations of oxygen ($P_{O_2} = 0.005$ atm), added NOM (2 mg C/l) was found to increase the rate of Fe(II) oxidation by a factor of five over rates observed in the absence of added NOM. In contrast, NOM showed little effect on oxidation rates of Fe(II) at high partial pressures of oxygen ($P_{O_2} = 0.2$ atm). NOM apparently serves as a mediator of electron transfer between O_2 and iron(II) which results in an enhancement of iron oxidation rates that is significant at low levels of dissolved O_2. The mechanism of this enhancement, however, has not been elucidated. It is perhaps mechanistically significant that Liang et al. (1993) also found that total concentrations of Fe(III) species in solution were enhanced by Fe(III)-NOM complex formation and stabilization of colloidal Fe(III) hydroxides by adsorption of NOM.

Recent observations in our laboratories (Peiffer & Macalady, unpubl. data) also indicate that NOM has a significant influence on metal redox reactions in the presence of varying amounts of sulfide and oxygen. Reduction of ferric iron, present as non-filterable (0.2 micrometer) ferric-NOM complexes, in an iron and NOM rich ground water from a shallow aquifer beneath a wetland, proceeded rapidly upon the

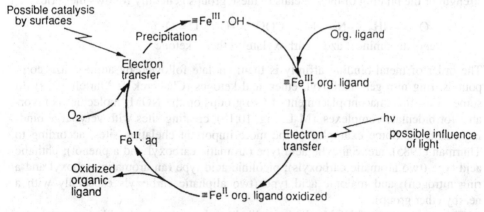

Figure 10.17. Schematic representation of the aquatic redox cycling of iron. Triple lines represent the lattice surface of an iron(III) (hydr)oxide (from Stumm 1992).

introduction of millimolar quantities of sulfide. Upon removal of the sulfide (by N_2 sparging), and the introduction of air, the iron was quickly re-oxidized. However, the NOM was apparently re-oxidised much more slowly. After exposure to air for periods of up to several days, the NOM retained the ability to reduce the iron back to the ferrous state after the solution was sparged with N_2 to remove oxygen. This suggests that NOM reduced by sulfide is reoxidised very slowly by molecular oxygen, at least at near-neutral pH's, providing a kind of redox-buffering of kinetic origin. The extent to which these observations are related to sulfide-NOM interactions as a unique form of redox reactions remains to be demonstrated.

10.3.5 *Natural organic matter and metal ion transport*

The binding of a metal ion to an organic complexing agent is generally expressed in terms of a *binding constant*. Binding constants can take a variety of forms depending upon the protonation state of the complexing agent used in the expressions, but they generally represent (for monodentate complexes) the equilibrium ratio of bound metal (metal-organic complex, ML) concentration (or activity) to the product of the concentrations (activities) of unbound ligand (L) and free aqueous metal ion (M), e.g.:

$$HL + M^{n+} \Leftrightarrow ML^{(n-1)+} + H^+ \tag{10.4}$$

$$K_b = [ML^{(n-1)+}] \cdot [H^+] / [M^{n+}] \cdot [HL] \tag{10.5}$$

where K_b is the binding constant and the square brackets represent concentrations, or more accurately, activities.

Because of the multiplicity of sites for metal complex formation on NOM 'molecules', precisely defined binding constants for NOM or its defined fractions are conceptually difficult. Nevertheless, the binding constant concept can be used in a general way to discuss the relative strengths of metal-binding interactions of NOM in aqueous systems. Binding constants can be used as a surrogate for binding strength.

As previously discussed, the presence of functional groups which may form bonds with metals suggests NOM is likely to play a major role in metal transport. The strength of the binding of most metals to these groups generally follows the series:

$$-O^- > -NH_2 > -N = N- > -COO^- > -O- > C = O$$
enolate   amine   azo   carboxylate   ether   ketone

The order of metal-binding affinity is then: enolate followed by amines, azo compounds, ring nitrogen, carboxyl, ether, and ketones (Chaberek & Martell 1959). In some cases, the adjacent placement of two groups on the NOM 'molecule' is favorable for bidentate complexes, (M_2L, Fig. 10.18), creating sites with very large binding constants. Some examples of the more important chelation sites, according to Thurman (1985), are: Salicylic acid type (aromatic carboxyl and a phenol); phthalic acid type (two aromatic carboxyls); picolinic acid type (an aromatic carboxyl and a ring nitrogen); and malonic acid type (two aliphatic carboxyls especially with a nearby ether group).

The components of NOM which are likely to be most responsible for metal mobility are: humic and fulvic acids, hydrophilic acids, pigments, and amino acids. However, humic and fulvic acids account for roughly 50-90% of dissolved NOM, so

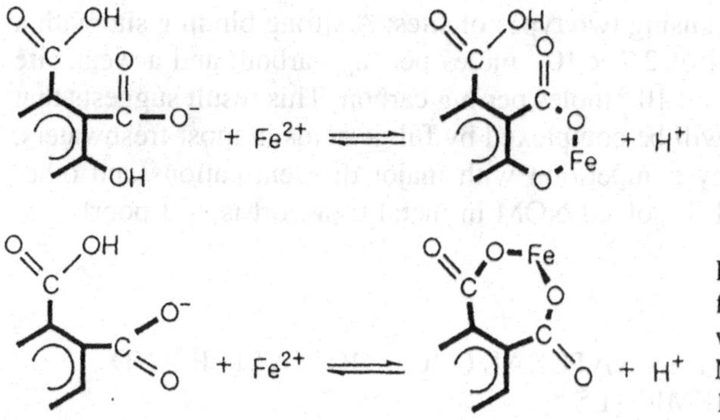

Figure 10.18. An example of the formation of a bidentate complex with a metal ion by portions of an NOM molecule (from Liang et al. 1993).

they are often considered to be the major complexers of metals in most natural waters. Certain other components of NOM, however, such as pigments (e.g. chlorophyll), other porphyrins, and low-molecular-weight polyfunctional organic acids such as oxalate and citrate are known to have very strong binding constants, perhaps several orders of magnitude greater than humic substances. Though these components are generally present at very low concentrations in natural waters, their extremely high binding constants may impart a measurable impact on metal mobility in certain systems. For example, the upper leaf litter zone of forest floor soils, where pigments may influence metal transport. Amino acids constitute about 3-5% of dissolved NOM. Their binding constants are similar to that of humic substances (Tuschall & Brezonik 1980). Simple acids such as acetate, formate, etc. have been identified in water and their metal binding constants are generally well known (Martel & Smith,1977, 1982). Little is known however, about the more complex components of the hydrophilic acid class, which can account for up to 50% of dissolved NOM. McKnight et al. (1983) determined copper binding constants for hydrophilic acids and found them similar to those of fulvic acid.

Schnitzer & Khan (1978) reviewed the early work on metal binding by pedogenic humic substances. Most workers found that in general the order of metal binding constants follow the Irving-Williams series (1948, as quoted by Thurman 1985, page 415). Part of the series found for pedogenic humic substances at neutral pH is:

$$Pb^{2+} > Cu^{2+} > Ni^{2+} > Co^{2+} > Zn^{2+} > Cd^{2+} > Fe^{2+} > Mn^{2+} > Mg^{2+}$$

and at pH = 3

$$Fe^{3+} > Al^{3+} > Cu^{2+} > Ni^{2+} > Co^{2+} > Pb^{2+} > Ca^{2+} > Zn^{2+} > Mn^{2+} > Mg^{2+}$$

For aquogenic humic substances Mantoura et al. (1978) found:

$$Hg^{2+} > Cu^{2+} > Ni^{2+} > Co^{2+} > Mn^{2+} > Cd^{2+} > Ca^{2+} > Mg^{2+}$$

In addition to the magnitude of the binding constants, the number of sites present on dissolved NOM is important to the mobility of metals. The estimates of the number and strength of sites varies considerably based on the fraction of NOM studied, the environment of isolation, and the metal of interest. As an example, McKnight & Wershaw (1989) studied Cu^{2+} binding by aquatic fulvic acid. They found that the

Cu^{2+} binding could be modeled using two types of sites: A strong binding site with a log $K_b = 8.1$, present at a level of 2.7×10^{-7} moles per mg carbon; and a weak site with log $K_b = 5.9$, present at 1.6×10^{-6} moles per mg carbon. This result suggests that a few micromoles/l of copper will be complexed by fulvic acids in most freshwaters. This will also be influenced by competition with major divalent cations and other trace metals. The exact role of dissolved NOM in metal transport is still poorly understood.

10.4 INTERACTIONS BETWEEN NATURAL ORGANIC MATTER AND ANTHROPOGENIC CHEMICALS

10.4.1 *Transport of pollutant metals as dissolved natural organic matter complexes*

NOM in an aquifer exists as dissolved or colloidal NOM in the aqueous phase and as adsorbed organic coatings on the aquifer matrix. Consequently the role of NOM as either a sink for metal contaminants or a facilitator of transport is complex and difficult to ascertain. It is likely that NOM may act in both roles with different NOM components contributing in different ways. Consider for example the role of NOM in the migration of radionuclides. Righetto et al. (1991) studied the sorption of actinides to mineral oxides in the presence of humic acid (Fig. 10.19). The humic acid was strongly sorbed to the oxides and significantly enhanced the actinide sorption. Thus the humic acid component of NOM can act as a sink for actinides through adsorption. Conversely, dissolved NOM may solubilize actinides, for example McKnight et al. (in prep) observed that 70-85% of the Pu present in surface and ground water at a nuclear weapons facility was associated with fulvic acid. Their results further suggested that fulvic acid contains very strong binding sites for Pu. Marley et al. (1993) performed a forced-gradient tracer injection using surface water NOM and monitored

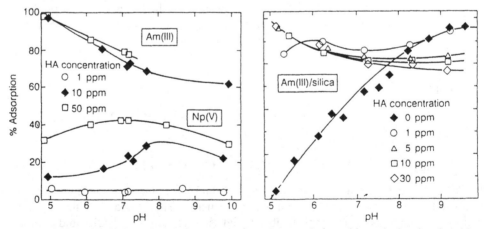

Figure 10.19. Left) pH dependence of Am(III) and Np(V) association with particulate humic acids. Right) Adsorption behavior of Am(III) in the amorphous silica (1200 pp.m)-humic acid system as a function of pH and humic acid concentration (from Righetto et al. 1991).

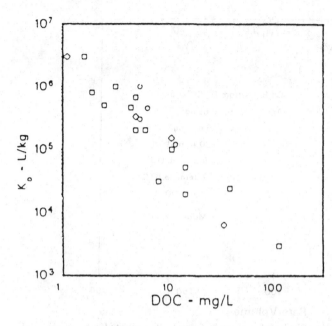

Figure 10.20. Variation of K_d (see Section 10.3.5) of reduced plutonium as a function of total DOC (from Nelson et al. 1985).

the breakthrough of NOM and actinides. Their NOM consisted of 81-98% fulvic acid and only 2-19% humic acid. Their results indicated that Am was transported by lower molecular weight components which were less sorbed than the more hydro-phobic components. These studies and others (Fig. 10.20) suggest that the fulvic acid component of NOM may provide a means of facilitated transport of actinides.

Another example of metal transport by NOM complexes was the work of Dun-nivant et al. (1992). They performed column experiments using aquifer sediments and NOM collected from a stream near a peat deposit to study cadmium transport. At an NOM concentration of 5.2 mg C/l the breakthrough volume ($C/C_0 = 0.5$; where C = the eluent concentration and C_0 = the influent concentration) was reduced by roughly 25% over that when NOM was absent. Increasing the NOM concentration further enhanced the breakthrough although the relative effect on the breakthrough curves (BTCs) was not as great per mg C/l (Fig. 10.21).

These two examples, along with numerous other literature sources and the large body of evidence on metal-NOM complexes in surface waters, suggests that these complexes are a very important if not the predominant transport mechanism.

10.4.2 *Effects of sorption/partitioning to natural organic matter on the transport of organic chemicals*

Hydrophobic interactions between soil or sediment organic matter and sparingly-soluble, non-ionic organic chemicals has been characterized in terms of a partitioning (or sorption) of the organic chemical into the NOM fraction of the soil or sediment. In a series of papers, (Karickhoff et al. 1979; Karickhoff 1980, 1981, 1984; Chiou et al. 1983; Voice & Weber 1983), this partitioning was shown to be adequately de-scribed in terms of an equilibrium between the dissolved and particulate phases. The partition coefficient, K_p, defines the ratio of the concentration of the organic con-

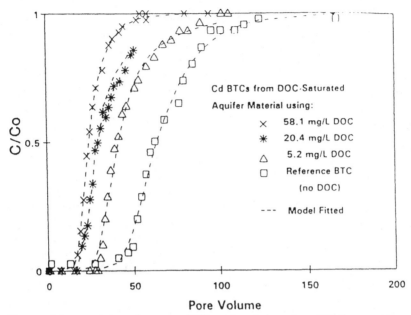

Figure 10.21. Breakthrough curves (BTCs) of cadmium from DOC-saturated aquifer material (0.058% C by weight) in the presence of four different DOC concentrations. Dashed lines represent model-fitted curves (from Dunnivant et al. 1992).

taminant in the solid soil or sediment phase, C_s expressed on a weight or mole of contaminant per kilogram of solid basis to the concentration in the aqueous phase, C_w expressed on a weight or mole per liter basis, or:

$$K_p = C_s/C_w \qquad (10.6)$$

(In some literature, K_p is called K_d, the solid-water distribution coefficient). This relationship is valid over a wide range of conditions for dilute (C_s less than 10^{-5} M or ½ the aqueous solubility) solutions of hydrophobic contaminants. The value of K_p for a particular soil or sediment and a particular contaminant is of course related to both the characteristics of the solid phase and the nature of the contaminant organic chemical. For most soils or sediment samples (fraction of organic carbon, f_{oc}, greater than about 0.002, Schwarzenbach et al. 1993, Abdul et al. 1987), the only relevant soil characteristic is its NOM fraction, and, for a given contaminant and a series of soils or sediments,

$$K_p = K_{oc} (f_{oc}) \qquad (10.7)$$

where f_{oc} represents the NOM fraction of the soil or sediment expressed as a fraction of the soil mass which is organic carbon (Fig. 10.22). K_{oc} is a characteristic of the contaminant chemical and represents its hypothetical partitioning to the organic carbon fraction of the soil or sediment. For soils with a dominant sand size fraction, the NOM content of the finer size fractions (< 50 micrometers) has been shown to be most useful for predictions of solid/water partitioning. For most soils and sediments, whole soil f_{oc} values can be used without substantial error (Karickhoff 1981).

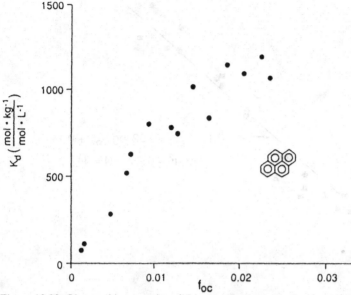

Figure 10.22. Observed increase in solid-water distribution ratios for a hydrophobic compound, pyrene, as a function of the NOM content of the solid (measured as organic carbon) in a variety of soils and sediments (from Schwarzenbach et al. 1993).

Values of K_{oc} for contaminant organic chemicals have been in turn shown to be related to a variety of molecular parameters which attempt to quantify the hydrophobicity of the organic substance. The most widely discussed of such parameters is the octanol/water partitioning coefficient, K_{ow}, which has been measured for a large number of chemicals of interest in studies of contaminant behavior (Pomona College 1984). Measured values of K_{ow} for pollutant organic chemicals range over at least seven orders of magnitude. Many empirical and semi-empirical relationships between K_{ow} values and other properties such as water solubility and molecular size (Karickhoff 1981 and others), and molecular size and topology indices such as molecular connectivity (Bahnick & Doucette 1988) have been developed. The most widely used relationships for the estimation of K_{oc} values involve empirical relationships of the form:

$$\log K_{oc} = a \log K_{ow} + d \tag{10.8}$$

A large number of such relationships have been reported in the literature (for example, Figure 10.23 and Brown & Flagg 1981; Schwarzenbach & Westall 1981; Karickhoff 1981; Hassett et al. 1983; Kenaga & Goring 1978; Chiou et al. 1983). Baker & Mihelcic (1995) recently developed a relationship valid for $\log K_{ow}$ values between 1.59 and 7.32 (spanning a wide variety of chemical classes) which gives a value of 1.015 for a and –0.404 for d, and which accounts for 90% of the variability in the data. Such correlations provide an adequate approximation for the prediction of the equilibrium partitioning of a wide variety of hydrophobic organic chemicals to soil or sediment NOM, but more precise predictions of K_{oc} are not possible using existing data and models.

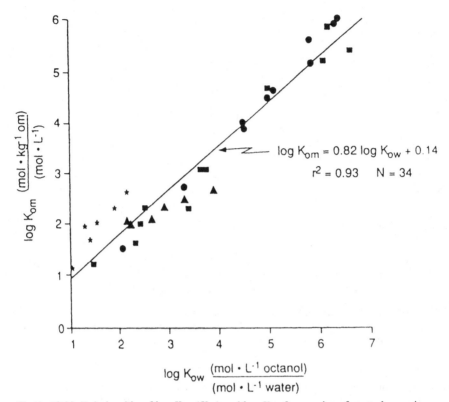

Figure 10.23. Relationship of log K_{om} (K_{oc}) and log K_{ow} for a series of neutral organic compounds: (•) Aromatic hydrocarbons; (■) Chlorinated hydrocarbons; (▲) Chloro-S-triazines; and (*) Phenyl ureas (from Schwarzenbach et al. 1993; data compiled by Karickhoff 1981).

10.4.3 *Effects of natural organic matter on hydrolytic reactions*

The degradation chemistry of many anthropogenic organic compounds is dominated by abiotic or biologically mediated hydrolysis reactions. Hydrolysis refers to nucleophilic attack on a molecule by water, resulting in the breakdown of the molecule into smaller, more hydrophilic products. Examples of hydrolysis reactions relevant to considerations of ground and surface water pollution are shown in Figure 10.24.

Specific examples and details of NOM-mediated abiotic hydrolysis reactions are reviewed by Macalady & Wolfe (1984) and Perdue & Wolfe (1982), for sorbed and aqueous pesticides respectively.

10.4.4 *Oxidation/reduction reactions facilitated by natural organic matter*

Oxidation/reduction (redox) reactions represent another important mode of transformation of anthropogenic organic chemicals in aqueous systems. These reactions can be defined as electron-transfer reactions, in which electrons are exchanged between components of an aqueous system and the contaminant molecule. As a consequence of a redox reaction, both dramatic and relatively subtle changes in the characteristics

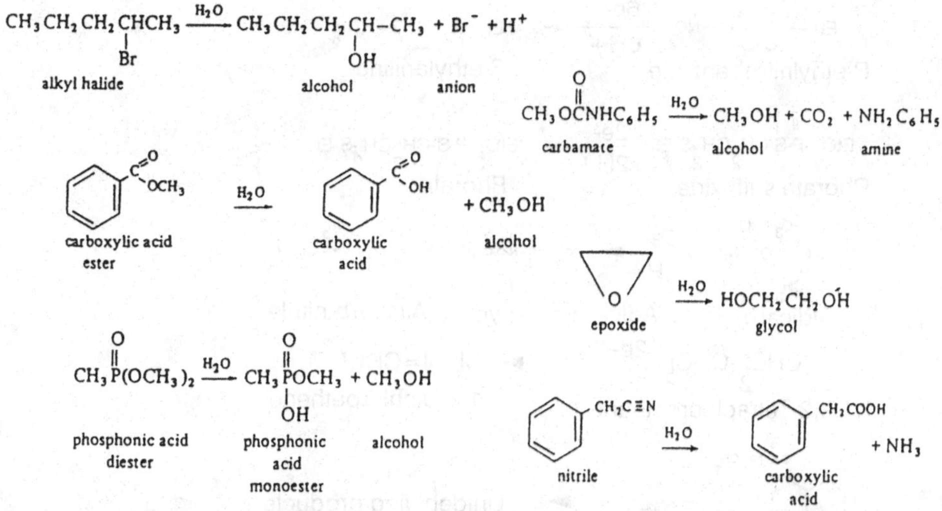

Figure 10.24. Some examples of hydrolysis reactions and reaction products (from Harris 1983).

of organic contaminants can be realized. For example, biologically mediated oxidations of a wide variety of organic substrates, both natural and anthropogenic, result in the conversion of the carbon/hydrogen framework of organic molecules to carbon dioxide and water, a process often referred to as 'mineralization'.

Less complete oxidations also occur, as in conversions of carbohydrates to simple organic alcohols or acids, thioethers to sulfoxides and sulfones, or amines to the corresponding nitroso or nitrate derivatives. Reduction reactions generally produce less dramatic changes, including the important class of reductive dehalogenation reactions and the inverse reactions of the partial oxidation processes listed above. Comprehensive reviews of redox reactions of anthropogenic organic chemicals in aquatic systems have been provided by Macalady et al. (1986) and Wolfe & Macalady (1992).

Because oxidation of organic molecules is the primary activity of a diverse and widespread group of biological systems, and because many organic molecules of concern in the pollution of aqueous systems are intentionally designed to be resistant to oxidation, reductive processes are of considerably more interest in considerations of the redox transformations of anthropogenic organic chemicals. A wide variety of reductive transformations of anthropogenic organic chemicals have been observed in natural systems. A few examples are illustrated in Figure 10.25.

Considerations of the role of NOM in reductive transformations have only been studied in detail relatively recently. In some senses the present understanding of the role of NOM in reductive transformations began with the observation of the extremely rapid reduction of the nitro-aromatic group on the pesticide methyl parathion to the corresponding aromatic amine. Experiments in our laboratories at the Colorado School of Mines with sediment/water slurries sampled in a variety of local ponds and reservoirs demonstrated first-order reductive transformations with half-lives from several hours to as fast as 15 seconds. Reactions proceeded in open beakers of unmodified sediment slurries, with no precautions to exclude atmospheric oxygen. Re-

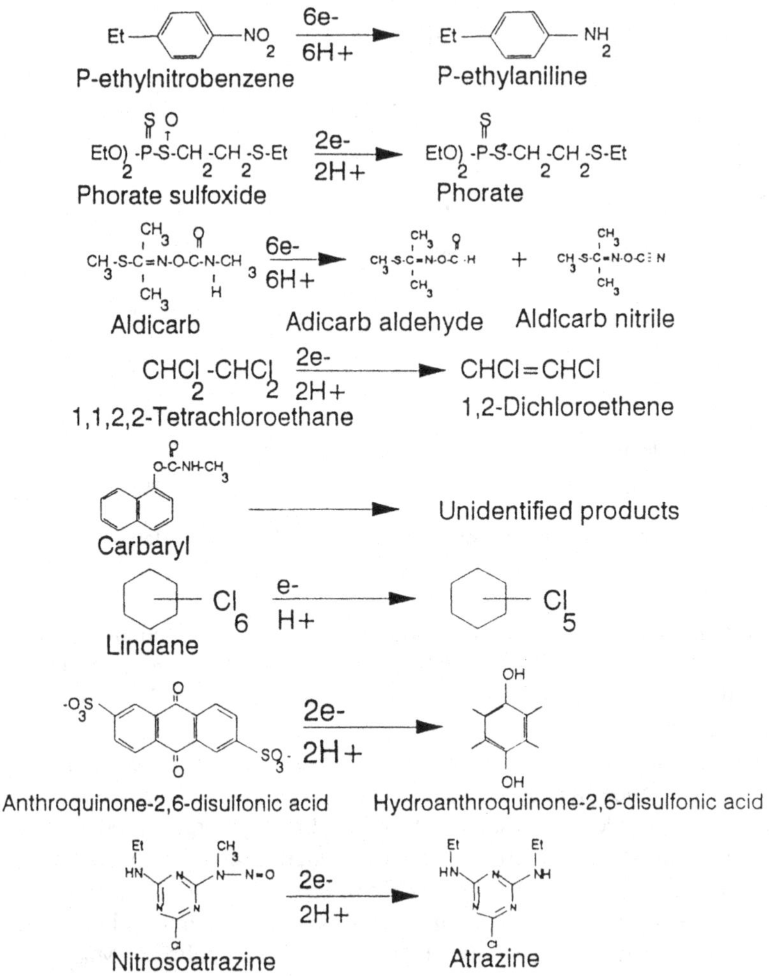

Figure 10.25. Examples of reductions which have been shown to occur in environmental samples under laboratory conditions (from Wolfe & Macalady 1992).

lated studies at the US. Environmental Protection Agency (USEPA) laboratories subsequently demonstrated rapid reduction of methyl parathion in additional sediment/water samples. Efforts to correlate reduction rates with sediment parameters revealed a direct relationship of the rate of these reductions to the NOM content of the sediments, with reaction rates also directly proportional to the sediment to water ratio in the slurries (Wolfe et al. 1987)

Similar results were obtained for reductions of azo compounds (Weber & Wolfe 1987) and halogenated organic compounds (Peijnenburg et al. 1992). Direct involvement of microbial systems in these reactions seemed improbable due to the extremely fast reaction rates observed in some systems. The emerging hypothesis is that NOM plays a key role in the observed transformations.

Other efforts to characterize the nature of the involvement of NOM in reductions of nitroaromatic and halogenated organic compounds have focused directly on reactions mediated by particulate NOM in sediments. Continuing work in the laboratories of N. Lee Wolfe at the USEPA Laboratories in Athens, GA has resulted in isolation and characterization of a several robust sediment enzymes which facilitate reductions of anthropogenic chemicals. Immunospecific assay techniques have been developed which provide a method to identify the sources of biological redox proteins in soils, sediments and aquifer materials (Carreira & Wolfe 1995). Proteins isolated from plant materials around soils and ponds have been shown to exhibit reductive activity which can account for a substantial fraction of the redox activities observed in sediment slurries. Thus, the reductive activity of sediment NOM may be largely due to biological components such as proteins.

Clearly, NOM can facilitate the rapid reduction of a variety of anthropogenic chemicals in sediment/water and homogeneous aqueous systems. It is equally clear that we have an incomplete understanding of the nature and scope of such redox activities.

10.4.5 *Colloidal natural organic matter and facilitated transport*

The importance of colloids, particularly colloidal NOM, in the transport of contaminants in the environment has recently received a great deal of attention. Part of this interest arose from 'anomalous' observations obtained in laboratory investigations of hydrophobic organic contaminant partitioning. It was observed that as the total solids content increased the observed partitioning coefficient, which should be invariant, decreased. This variation was termed the 'solids effect' and has since been explained by the presence of colloids which were not efficiently separated from the aqueous phase by standard techniques (i.e. filtration, centrifugation) due to their small size and/or low density (Morel & Geschwend 1987). These results highlighted the potential for colloidal influences on contaminant distributions.

Further evidence of the role of colloids in contaminant transport came from investigations of the distribution of radionuclides in the environment. For example, Penrose et al. (1990) found elevated levels of Pu and Am in ground water more than 3 km from a radioactive liquid waste outfall. Laboratory partitioning studies with soils from the site indicated the plutonium should be highly retarded by sorption to the soil and should not have traveled more than a few meters. Ultrafiltration studies showed that Pu was associated with material in the size range of 0.025-0.45 microns. This process of enhanced transport (over model predictions) of strongly sorbing contaminants has been termed 'facilitated' transport.

In the traditional view of partitioning in ground water, a solute exists as partitioned to an immobile solid phase and dissolved in the mobile aqueous phase. It has become clear that a third phase must be considered. Facilitated transport occurs as the result of the presence of this third mobile sorbing phase. This phase shares the property of the immobile solid phase in that it binds the solute. However, due to its small particle size (generally sub-micron) and/or hydrophilic character, it is mobile and transported in the aqueous phase. The transport of the contaminant in the environment thereby becomes linked to the transport of the colloidal phases. Colloid transport in surface waters is mainly influenced by aggregation followed by settling. In ground waters colloid transport is influenced by straining of larger particles by pores in the porous media and by interception and capture of smaller colloids by the surfaces of the aquifer particles. These processes are all dependent on colloid size and surface charge.

Colloidal NOM is particularly well suited to provide a means of facilitated transport. Colloidal NOM, although not well-characterized, has ionizable functional groups of the same type as dissolved NOM (primarily carboxyl) which may bind metals and other positively charged species. Some of these groups are in favorable positions to allow bidentate chelate formation and therefore have large binding constants for metals (Schnitzer & Kahn 1972). Other minor components of colloidal NOM, nitrogen in particular, may provide additional sites for metal interactions. Electrostatic repulsion between the negatively charged colloidal NOM and soil minerals, which are predominantly negatively charged aluminosilicates (clays) and quartz, promotes the transport of the colloidal NOM. Adsorption of NOM to iron(III) oxides will inhibit transport, but under anoxic, reducing conditions common in contaminated ground waters, NOM has been shown to stabilize iron oxide colloids and thereby promote transport (Liang et al. 1993). Also present on colloidal NOM are regions of relatively hydrophobic character (i.e. aromatic rings, aliphatic chains) which can bind hydrophobic contaminants. It is suspected that colloidal NOM, in part due to its higher apparent molecular weight, should provide a greater amount of hydrophobic character than dissolved NOM and should therefore exert a greater influence on the transport of sorbed hydrophobic contaminants.

Considerable evidence demonstrates the importance of the association of organic pollutants with colloidal NOM. Less is known about the specific role of colloidal NOM in enhancing the transport of contaminants in aqueous systems, especially ground water. A number of studies have demonstrated that colloidal NOM may adsorb nonpolar organic pollutants in a manner similar to soil and sediment organic matter. Chiou et al. (1986) demonstrated an increase in the 'apparent solubility' of DDT and poly-chlorinated biphenyls (PCBs) in the presence of humic colloids (Fig. 10.26). Estuarine colloids have been shown to be 10-35 times better than soil or sediment organic matter at adsorbing the herbicides atrazine and linuron (Means & Wijayaratne,1982) and 10 times greater for poly-aromatic hydrocarbons (PAH's) (Wijayaratne & Means 1984). Accounting for the presence of colloidal NOM improved the modeling of the partitioning of PCB's in Lake Superior (Baker et al. 1986). This study indicated that a three-phase model, including non-filterable colloidal NOM, best explained the data and also suggested that colloid-associated PCB's may be the dominant species in surface waters. In addition to hydrophobic partitioning, which is related to the organic carbon content of the colloid and the hydrophobicity of the pollutant (Karikhoff et al. 1979), specific interactions between certain types of pollutants are also important. Interaction of pollutants with NOM can occur by various mechanisms including hydrogen bonding, charge transfer complexes through aromatic pi-electrons, cation exchange, and conjugate formation through biochemical processes (Leenheer 1991). As an example, the sorption of benzidine and toluidine onto estuarine colloids was enhanced over that explained by hydrophobic partitioning (Means & Wijayaratne 1989). Under pH conditions favoring the cationic form of the amine, sorption was enhanced, presumably by cation exchange.

Despite overwhelming evidence showing interaction of NOM with organic and inorganic pollutants, the role of NOM in enhancing transport, especially in groundwaters, is poorly known.

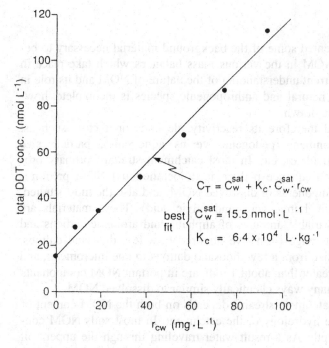

Figure 10.26. Plot of apparent DDT concentration in the aqueous phase as a function of the concentration of humic colloids, r_{cw} (from Chiou et al. 1986)

The equation shown in the figure:

$$C_T = C_w^{sat} + K_c \cdot C_w^{sat} \cdot r_{cw}$$

$$\text{best fit} \begin{cases} C_w^{sat} = 15.5 \text{ nmol} \cdot L^{-1} \\ K_c = 6.4 \times 10^4 \text{ L} \cdot \text{kg}^{-1} \end{cases}$$

10.4.6 *Natural organic matter... sink for pollutants or facilitator of transport? Mysteries and research questions*

NOM is a complex mixture of components, with a wide range of molecular weight and functional group contents. As scientists, we seek to understand how this complex mixture of chemical and physical properties determines NOM's distribution among various environmental compartments (i.e. water, soil, sediment), its role in soil development and weathering processes, its ability to bind metals and moderate hydrolysis reaction rates and its role in inorganic and organic redox reactions. It should be clear from the discourse above that our present understanding of relationships between the properties of NOM, the time, source and spatial variations in these properties, and the role of NOM in geochemical processes is superficial and incomplete. Consequently research to deepen and broaden the role of NOM in these processes must proceed along a wide variety of pathways.

The research outlined above provides a confusing array of conclusions which indicate that, in various soil and aqueous environments, NOM can act either as an agent for retardation of contaminant transport and transformation or as facilitator of such processes. If we have as our goal the ability to predict the roles of NOM in geochemical processes, both natural and those induced by anthropogenic disturbances, a considerable effort must continue to define in a fundamental way the nature of the chemistry and physics of NOM. There is no lack of mysteries, and the attempt to obtain a coherent picture of the interplay of NOM with its surrounding chemical and geological environments will continue to provide exciting and important research challenges.

10.5 CONCLUSIONS

In this chapter we have presented some of the background material necessary to begin to evaluate the role of NOM in the various mass balances which take place in catchments. Although our current understanding of the nature of NOM and its role in the mass balance of various natural and anthropogenic species is incomplete, however, some conclusions can be drawn.

The nature of NOM, and therefore its reactivity, depends on factors such as: source material and environment (pedogenic versus aquogenic), particle size (dissolved, colloidal, or particulate), etc. In most catchment streams, primary productivity is low and as a consequence pedogenic (allochthonous) NOM predominates. The most abundant components of dissolved NOM, and also the most studied, are humic substances (humin, humic acid, and fulvic acid). These materials are polyelectrolytes containing variable amounts of aliphatic and aromatic carbons and have molecular weights that range from five hundred to a few thousand daltons. Colloidal NOM, ranging in size from a few thousand daltons to one micrometer, and particulate NOM, material greater than about 1 mm, are important NOM components in some systems and are in many ways chemically similar to dissolved NOM.

The amount of NOM in catchment streams depends on both the NOM content of the surrounding soils and the hydrology of the catchment. In most soils NOM concentrations decrease with depth. As a result water traveling through the upper soil horizons transports more NOM to streams than does water traveling through deeper soil zones. The relative importance of these different flow-paths often vary with season.

NOM contains functional groups which provide both acidic protons and metal chelating sites. Although not clearly understood, it is likely that NOM participates in weathering reactions of soil minerals by both acid dissolution and complex formation. One result of these processes is the development of soil profiles (podzolization). Dissolved NOM can contribute to the transport of elements through the soil and eventually into the catchment streams. These elements may be naturally occurring constituents of the soil and bedrock or anthropogenic contaminants such as heavy metals. Conversely, complexation of these species to NOM associated with the solid phases in soil may work to prevent mobilization and result in retention in the catchment.

The mobility of many species depend on their oxidation state. NOM plays an important role in the redox state of many species. Iron for example can be reduced by NOM from the immobile ferric state to the relatively more mobile ferrous state. The mass balance of redox-sensitive metals may therefore depend on both the redox and complexation characteristics of NOM.

NOM plays a significant role in the transport and transformation of anthropogenic organic compounds. Partitioning of anthropogenic compounds between aqueous and solid phases depends to a great extent on the hydrophobicity of the compound and the NOM content of the solid phase. Generally we think of anthropogenic compounds as sorbing to solid phase NOM and consequently being retained in the catchment soils. However dissolved and colloidal NOM may also partition these compounds and therefore facilitate their transport through the catchment. Further, sorbed contaminants may be transported by particulate NOM associated with suspended sediments. As a consequence, NOM may act as an agent of transport and/or retardation in catchments.

Finally, NOM participates in the transformation of many anthropogenic organic compounds. Processes including hydrolysis, oxidation, and reduction of organic compounds may involve NOM. In particular, NOM seems important in the reduction reactions of various organic compounds. The mass balance of organic compounds in catchments may therefore be influenced by NOM through processes other than transport.

10.6 FUTURE RESEARCH DIRECTIONS

Further characterization of the physical and chemical properties of NOM are needed, especially in relation to the influence of particle size (i.e. dissolved, colloidal, and particulate) and environment (surface water, soil, groundwater, etc.) on NOM properties. Field-scale studies of groundwater-surface water interactions, in relation to NOM transport, are needed. The role of organic-rich upper soil horizons versus deeper low-NOM groundwaters in transport of NOM to streams needs further investigation. Continued study of natural and contaminant species partitioning to and reaction with (hydrolysis, oxidation, reduction) NOM is needed. A synthesis of the results of NOM characterization, NOM transport studies, and NOM-contaminant interaction studies is needed to provide a mechanistic understanding of the importance of NOM for the mass balances of natural and contaminant species in catchments.

REFERENCES

Abdul, A.S., Gibson, T.L. & Rai, D.N. 1987. Statistical correlations for predicting the partition coefficents for non-polar organic contaminants between aquifer organic carbon and water. *Hazardous Waste and Hazardous Material* 4: 211-222.

Aiken, G.R., McKnight, D.M., Wershaw, R.L. & MacCarthy, P. 1985. An introduction to humic substances in soil, sediment and water. In: Aiken, G.R., McKnight, D.M., Wershaw, R.L. & MacCarthy, P. (eds), *Humic Substances in Soil, Sediment and Water: Geochemistry, Isolation, and Characterization*. John Wiley and Sons, New York. p. 6

Antweiler, R.C. & Drever, J.I. 1983. The weathering of a late Tertiary volcanic ash: importance of organic solutes. *Geochim. Cosmochim. Acta.* 47: 623-629.

Bahnick, D.A. & Doucette, W.J. 1988. Use of molecular connectivity indices to estimate soil sorption coefficents for organic chemicals. *Chemosphere*, 17. p. 1703.

Baker, W.E. 1973. The role of humic acids from Tasmanian podzolic soils in mineral degradation and metal mobilization. *Geochim. Cosmochim. Acta*, 37: 269-281.

Baker, J.R. & Mihelcic, J.R. 1995. Development and evaluation of models for predicting soil/water partition coefficents. (submitted for publication).

Baker, J.E., Capel, P.D. & Eisenreich, S.J. 1986. Influence of colloids on sediment-water partition coefficents of polychlorobiphenyl congeners in natural waters. *Environ. Sci. Technol*, 20: 1136-1143.

Bennet, P. & Siegel, D.I. 1987. Increased solubility of quartz in water due to complexing by organic compounds. *Nature* 326: 684-686.

Beven, K.J. & Kirkby, M.J. 1979. A physically based, variable contributing area model of basin hydrology. *Hydrological Sci. Bull.* 24: 43-69.

Bockheim J.G. 1979. Properties and relative ages of soils of southwestern Cumberland Peninsula, Baffin Island, N.W. T., Canada. *Arctic and Alpine Res.* 11: 289-306.

Brown, D.S. & Flagg, E.W. 1981. Emprical prediction of pollutant sorption in natural sediments. *J. Environ. Qual.* 10: 382-386.

Buffle, J. 1984, Natural Organic Matter and Metal-Organic Interactions in Aquatic Systems. In: H. Siegel (ed.), *Metal Ions in Biological Systems,* Marcel Dekker. New York. pp 154-221.

Buffle, J., Perret, D. & Newman, M. 1992. The use of filtration and ultrafiltration for size fractionation of aquatic particles, colloids, and macromolecules. In: Buffle, J. & van Leeuwen, H.P. (eds), *Environment Particles. Volume 1.* Lewis Publishers. Boca Raton. pp 171-230.

Buffle, J., DeVitre, R.R., Perret, D. & Leppard, G.G. 1989. Physico-chemical characteristics of a colloidal iron phosphate species formed at the oxic-anoxic interface of a eutrophic lake. *Geochim. Cosmochim. Acta.* 53: 399-408.

Carreria, L.H. & Wolfe, N.L. 1995. Use of field immunoassays to identify redox proteins in natural systems. (submitted for publication).

Chaberek, S. & Martell, A.E. 1959. *Organic Sequestering Agents.* John Wiley & Sons. New York. 616 p.

Chesworth, W. & Macias-Vasquez, F. 1985. Pe, pH, and podzolization. *Am. J. Sci.* 285: 128-146.

Chiou, C.T., Porter, L.J. & Schmedding, D.W. 1983. Partition equilibria of non-ionic organic compounds between soil organic matter and water. *Environ. Sci. Technol.* 17: 227-231.

Chiou, C.T., Malcolm, R.L., Brinton, T.I. & Kile, D.E. 1986. Water solubility enhancement of some organic pollutants and pesticides by dissolved humic and fulvic acids. *Environ. Sci. Technol.* 20: 502-508.

Chittleborough, D.J., Smettem, K.R.J., Cotsaris, E. & Leaney, F.W. 1992. Seasonal changes in the pathways of dissolved organic carbon through a hillslope soil (Xeralf) with contrasting texture. *Aust. J. Soil. Res.* 30: 465-76.

Cronan, C.S. & Aiken, G.R. 1985. Chemistry and transport of soluble humic substances in forested watersheds of the Adirondack Park, New York. *Geochim. Cosmochim. Acta.* 49: 1697-1705.

Cuevas, E. & Medina, E. 1988. Nutrient dynamics within Amazonia forests. II. Fine root growth, nutrient availablity and leaf litter decomposition. *Oecologia.* 76: 222-235.

Cuffney, T.F. 1988. Input, movement and exchange of organic matter within a subtropical coastal blackwater river-floodplain system. *Freshwater Biol.* 19: 305-320.

Dethier, D.P., Jones, S.B., Fiest, T.P. & Ricker, J.E. 1988. Relations among sulfate, aluminum, iron, dissolved organic carbon, and pH in upland forest soils of northwestern Massachusstts. *Soil Sci. Soc. Amer. J.* 52: 506-512.

Dunnivant, F.M., Jardine, P.M., Taylor, D.L. & McCarthy, J.F. 1992. Co-transport of cadmium and hexachlorobiphenyl by dissolved organic carbon through columns containing aquifer material. *Environ. Sci. Technol.* 26: 360-368.

Edwards, R.T. & Meyer, J.L. 1987. Metabolism of a sub-tropical low gradient blackwater river. *Freshwater Biol.* 17: 251-263.

Esser, G., Aselmann, I. & Lieth, H. 1982. Modeling the carbon reservoir in the system compartment 'litter'. In: *Mitteilungen aus dem Geologish-Palantologischen Institut der Universitat Hamburg. Vol. 52,* University of Hamburg, Germany. pp 39-58.

Fisher, S.G. & Likens, G.E. 1973. Energy flow in Bear Brook, New Hampshire: An integrative approach to stream ecosystem metabolism. *Ecological Monographs* 43: 421-439.

Greenland, D.J. 1971. Interactions between humic and fulvic acids and clays. *Soil Sci.* 111: 34-41.

Harris, J.C. 1983. Rate of hydrolysis. In: Lyman, W., Reehl, W. & Rosenblatt, D. (eds), *Handbook of Chemical Property Estimation Methods.* McGraw-Hill, New York. Chapter 7.

Hassett, J.J., Banwart, W.L. & Griffen, R.A. 1983. Correlation of compound properties with sorption characteristics of non-polar compounds by soils and sediments: Concepts and limitations. In: C.W. Francis & S.I. Aurebach (eds), *Environmental and Solid Waste Characterization, Treatment and Disposal.* Butterworth Publishers, MA.

Hayes, M.H.B., MacCarthy, P., Malcom, R.L. & Swift, R.S. 1989. Structures of humic substances: the emergance of 'forms'. In: M.H.B. Hayes, P. Macarthy, R.L. Malcom & R.S. Swift (eds). *Humic Substances II. In Search of Structure.* John Wiley and Sons. New York. pp 690-733.

Hedges, J.I. & Lee, C. 1992. Measurement of dissolved organic carbon and nitrogen in natural waters. *Marine Chem.* 41, 290 pp

Hornberger, G.M., Bencala, K.E. & McKnight, D.M. 1994. Hydrological controls on dissolved organic carbon during snowmelt in the Snake River near Montezuma, Colorado. *Biogeochemistry* 25: 147-165.

Huang, W.H. & Keller, W.D. 1971. Dissolution of clay minerals in dilute organic acids at room temperature. *Am. Mineral.* 56: 1082-1095.

Hunter, K.A. 1980. Microelectrophoretic properties of natural surface-active organic matter in coastal seawater. *Limno. Oceanogr.* 25: 807-822.

Jardine, P.M., Wilson, G.V., Luxmoore, R.J. & McCarthy, J.F. 1989. Transport of inorganic and natural organic tracers through an isolated pedon in a forest watershed. *Soil Sci. Soc. Am. J.* 53: 317-323.

Jardine, P.M., Wilson, G.V., McCarthy, J.F., Luxmoore, R.J., Taylor, D.L. & Zelazny, L.W. 1990. Hydrogeochemical processes controlling the transport of dissolved organic carbon through a forested hillslope. *J. Contam. Hydrol.* 6: 3-19.

Karickhoff, S.W. 1980. Sorption kinetics of hydrophobic pollutants in natural sediments. In: R.A. Baker (ed.), *Contaminants and Sediments. Vol. 2: Analysis, Chemistry, and Biology.* Ann Arbor Science, Ann Arbor. pp 193-205.

Karickhoff, S.W. 1981. Semi-empirical estimation of sorption of hydrophobic pollutants on natural sediments and soils. *Chemosphere* 10: 833-846.

Karickhoff, S.W. 1984. Organic pollutant sorption in aquatic systems. *J. Hydraulic Eng.* 110: 707-735.

Karickhoff, S.W., Brown, D.S. & Scott, T.A. 1979. Sorption of hydrophobic pollutants on natural sediments. *Water Res.* 13: 241-248.

Kenaga, E.E. & Goring, C.A.I. 1978. Relationship between water solubility, soil sorption, octanol-water partitioning and bioconcentration of chemicals in biota. *Proceedings of the American Society for Testing and Materials, 3rd Aquatic Toxicology Symposium. No. STP 707.* pp 78-115.

Lee, C.M. & Macalady, D.L. 1989. Towards a standard method for the measurement of organic carbon in sediments. *Intern. J. Environ. Anal. Chem.* 35: 219-225.

Leenheer, J.A. 1991. Organic substance structures that facilitate contaminant transport and transformations in aquatic sediments. In: R.A. Baker (ed.), *Organic Substances and Sediments in Water. Vol 1: Humics and Soils.* Lewis Publishers. Chelsea. MI. pp 3-22.

Leenheer, J.A., Malcom, R.L., McKinley, P.W. & Eccles, L.A. 1974. Occurence of dissolved organic carbon in selected groundwater samples in the United States. *US Geol. Surv. J. Res.* 2: 361-369.

Leppard, G.G. 1992. Evaluation of electron microscope techniques for the description of aquatic colloids. In: Buffle, J. & van Leeuwen, H.P. (eds), *Environmental Particles, Vol. 1.* Boca Raton, FL, Lewis Publishers, pp 231-290.

Leppard, G.G., Buffle, J. & Baudat, R. 1986. Description of the aggregation properties of aquatic pedogenic fulvic acids-Combining physico-chemical data and microscopical observations. *Water Res.* 20: 185-196.

Lewis, W.M. 1988. Primary production in the Orinoco River. *Ecology* 69: 679-692.

Liang, L., McNabb, J.A., Paulk, J.M., Gu, B. & McCarthy, J.F. 1993. Kinetics of Fe(II) oxygenation at low partial pressure of oxygen in the presence of natural organic matter. *Environ. Sci. Technol.* 27: 1864-1870.

Lonsdale, W.M. 1988. Predicting the amount of litterfall in forests of the world. *Annals of Botany* 61: 319-324.

Loughnan, F.C. 1969. *Chemical Weathering of Silicate Minerals.* Elsevier, New York, 154 pp.

Macalady, D.L. & Wolfe, N.L. 1984. Abiotic hydrolysis of sorbed pesticides. In: R.F. Krueger & J.N. Seiber (eds), *Treatment and disposal of pesticide wastes.* Advances in Chemistry Series 259, American Chemical Society, Washington, D.C. pp 221-244.

Macalady, D.L., Tratnyek, P.G. & Grundl, T.J. 1986. Abiotic reduction reactions of anthropogenic chemicals in anaerobic systems: A critical review. *J. Contam. Hydrol* 1: 1-28.

Macalady, D.L., Tratnyek, P.G. & Wolfe, N.L. 1989. Influence of natural organic matter on the abiotic hydrolysis of organic contaminants in aqueous systems. In: I.H. Suffet & P. MacCarthy (eds), *Aquatic Humic Substances: Influence on Fate and Treatment of Pollutants*. Advances in Chemistry Series 219, American Chemical Society, Washington, D.C., pp 323-332.

MacCarthy, P., Peterson, M.J., Malcom, R.L. & Thurman, E.M. 1979. Separation of humic substances by pH gradient desorption from a hydrophobic resin. *Anal. Chem.* 51: 2041-2043.

Malcolm, R.L. 1985. Humic substances in rivers and streams. In: G.A. Aiken, D.M. McKnight & R.L. Wershaw (eds), *Humic Substances in Soil, Sediment, and Water: Geochemistry, Isolation, and Characterization*. Wiley-Interscience. New York, NY, 692 pp.

Mantoura, R.F.C., Dickson, A. & Riley, J.P. 1978. The complexation of metals with humic materials in natural waters. *Estuarine and Coastal Marine Sci.* 6: 387-408.

Marley, N.A., Gaffney, J.S., Orlandini, K.A. & Cunningham, M.M. 1993. Evidence for radionuclide transport and mobilization in a shallow, sandy aquifer. *Environ. Sci. Technol.* 27: 2456-2461.

Martell, A.E. & Smith, R.M. 1977. *Critical Stability Constants*. Vol. 3; Plenum Press, New York.

Martell, A.E. & Smith, R.M. 1982. *Critical Stability Constants*. Vol. 5; Plenum Press, New York.

Mast, M.A. & Drever, J.I. 1987. The effect of oxalate on the dissolution rates of oligoclase and tremolite. *Geochim. Cosmochim. Acta.* 51: 2559-2568.

McDowell, W.H. & Wood, T. 1984. Podzolization: Soil processes control dissolved organic carbon concentrations in stream water. *Soil Sci.* 137: 23-32.

McKnight, D.M. & Wershaw, R.L. 1989. Complexation of copper by fulvic acid from the Suwannee River-Effect of counter-ion concentrations. In: R.C. Averett, J.A. Leenheer, D.M. McKnight & K.A. Thorn (eds), *Humic Substances in the Suwannee River, Georgia: Interactions, Properties, and Proposed Structures. US Geol. Survey, Open-File Report* 87-557: 59-80.

McKnight, D.M., Feder, G.L., Thurman, E.M., Wershaw, R.L. & Westall, J.C. 1983. Complexation of copper by aquatic humic substances from different environments. In: R.E. Wildung & E.A. Jenne (eds), *Biological Availability of Trace Metals*. Elsevier, Amsterdam. pp 65-76.

McKnight, D.M., Bencala, K.E., Zellweger, G.W., Aiken, G.R., Feder, G.L. & Thorn, K.A. 1992. Sorption of dissolved organic carbon by hydrous aluminum and iron oxides occuring at the confluence of Deer Creek with the Snake River, Summit County, Colorado. *Environ. Sci. Technol.* 26: 1388-1396.

Means, J.C. & Wijayaratne, R. 1982. Role of natural colloids in the transport of hydrophobic pollutants: *Science* 215: 968-970.

Means, J.C. & Wijayaratne, R. 1989. Sorption of benzidine, toluidine, and azobenzene on colloidal organic matter. In: I.H. Suffet & P. MacCarthy (eds), *Aquatic Humic Substances: Influence on Fate and Treatment of Pollutants*. Advances in Chemistry Series 219, American Chemical Society, Washington, D.C., pp 209-222.

Meybeck, M. 1981. River transport of organic carbon to the ocean. In: G.E. Likens (ed.), *Flux of Organic Carbon by Rivers to the Ocean*. US Department of Energy. NTIS report # CONF-8009140, UC-11, Springfield, VA. pp 219-269.

Meybeck, M. 1982. Carbon, nitrogen, and phosphorous transport by world rivers. *Amer. J. Sci.* 282: 401-450.

Meyer, J.L., Likens, G.E. & Sloane, J. 1981. Phosphorous, nitrogen, and organic carbon flux in a headwater stream. *Archiv. fuer. Hydrobiologie* 91: 28-44.

Morel, F.M.M. & Gschwend, P.M. 1991. The role of colloids in the partitioning of solutes in natural waters. In: W. Stumm (ed.), *Aquatic Surface Chemistry*. John Wiley & Sons. New York. 508 pp.

Nelson, D.M., Penrose, W.R., Karttunen, J.O. & Melhoff, P. 1985. Effects of dissolved organic carbon on the adsorption properties of plutonium in natural waters. *Environ. Sci. Technol.* 19: 127-131.

O'Brien, B.J. & Stout J.D. 1978. Movement and turnover of soil organic matter as indicated by carbon isotope measurements. *Soil Biology and Biochemistry* 10: 309-317.

Olson, J.S. 1963. Energy storage and the balance of producers and decomposers in ecological systems. *Ecology* 44: 322-331.

O'Melia, C.R. & Tiller, C.L. 1993. Physicochemical aggregation and deposition in aquatic environments. In: J. Buffle & H.P. van Leeuwen (eds), *Environmental Particles. Volume 2.* Lewis Pubs. Boca Raton. pp 353-386.

Pedro, G., Jamagne, M. & Begon, J.C. 1978. Two routes in the genesis of strongly differentiated acid soils under humid, cool-temperate conditions. *Geoderma* 20: 173-189.

Peijnenburg, W.J.G.M., 't Hart, M.J., den Hollander, H.A., van de Meent, D., Verboom, H.H. & Wolfe, N.L. 1992. Reductive transformations of halogenated aromatic hydrocarbons in anaerobic water-sediment systems: Kinetics, Mechanisms, and Products. *Environ. Toxicol. and Chem.* 11: 289-300.

Penrose, W.R., Polzer, W.L., Essington, E.H., Nelson, D.M. & Orlandini, K.A. 1990. Mobility of plutonium and americium through a shallow aquifer in a semiarid region. *Environ. Sci. Technol.* 24: 228-234.

Perdue, E.M. & Wolfe, N.L. 1982. Modification of pollutant hydrolysis kinetics in the presence of humic substances. *Environ. Sci. Technol.* 16: 847-852.

Pomona College 1984. *Log P and Parameter Database.* Pomona College, Claremont, CA. Seaver Chemistry Labortory, Medicinal Chemistry Project. Technical Database Services Inc. New York.

Ranville, J.F., Harnish, R.A. & McKnight, D. 1991. Particulate and colloidal organic material in Pueblo Reservoir, Colorado: Influence of autochthonous source on chemical composition. In: R.A. Baker (ed.), *Organic Substances and Sediments in Water. Volume 1.* Lewis Publishers, Chelsea. pp 47-74.

Righetto, L., Bidoglio, G., Azimonti, G. & Bellobono, I.R. 1991. Competitive acitinde interactions in colloidal humic acid-mineral oxide systems. *Environ. Sci. Technol.* 25: 1913-1919.

Robert, M. & Berthelin, J. 1986. Role of biological and biochemical factors in soil mineral weathering. In: P.M. Huang & M. Schnitzer (eds), *Interactions of Soil Minerals with Natural Organics and Microbes.* Soil Science Society of America (SSSA) Special Publication Number 17. Madison. pp 453-496.

Romell, L.G. 1935. Ecological problems of the humus layers in the forest. *Cornell University, Agricultural Experiment Station Memoir 170,* Ithaca, New York.

Schalscha, E.B., Appelt, H. & Schatz, A. 1967. Chelation as a weathering mechanism-I. Effect of complexing agents on the solubilization of iron from minerals and granodiorite. *Geochim. Cosmochim. Acta.* 31: 587-596.

Schlesinger, W.H. 1977. Carbon balance in terrestrial detritus. *Annual Review of Ecology and Systematics* 8: 51-81.

Schlesinger, W.H. 1991. *Biogeochemistry: An analysis of Global Change.* Academic Press, San Diego, 443 pp.

Schnitzer, M. & Kahn, S.U. 1972. *Humic Substances in the Environment.* Marcel Dekker, New York.

Schnitzer, M. & Kahn, S.U. 1978. *Soil Organic Matter.* Elsevier. Amsterdam. 319 pp.

Schnitzer, M. & Kodama, H. 1976. The dissolution of micas by fulvic acid. *Geoderma* 15: 381-391.

Schwartzenbach, R.P. & Westall, J. 1981. Transport of non-polar organic compounds from surface water to groundwater: Laboratory sorption studies. *Environ. Sci. Technol.* 15: 1360-1367.

Schwarzenbach, R.P., Gschwend, P.M. & Imboden, D.M. 1993. *Environmental Organic Chemistry.* Wiley-Interscience. New York, NY, 681pp.

Senesi, N. & Steelink, C. 1989. Application of ESR spectroscopy to the study of humic substances. In: M.H.B. Hayes, P. Macarthy, R.L. Malcom & R.S. Swift (eds), *Humic Substances II. In Search of Structure.* John Wiley and Sons. New York. pp 373-408.

Stevenson, F.J. 1982. *Humus Chemistry.* John Wiley and Sons. New York. 443 pp.

Stevenson, F.J. 1985. Geochemistry of soil humic substances. In: G.A. Aiken, D.M. McKnight & R.L. Wershaw (eds), *Humic Substances in Soil, Sediment, and Water: Geochemistry, Isolation, and Characterization.* Wiley-Interscience. New York, NY, pp 13-52.

Stevenson, F.J. 1986. *Cycles of Soil*. John Wiley and Sons. New York. 380 pp.

Stumm, W. 1992. *Chemistry of the Solid-Water Interface*. John Wiley & Sons.New York, 428 pp.

Swift, M.J., Heal, O.W. & Anderson, J.M. 1979. *Decomposition in Terrestrial Ecosystems*. University of California Press, Berkeley, CA. 372 pp.

Tan, K.H. 1980. The release of silicon, aluminum, and potassium during decomposition of soil minerals by humic acid. *Soil Sci.* 129: 5-11.

Thorn, K.A., Arterburn, J.B. & Mikita, M.A. 1992. ^{15}N and ^{13}C NMR invstigation of hydroxylamine-derivatized humic substances. *Environ. Sci. Technol.* 26: 107-116.

Thurman, E.M. 1985. *Organic Geochemistry of Natural Waters*. Martinus Nijhoff/Dr. W. Junk Pubs. Dordrecht, 497 pp.

Thurman, E.M. & Malcom, R.L. 1981. Preparative isolation of aquatic humic substances. *Environ. Sci.Technol.* 15: 463-466.

Tiller, C.L. & O'Melia, C.R. 1993. Natural organic matter and colloidal stability: models and measurements. *Colloids and Surfaces A: Physiochem. and Eng. Aspects.* 73: 89-102.

Tipping, E. & Cooke, D. 1982. The effects of adsorbed humic substances on the surface charge of goethite (a-FeOOH) in freshwaters. *Geochim. Cosmochim. Acta.* 46: 75-80.

Tuschall, J.R. & Brezonik, P.L. 1980. Characterization of organic nitrogen in natural waters: Its molecular size, protein content, and interactions with heavy metals. *Limnol. Oceanog.* 25: 495-504.

Ugolini, F.C. 1968. Soil development and alder invasion in a recently deglaciated area of Glacier Bay, Alaska. In: J.M. Trappe, J.F. Franklin, R.F. Tarrant & G.M. Hansen (eds), *Biology of Alder*. US Forest Service, Pacific Northwest Forest and Range Experiment Station, Portland, Oregon. pp 115-140.

Ugolini, F.C., Minden, R., Dawson, H. & Zachara, J. 1977. An example of soil processes in the *Abies amabilis* zone of central Cascades. Washington. *Soil Sci.* 124: 291-302.

Voice, T.C. & Weber, W.J. 1983. Sorption of hydrophobic compounds by sediments, soils, and suspended solids. I. Theory and background. *Water Res.* 17: 1433-1441.

Weber, E.J. & Wolfe, N.L. 1987. Kinetic studies of the reduction of aromatic azo compounds in anaerobic sediment/water systems. *Environ. Toxicol. Chem.* 6: 911-919.

Wijayaratne, R.D. & Means, J.C. 1984. Sorption of polycylic aromatic hydrocarbons by natural estuarine colloids. *Mar. Environ. Res.* 11: 77-89.

Wilson, G.V., Jardine, P.M., Luxmoore, R.J. & Jones, J.R. 1990. Hydrology of a forested watershed during storm events. *Geoderma* 46: 119-138.

Wolfe, N.L. & Macalady, D.L. 1992. New perspectives in aquatic redox chemistry: Abiotic transformations of pollutants in groundwater and sediments. *J. Contam. Hydrol.* 9: 17-34.

Wolfe, N.L., Macalady, D.L, Kitchens, B.E. & Grundl, T.J. 1987. Physical and chemical factors that influence the anearobic degradation of methyl parathion in sediment systems. *Environ. Toxicol. Chem.* 6: 827-837.

Zunino, H. & Martin, J.P. 1977. Metal-binding organic macromolecules in soils. *Soil Sci.* 123: 65-76.

SUGGESTED READINGS

Aquatic Humic Substances: Influence on Fate and Treatment of Pollutants. 1989. I.H. Suffet & P. MacCarthy (eds), Advances in Chemistry Series 219, American Chemical Society, Washington, D.C., 864 pp.

Biogeochemistry: An Analysis of Global Change. 1991. Schlesinger, W.H., Academic Press. San Diego, CA, 443 pp.

Environmental Organic Chemistry. 1993. Schwarzenbach, R.P., Gschwend, P.M. & Imboden, D.M., Wiley-Interscience. New York, NY, 681 pp.

Geochemical Processes: Water and Sediment Environments. (1979) Lerman, A., John Wiley and Sons, New York, 481 pp.

Humic Substances in Soil, Sediment, and Water: Geochemistry, Isolation, and Characterization. 1985. G.A. Aiken, D.M. McKnight & R.L. Wershaw (eds), Wiley-Interscience. New York, NY, 692 pp.

Humic Substances II: In Search of Structure. 1989. M.H.B. Hayes, P. MacCarthy, R.L. Malcolm & R.S. Swift (eds), Wiley-Interscience. Chichester, UK, 764 pp.

Humus Chemistry: Genesis, Composition, Reactions. 1982. Stevenson, F.J., Wiley-Interscience. New York, NY, 443 pp.

Interactions of Soil Minerals with Natural Organics and Microbes. 1986. P.M. Huang & M. Schnitzer (eds), SSSA Special Publication Number 17, Soil Science Society of America, Inc., Madison, WI, 606 pp.

Organic Geochemistry of Natural Waters. 1985. Thurman, E.M., Martinus Nijhoff/Dr. W. Junk Publishers. Dordrecht, 497 pp.

Organic Substances and Sediments in Water. Volume 1: Humics and Soils. 1991. R.A. Baker (ed.), Lewis Publishers Chelsea, MI, 392 pp.

The Geochemistry of Natural Waters. 1988. Drever, J.I., 2nd Ed., Prentice Hall. Englewood Cliffs, N.J., 437 pp.

CHAPTER 11

Relationship between rock, soil and groundwater compositions

JOHN MATHER
Royal Holloway, University of London, Egham, Surrey, UK

11.1 INTRODUCTION

Much of the chemical character of groundwater is established within the soil and the unsaturated zone. This is a region where there are rapid changes in water chemistry. In addition, it provides an important buffer, which protects deeper groundwater from pollutants derived from both agriculture and industry. On a catchment scale groundwater is important in the overall water balance. In upland areas, topographic differences in the landscape mean that groundwater can emerge as the baseflow component of streams and in lowland regions such baseflow is often the main component of rivers in drier summer months.

There are probably few hydrological systems that are not influenced by the activities of man. The presence of tritium derived from thermonuclear tests, trace levels of synthetic organic compounds and acidification in remote mountain areas are clear evidence of this. Geochemical inputs arise from the atmosphere, soil, vegetation and the bedrock and give rise to varying concentrations of different chemical elements and species. In order to understand the impact of such contamination it is necessary to understand the interactions which contribute to the natural chemistry of groundwaters. The objective of this chapter is to examine these interactions with particular reference to how they affect the geochemistry of groundwaters. Finally anthropogenic inputs are briefly reviewed to provide an overview of all the influences and processes involved. A recent treatment on the main factors affecting the composition of natural waters is presented by Bricker & Jones (1995).

11.2 THE SOURCE TERM

The ultimate source of most groundwater is atmospheric deposition, rain and snow melt. Deep groundwater may reflect contributions from additional sources such as fluid inclusions or waters which existed in the aquifer when the materials forming the aquifer were deposited (known as connate waters). The atmospheric or meteoric

305

water infiltrates through the soil and the unsaturated zone before entering the saturated zone.

The composition of this source term is controlled by the dissolution of atmospheric gases particularly carbon dioxide, and by the wash out of components derived from the sea, the land and pollution sources. Thus rain is essentially a dilute solution of carbonic acid and a sea-salt aerosol plus a variable mixture of sulphuric, nitric and hydrochloric acids and usually some ammonium ion. In the UK, the average pH of rainfall varies from 5.0 to 4.4 in the west to 4.3 to 4.1 in the east. This latter figure is comparable to that for rainfall in parts of Eastern Europe and Scandinavia that have been affected by industrial pollution.

In areas near the sea, rainfall has a variable load of major sea-salt components such as Cl, Na and Mg; further inland both the variability and mean concentration of these components decline. The concentrations of most minor elements are higher than the level which can be attributed to a marine origin. This is probably due to the partial dissolution of particles of terrestrial and industrial origin by acidic rain. However, the concentration of dissolved material is generally low and frequently below 20 mg/l. Figure 11.1 shows a typical rainwater with a total dissolved solids of 7.1 mg/l and Figure 11.2 demonstrates how the chloride content of rainwater decreases with increasing distance inland from the coast (from Raiswell et al. 1980). The fact that different influences control the composition of precipitation at inland sites is illustrated by the data from England and Australia in Figure 11.2 which show a secondary increase in chloride at some distance from the coast.

Vegetation can have a significant effect on the composition of rainfall, depending on the type of vegetation, meteorological conditions and even the presence or absence of daylight. Studies with forest trees have shown that there is an increase in the concentration of most solutes in the throughfall and stemflow and an increase by a factor of two or three is typical (Kinniburgh & Edmunds 1984). Some of this may be due to enrichment of solutes by evaporation from the surface of the vegetation, but some, at least, is likely to be derived from the vegetation itself.

It is important to recognise the role of evaporation in concentrating solutes. Thus, to look at it very simplistically, if the chloride content of precipitation is 10 mg/l and 75% of this precipitation is lost by evaporation and transpiration, the chloride concentration in the water interacting with the soil could be increased to 40 mg/l.

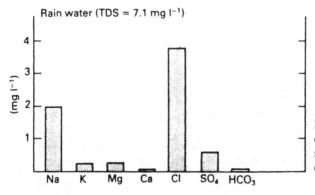

Figure 11.1. Typical composition of rainwater with a total dissolved solids (TDS) of 7.1 mg/l (Raiswell et al. 1980).

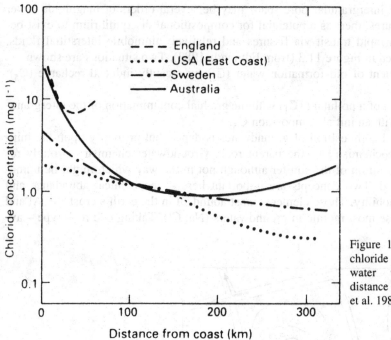

Figure 11.2. Decrease in chloride content of rain-water with increasing distance inland (Raiswell et al. 1980).

11.3 FACTORS AFFECTING GROUNDWATER CHEMISTRY

A number of natural processes can be identified by which groundwater gains its intrinsic properties. Most of the chemical properties of groundwater are attained rapidly, during movement through the soil and the unsaturated zone, and an understanding of the reactions that occur there are of particular importance in hydrogeochemical studies. Although there may be little change in total mineralisation within the saturated zone it is here that exchange reactions and other processes slowly modify groundwater chemistry. It is here also that there is an opportunity for the removal or modification of harmful substances by adsorption, precipitation or oxidation/reduction reactions.

However, before considering the geochemical reactions themselves in more detail it is necessary to consider the physical factors which are important with respect to the chemical composition of groundwater. These are residence time and the pathways or routes along which water moves through the system. A longer residence time provides an opportunity for reactions to occur between water and the materials with which it is in contact, while the pathways along which water moves determine the materials water contacts during its passage. In general, water that follows shallow pathways contacts more weathered and consequently less reactive materials than water moving along deeper pathways.

The flow system is very important as it brings in new reactants and removes reaction products. Flow in fractures is quantitatively more important than intergranular flow in the majority of aquifers. However, fissure storage is commonly only a small fraction of the total storage in most sedimentary aquifers. Because the rate of solute transport in the intergranular pore space may be several orders of magnitude lower than in the fissures, there is a potential for compositional disequilibrium to exist between water in rapid transit via fissures and relatively immobile interstitial fluids. This is illustrated in Figure 11.3 (from Edmunds 1983). Two situations are shown:

a) Displacement of old formation water (C_1) by recently induced recharge (C_5); and

b) Dispersion of a pollutant (C_1) with the gradual contamination of the intergranular pore fluid with an initial composition C_5.

The chemical composition of groundwater is dependent primarily upon the mineralogy and geochemistry of the parent rock. Groundwater chemistry strongly reflects the composition of the aquifer although not in the way which might be immediately supposed. Two concepts are important here – geochemical abundance and geochemical mobility. Those elements most abundant in the earth's crust (Si, Al and Fe) are not those most mobile in groundwaters (Na, Cl). Taking one rock type – av-

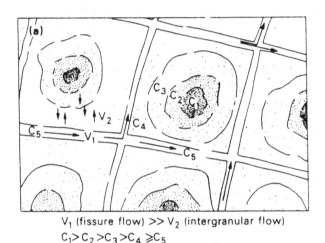

V_1 (fissure flow) >> V_2 (intergranular flow)
$C_1 > C_2 > C_3 > C_4 \geqslant C_5$

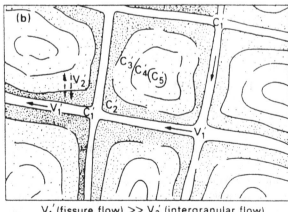

V_1' (fissure flow) >> V_2' (intergranular flow)
$C_1' \geqslant C_2' > C_3' > C_4' > C_5'$

Figure 11.3. Schematic representation of the effects of heterogeneous groundwater flow on water quality distribution in the fissure and intergranular pore spaces of an aquifer (Edmunds 1983). For definition of symbols see text.

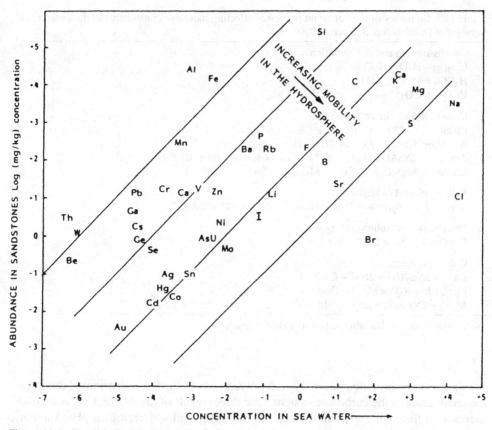

Figure 11.4. Geochemical abundance of the elements, based on their abundance in sandstone compared with sea water (Edmunds 1986).

erage sandstone – and comparing elemental abundance with that in sea water gives an approximate summary of geochemical mobility for the most important elements (Fig. 11.4, from Edmunds 1986).

11.4 REACTIONS IN THE UNSATURATED ZONE

As meteoric waters infiltrate downwards through the soil and the unsaturated zone chemical evolution takes place as they react with the soil and aquifer minerals. The principal processes affecting the inorganic constituents are summarised in Table 11.1 (from Domenico & Schwartz 1990).

11.4.1 *Gas dissolution and redistribution*

The dissolution and redistribution of CO_2 (gas) are very important soil zone processes (see Table 11.1). Rainwater is acidic, has low dissolved solids and rapidly dissolves CO_2 which occurs in the soil as a result of root and microbial respiration and

Table 11.1. Summary of the important processes affecting inorganic constituents in the unsaturated zone (after Domenico & Schwartz 1990).

---

1   Gas dissolution and redistribution
$$CO_2(g) + H_2O = H_2CO_3$$
$$H_2CO_3 = HCO_3^- + H^+$$
$$HCO_3^- = CO_3^{2-} + H^+$$

2   Carbonate and silicate dissolution
Calcite: $CaCO_3(s) + H^+ = Ca^{2+} + HCO_3^-$
Anorthite: $CaAl_2Si_3O_8(s) + 2H^+ + H_2O = kaolinite(s) + Ca^{2+}$
Albite: $2NaAlSi_3O_8(s) + 2H^+ + 5H_2O = kaolinite(s) + 4H_2SiO_3 + 2Na^+$
Enstatite: $MgSiO_3(s) + 2H^+ = Mg^{2+} + H_2SiO_3$

3   Sulfide mineral oxidation
$$4FeS_2(s) + 15O_2(g) + 14H_2O = 4Fe(OH)_3(s) + 16H^+ + 8SO_4^{2-}$$

4   Precipitation-dissolution of gypsum
$$CaSO_4 \cdot 2H_2O(s) = Ca^{2+} + SO_4^{2-} + 2H_2O$$

5   Cation exchange
$$Ca^{2+} + 2Na\text{-}X(s) = 2Na^+ + Ca\text{-}X(s)$$
$$Ca^{2+} + Mg\text{-}X(s) = Mg^{2+} + Ca\text{-}X(s)$$
$$Mg^{2+} + 2Na\text{-}X(s) = 2Na^+ + Mg\text{-}X(s)$$

Note: where Na-X is Na adsorbed on to a clay mineral.

the oxidation of organic matter. This dissolved CO_2 is further redistributed amongst the weak acids of the carbonate system. One direct result of dissolving CO_2 is a rapid increase in the total carbonate content of the water and a decrease in pH. Another important soil zone process is the dissolution of O_2 (gas). The resulting concentrations of dissolved oxygen control the redox chemistry in shallow groundwaters.

11.4.2 *Carbonate and silicate dissolution*

The CO_2-charged water is effective in dissolving minerals. The most common reactions involve the weak acids of the carbonate and silicate systems and strong bases from the dissolution of carbonate, silicate and aluminosilicate minerals (examples are given in Fig. 11.5). The dissolution of calcite is probably the most important reaction in the unsaturated zone and a model involving the dissolution of CO_2 gas, accompanied by calcite dissolution, adequately describes the geochemistry of groundwater in the unsaturated zone in carbonate terrains.

When there are no carbonates present, the ion chemistry will be controlled by the dissolution of silicate and aluminosilicate minerals. The relatively low solubility of these minerals, however, means that the concentration of most ions is low. Figure 11.5 (from Bricker 1993) shows theoretical groundwater compositions resulting from carbonic acid weathering of some common primary silicate minerals to form kaolinite. The compositions are normalised with respect to bicarbonate.

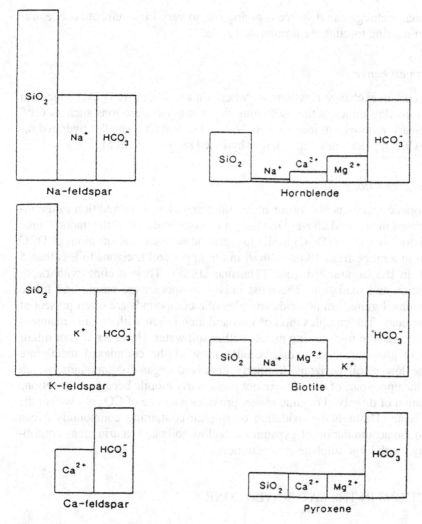

Figure 11.5. Water compositions resulting from carbonic acid weathering of some common primary silicate minerals to form kaolinite (Bricker 1993).

11.4.3 *Sulphide oxidation*

In the unsaturated zone minerals like pyrite or marcasite are oxidized to produce $Fe(OH)_3$(solid). This is one of the most important acid producing reactions in geological systems and is the cause of serious acid drainage problems in mining areas.

11.4.4 *Gypsum precipitation and dissolution*

Cyclical precipitation and dissolution of gypsum can be important in arid areas. If evaporation generally exceeds precipitation, water that infiltrates in normal years evaporates and deposits a small quantity of gypsum. This gypsum may accumulate and exceptional recharge can dissolve it giving rise to very high sulphate concentrations in water moving through the unsaturated zone.

11.4.5 *Cation exchange*

The most important exchange reactions are where Ca and Mg in water exchange with sorbed Na from clay minerals thus softening the water. Divalent ions such as Ca^{2+} and Mg^{2+} replace monovalent ions such as Na^+ and ions with a smaller hydrated radius such as Ca^{2+} replace ones with a larger hydrated radius such as Mg^{2+}.

11.4.6 *Organic reactions*

Important organic reactions also occur in the unsaturated zone in addition to the inorganic reactions mentioned above. Dissolution of organic debris is the major source of dissolved organic carbon (DOC) in shallow groundwaters. Concentrations of DOC typically fall in a range from 10 to 50 mg/l in the upper soil horizons to less than 5 mg/l deeper in the unsaturated zone (Thurman 1985). They decline with depth through sorption and oxidation. The most important species are humic and fulvic acids but tannins, lignins, amino acids and phenolic compounds are often present at low concentrations. The complexation of iron and aluminium with organic matter is an important process in transporting metals within soil water. However, it is of minor significance in groundwater systems, because most of the complexed metals are sorbed in the lower soil horizons. Similarly dissolved organic compounds, which originate in the upper part of the soil, are not particularly mobile because of sorption.

The oxidation of dissolved organic matter provides a source of CO_2 gas within the unsaturated zone. Through the oxidation of sulphur-containing compounds it can also result in the accumulation of gypsum in shallow soils and in arid areas contribute to recharge with large sulphate concentrations.

11.5 REACTIONS IN THE SATURATED ZONE

Processes in the saturated zone are more complex than in the unsaturated zone because of the potential for dispersive mixing and more complicated geological environments. However, most of the same processes are at work and are summarised in Table 11.2 (after Domenico & Schwartz 1990).

11.5.1 *Carbonate and silicate dissolution*

If groundwater is not yet in equilibrium with carbonate, silicate and aluminosilicate minerals, they will continue to dissolve in the saturated zone (e.g. Table 11.2). Basically the groundwater composition will simply proceed towards equilibrium with those minerals available for dissolution. The extent to which equilibrium is established with respect to particular minerals, such as calcite, dolomite and gypsum can be calculated by specific computer programmes. Calculated ion activity products (K_{IAP}) can be compared with theoretical equilibrium constants ($K_{mineral}$) to derive a mineral saturation index $SI_{mineral}$.

Table 11.2. Summary of the important processes affecting ion concentrations in the saturated zone (after Domenico & Schwartz 1990).

---

1   Carbonate and silicate dissolution
    Carbonate minerals + H^+ = cations + HCO_3^-
    Silicate minerals + H^+ = cations + H_2SiO_3
    Alumino-silicate minerals + H^+ = cations + H_2SiO_3 + secondary minerals (for examples, clay minerals)

2   Dissolution of soluble salts, for example,
    Halite: $NaCl(s) = Na^+ + Cl^-$
    Anhydrite: $CaSO_4(s) = Ca^{2+} + SO_4^{2-}$
    Gypsum: $CaSO_4 \cdot 2H_2O(s) = Ca^{2+} + SO_4^{2-} + 2H_2O$
    Carnalite: $KCl \cdot MgCl_2 \cdot 6H_2O(s) = K^+ + Mg^{2+} + 3Cl^- + 6H_2O$
    Kieserite: $MgSO_4 \cdot H_2O(s) = Mg^{2+} + SO_4^{2-} + H_2O$
    Sylvite: $KCl(s) = K^+ + Cl^-$

3   Redox reactions, for example,
    $1/4\ O_2(g) + H^+ + e^- = 1/2\ H_2O$
    $1/2\ Fe_2O_3(s) + 3H^+ + e^- = Fe^{2+} + 3/2\ H_2O$
    $1/2\ MnO_2(s) + 2H^+ + e^- = 1/2\ Mn^{2+} + H_2O$
    $1/8\ SO_4^{2-} + 9/8\ H^+ + e^- = 1/8\ HS^- + 1/2\ H_2O$
    $1/8\ CO_2(g) + H^+ + e^- = 1/8\ CH_4(g) + 1/2\ H_2O$
    $1/4\ CO_2(g) + H^+ + e^- = 1/4\ CH_2O + 1/4\ H_2O$

4   Cation exchange, for example,
    Ca^{2+} Ca
    $Mg^{2+} + 2Na\text{-}clay(s) = 2Na^+ + Mg\text{-}clay(s)$
    Fe^{2+} Fe

The dissolution of calcite provides a good illustrative example. Groundwater is saturated with respect to calcite if $SI_{calcite} = \log K_{IAP}/\log K_{calcite}$ is positive and undersaturated if it is negative. Figure 11.6 shows data from a suite of groundwaters from the Triassic Sandstones of the English East Midlands where the degree of apparent saturation with respect to the carbonate minerals is related to the various aquifers (Edmunds & Morgan-Jones 1976). Most groundwaters from the Keuper Sandstone, which has a relatively abundant carbonate cement, show calcite supersaturation and a few are saturated with respect to dolomite in contrast to groundwaters from the underlying carbonate-poor Bunter Sandstone. The extent of attainment of a particular mineral equilibrium will indicate whether solution or precipitation is occurring in the aquifer which is relevant to problems such as encrustation in boreholes.

11.5.2 *Dissolution of soluble salts*

When groundwater encounters soluble salts in the saturated zone the impact on groundwater chemistry can be considerable. Common dissolution reactions are given in Table 11.2. Where active groundwater flow encounters evaporites, these can dissolve to produce a brine whose composition depends on the particular minerals present. For example in Cumbria (UK), where a major hydrogeological investigation of a potential radioactive waste disposal site is currently underway (UK Nirex 1993),

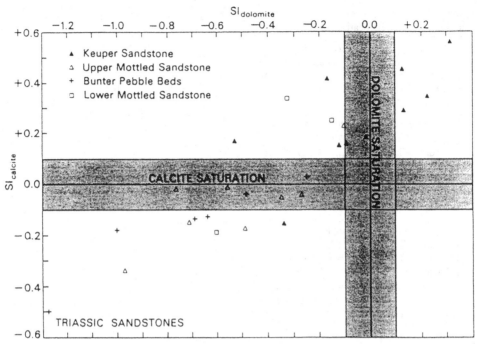

Figure 11.6. Plot showing the extent of attainment of equilibrium with respect to calcite and dolomite by groundwaters in the various Triassic aquifers of the English West Midlands (Edmunds & Morgan-Jones 1976).

high salinities, up to three times that of seawater, are thought to be derived from the dissolution of halite associated with Permo-Triassic evaporites.

11.5.3 *Redox reactions*

Redox reactions encountered along a flow system are important in controlling the hydrochemistry of metals such as iron and manganese and carbon, nitrogen and sulphur species. It is often possible to define redox zones in a flow system and an idealized example based on a limestone aquifer is given in Figure 11.7.

It is possible that many redox reactions are microbiologically mediated. In most groundwater flow systems bacteria are likely to be present, although exact conditions for their active participation may not be present. Thus to demonstrate the presence of sulphate-reducing bacteria is one thing, but to show that they are actively reducing sulphate is another (Edmunds 1986).

11.5.4 *Cation exchange*

The most important reactions are those which take Ca^{2+} and Mg^{2+} out of solution and replace them with Na^+. The main requirement is a large reservoir of exchangeable Na which is usually provided by clay minerals deposited in a marine environment. However, nearly all minerals have measurable exchange capacities as a result

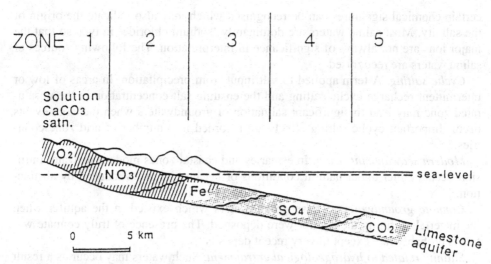

ZONE 1 2 3 4 5

Solution CaCO3 satn.

Zone 1. Oxygen reduction; nitrate present; no iron or $H_2S(g)$ in solution.
Zone 2. Nitrate reduction; no oxygen, iron or $H_2S(g)$ in solution.
Zone 3. Iron reduction; no oxygen, nitrate or $H_2S(g)$ in solution; iron present in solution.
Zone 4. Sulphate reduction; no oxygen or nitrate in solution; iron precipitated as FeS_2; $H_2S(g)$ in solution.
Zone 5. $CO_2(g)$ reduction/decarboxylation; methane ($CH_4(g)$) in solution.
Figure 11.7. Redox zonation, typical of a limestone aquifer, as groundwater moves from outcrop into confined conditions.

of crystal imperfections or impurities. For most conditions an affinity series may be defined where an ion with a larger hydrated radius will tend to be displaced from an adsorption site by an ion with a smaller hydrated radius:

$$Cs^+ > Rb^+ > K^+ > Na^+ > NH_4^+ > Li^+ \text{ (monovalent)}$$

$$Ba^{2+} > Sr^{2+} > Ca^{2+} > Mg^{2+} \text{ (divalent)}$$

Thus K^+ has a tendency to displace Li^+ from adsorption sites and so on (Edmunds 1986). In a limestone aquifer one might expect the Mg^{2+}/Ca^{2+} ratio to increase along flow paths as Ca^{2+} displaces Mg^{2+} and for the Sr^{2+}/Ca^{2+} ratio to decrease. However, other factors can complicate the geochemistry and for example in the English Chalk both ratios increase, because of a gradual recrystallisation of calcite resulting from the adaption of a marine rock to a freshwater environment (Edmunds et al. 1987). In this process, the impurities in the Chalk are gradually released, while a fractionally purer low-magnesian calcite is precipitated.

11.6 SALINE GROUNDWATERS

Brackish and saline groundwaters frequently occur in hydraulic continuity with fresh groundwaters and can cause considerable constraints on the exploitation of the fresh water. Because of the various ways in which salinity is imparted to a groundwater, certain chemical signatures can be recognised which may also indicate the origin of the salinity. Most saline waters are dominantly 'sodium chloride' in type so that the major ions are not always of significance in interpretation. The following sources of saline waters are recognised.

Cyclic salting. A term applied to salt input from precipitation. In areas of low or intermittent recharge cyclic salting and the ensuing salt concentration in the unsaturated zone may lead to significant salination of groundwaters when recharge events occur. Important cyclic salting effects are recorded in a number of arid zone countries.

Modern seawater intrusion. In estuaries and coastal zones modern seawater intrusion occurs either under natural conditions or because of flows induced by abstraction.

Connate groundwaters. Those groundwaters which existed in the aquifer when the materials forming the aquifer were deposited. The presence of truly connate waters is probably rare except in very recent deposits.

Salinity related to hydrogeological entrapment. Such waters may occur as a result of the poor flushing of groundwater systems where saline waters have been entrapped as a result of some geological event such as glacial activity.

Diffusion salinity. Undoubtedly diffusion occurs in transition zones between fresh and saline groundwaters. However, flow-mixing is likely to be dominant and such diffusion effects are not easily recognisable.

11.7 GROUNDWATER COMPOSITIONS

Having examined the processes which contribute to the ultimate composition of a groundwater it is now of interest to examine the compositions of groundwaters which result in both carbonate and non-carbonate aquifers. In the UK compositions for a range of aquifers have been discussed by Edmunds et al. (1989). One of the areas sampled was the Peak District of Derbyshire in the English Midlands. Here an anticline exposes a limestone inlier (Carboniferous Limestone) beneath shales and sandstones (Millstone Grit). Groundwaters from the two formations have very different geochemical characteristics, which are shown in Figure 11.8.

The absence of carbonate in the Millstone Grit is reflected in the groundwaters with pHs in the range 4.7 to 6.9 and alkalinities of < 1 to 88 denoting that there is relatively little acid neutralising capacity. They are well below calcite saturation, with $SI_{calcite}$ between -1.6 and -5.3. In contrast the limestone groundwaters are alkaline and saturated with calcite. The geochemical environment of the Millstone Grit therefore favours the mobilization of metals, whereas the more alkaline conditions in the limestone should not give rise to high concentrations. Figure 11.8 shows the much higher values of Fe, Mn, Al and other metals occurring in Millstone Grit groundwaters.

Perhaps the most studied aquifer in the UK is the Chalk and distinct evolutionary changes can be recognised in the chemistry of its groundwaters as it moves from the outcrop beneath confining deposits. A comparison of the chemical composition of rainfall and groundwater in the outcrop groundwaters is shown in Table 11.3. Significant amounts of Na^+, Mg^{2+}, K^+, SO_4^{2-} Cl^- and NO_3^- in the infiltrating water may

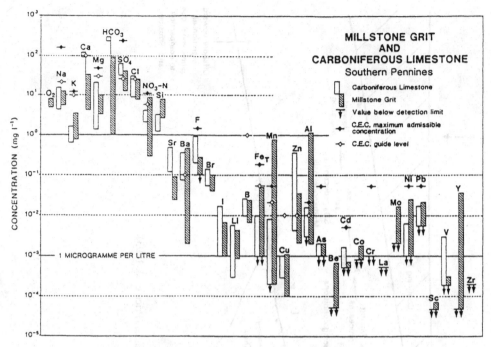

Figure 11.8. Element concentrations in the Millstone Grit and the Carboniferous Limestone in order of approximately decreasing abundance (Edmunds et al. 1989).

Table 11.3. Comparison of the chemical composition of rainfall and groundwater in the Chalk, in mg/l (after Price et al. 1993).

	Rainfall	Composition of the infiltrate*	Composition of groundwater at outcrop
Calcium	1.4	3.9	100
Magnesium	0.3	0.8	2
Sodium	1.8	5.0	10
Potassium	0.3	0.8	1
Bicarbonate	–	–	280
Sulphate	6.3	17.6	15
Chloride	3.1	8.7	15
Nitrate (as N)	0.4	11.2	20
Ammonia (as N)	0.5	–	–

* Following concentration by evaporation and transpiration

be derived from rainfall but some of the NO_3^-, SO_4^{2-}, and K^+, are probably derived from fertilizers (Price et al. 1993).

The sequence of chemical changes as the groundwater moves downgradient are shown in Figure 11.9. The dominant change is the conversion of the hard 'Ca-HCO$_3$'- or 'Ca-Mg-HCO$_3$'-type water at outcrop to a soft 'Na-HCO$_3$'-type water in the confined aquifer. At increasing distance downgradient mixing gradually

318   *John Mather*

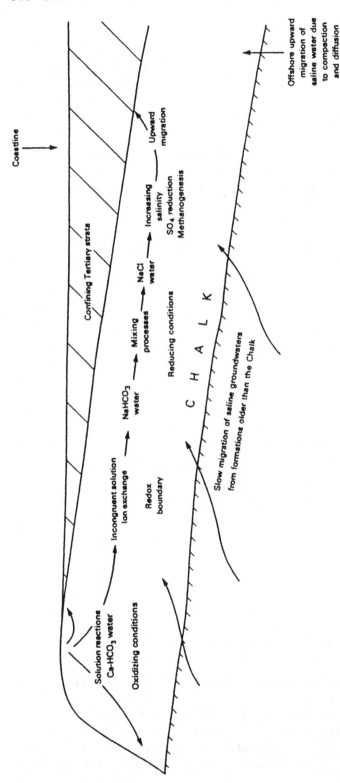

Figure 11.9. Schematic section of downgradient chemical changes in Chalk groundwaters (Price et al. 1993).

occurs with a 'Na-Cl' water that probably has a marine origin. A redox boundary is present within the confined zone and dissolved iron increases significantly in the aquifer following the onset of reducing conditions (Price et al. 1993).

The chemical hydrogeology of the Floridan limestone aquifer in the southeastern USA is also well known (Back & Hanshaw 1970 reviewed by Fetter 1994). As groundwater travels down the flow path from the recharge area it increases in total dissolved solids from 138 to 726 mg/l. All ions except bicarbonate show a progressive increase along the flowpath and computation of ion-activity products from the analyses show that both dolomite and calcite saturation increase. A trilinear plot of borehole samples shows that the chemical composition of the groundwater is changing along the flow path (Fig.11.10). This is a result of an increase in the Mg^{2+}/Ca^{2+} and the SO_4^{2-}/HCO_3^- ratios with increasing distance from the recharge area. The change in these ratios is due to the solution of gypsum ($CaSO_4 \cdot 2H_2O$) and dolomite ($CaMg(CO_3)_2$) along the flow path.

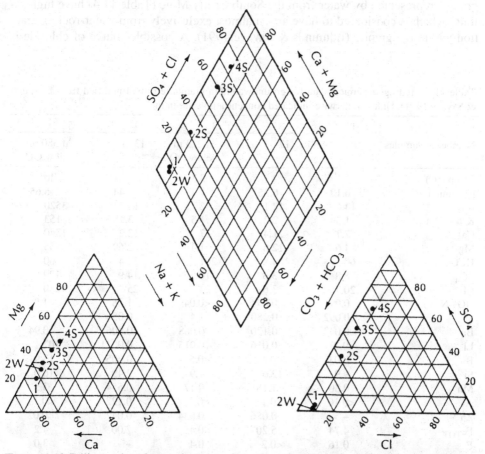

Figure 11.10 Trilinear plot of groundwater analyses from the Floridan aquifer of central Florida. Borehole 1 is at the southern edge of the recharge area with boreholes 2W, 2S, 3S and 4S at progressively greater distances along the flow path. Site 4S is about 100 km from site 1 (Back & Hanshaw 1970; Fetter 1994).

As much of the Norwegian landmass is classified as basement rocks it is important to briefly examine the geochemistry of groundwaters which are likely to occur. In terms of bulk composition, there may be little to distinguish such basement rocks from younger cover and indeed granitic basement is likely to be very similar in composition to many sandstones such as the Millstone Grit discussed above (Edmunds & Savage 1991).

Analyses from the Strath Halladale Granite in Northern Scotland are given in Table 11.4. Quite high alkalinities are found in the groundwaters which are related to the presence of secondary vein carbonates which, although a minor phase, dominate the hydrogeochemistry. Such dissolved carbonate derived from vein calcite is likely to be a common characteristic of groundwaters from basement terrains. In the shallower environments of the springs, the hydrogeological pathways are in weathered zones, decalcification has occurred and more acid waters are present.

In other areas, such as the Carnmenellis Granite in Cornwall, shallow groundwaters and baseflow have much lower pHs (Table 11.4). Deeper groundwaters in this granite, represented by water from the South Crofty Mine (Table 11.4) have high salinity which is considered to have arisen almost exclusively from water-rock interaction within the granite (Edmunds & Savage 1991). A possible source of chloride is

Table 11.4. Hydrogeochemical data for groundwaters in granite terrains (modified from Edmunds & Savage 1991). Unless otherwise indicated concentrations are in mg/l.

	1	2	3	4	5
Number of samples	16	17	At 223 m below OD	13	At 580 m below OD
Temp (°C)	–	–	7.3	–	40
pH (units)	6.19	6.35	7.4	5.44	6.65
Na	14.8	17.1	32	14.7	3520
K	1.38	2.2	1.5	3.18	153
Ca	7.7	21.4	37	12.2	1840
Mg	4.6	12.6	1.4	2.69	55
HCO_3	60	164	201	32.4	60
SO_4	8.8	3.6	10	14.9	129
Cl	20.7	23.6	24	25	9280
NO_3-N	0.09	0.09	<0.04	1.28	<1.0
Sr	0.092	0.280	1.4	0.056	30
Ba	0.043	0.120	0.068	0.006	0.94
Li	<0.2	0.016	0.015	0.0011	107
B	0.03	<0.5	<0.5	0.005	13.9
Si	9.9	12.6	7.9	2.49	16.0
Mn	0.30	1.18	0.17	0.554	3.5
Al	<0.10	<0.1	<0.1	0.122	<0.1
Zn	–	0.086	0.034	0.047	<0.02
Fe_{TOT}	4.74	5.20	0.45	2.00	4.2
F	0.16	<0.2	0.4	–	3.0
Total mineralisation	133	264	332	112	15250

1. Strath Halladale Springs, 2. Strath Halladale Shallow Boreholes, 3. Strath Halladale Deep Borehole, 4. Carnmenellis, 5. South Crofty Mine.

biotite, which in Cornwall may contain up to 1% chloride. Other suggested origins for high salinity in granite groundwaters include migration of sedimentary brines, marine transgressions, residual hydrothermal fluids and breakdown of fluid inclusions.

11.8 ANTHROPOGENIC INFLUENCES

No discussion of groundwater geochemistry would be complete without consideration of the major impacts which urban and industrial developments and modern agricultural practices are having. Incidents can be divided into two broad types on the basis of the source of pollutants. Point-source pollution occurs where pollutants are derived from a discrete, and often readily identifiable source such as a factory, leaking tank or waste disposal site. Diffuse pollution occurs where pollutant input is widely distributed and cannot be pin-pointed precisely. Examples are the widespread application of pesticides and fertilisers to arable land, which may involve the whole outcrop of the aquifer, and acid rain, which may be distributed on a regional or even national scale. In some urban areas, pollutants may leak into groundwater via so many discrete sources that the effect is identical to that of a diffuse pollutant. The major sources of groundwater pollution are outlined in Table 11.5 and the most important are briefly discussed in this section.

11.8.1 *Agricultural pollution*

Over the past few decades there has been a continuous increase in agricultural pro-

Table 11.5. Major sources of groundwater pollution (from Mather 1992).

Source of pollution	Type	Principal contaminants causing concern
Agricultural practices – including the spreading of sewage sludges on land	Diffuse	Nitrates, pesticides and herbicides, biological contaminants
Industry:		
a) Landfill	Point source	Range of organic and inorganic contaminants
b) Industrial development particularly associated with the engineering and chemical industries	Multi-point source	Hydrocarbons, chlorinated solvents
c) Mining and resource exploitation	Point source and diffuse	Chloride, sulphate, trace metals
d) Old gasworks	Point source	Phenol
Casual effluent spillages and discharges, including those from septic tanks	Point source	Range of contaminants depending on the source of the spillage
Soakaway drainage from paved areas, roads and airfields	Point source	Hydrocarbons, chloride
Acid rain	Diffuse	Trace metals

ductivity and production. One of the factors contributing to this has been the increasing use of fertilizer nitrogen. During this period the concentration of nitrate in groundwater has increased sharply and a correlation has been made between the two events (e.g. Foster & Crease 1974). Despite many years of research it is still not clear if groundwater nitrate increases are the result of the leaching of excess fertilizer or because fertilizers have built up the amount of organic matter that microbes can break down to produce nitrate. In the United Kingdom after 1940, substantial areas of old grassland were ploughed, stimulating the microbial population to break down the accumulated organic matter and release nitrate. Other sources of nitrogen include animal manures and slurries and digested sewage which are spread on farmland.

Steadily rising nitrate concentrations are now a feature of many unconfined aquifers and modelling studies predict that they will continue to increase for some time to come. For example in England and France well over 2 million people have been supplied with water from the Chalk containing more than 50 mg/l NO_3^- for periods of many years (Foster 1993). Of importance from the point of view of this chapter are the processes which might attenuate nitrate concentrations. In situ denitrification within the unsaturated zone seems rather unlikely, but there is evidence that denitrification occurs in confined conditions (Mariotti et al. 1988).

There is no doubt that pesticides and herbicides are a potential threat to groundwaters. According to Foster et al. (1991) concentrations in water-supply boreholes are unlikely to approach equilibrium with current pesticide leaching for many years. However, only time and further research will show whether or not impacts similar to those for nitrate will be produced.

11.8.2 *Industrial pollution*

The potential of landfill sites to pollute groundwater is well documented. The main components in the leachates which arise from such sites are major elements or species such as calcium, magnesium, iron, sodium, ammonium, carbonate, sulphate and chloride; trace metals such as manganese, chromium, nickel, lead and cadmium; a wide variety of organic components, measured as Total Organic Carbon (TOC) or Chemical Oxygen Demand (COD) and microbiological components.

Household waste is reasonably consistent in composition, as is the resulting leachate. Leachate compositions at sites which accept industrial wastes are likely to be much more variable and in general terms leachate composition will reflect the composition of the waste. However, it has been common practice in the UK to dispose of both household and industrial wastes, particularly those described as difficult wastes, at codisposal sites. In many cases the leachate obtained is comparable to that from mature household refuse and it is probable that landfill management and design is as important in determining the composition of leachate as the type of waste deposited (Johanson & Carlson 1976; Knox 1990). As a result of legislation, the management of landfill sites is now moving towards a total containment option where liners are designed to contain leachates and cappings are designed to reduce infiltration and the formation of leachates. However, many existing landfills still operate using the principle of dilute and disperse where leachates are allowed to migrate in the expectation that natural processes will attenuate leachate components before they reach groundwater abstraction zones.

It is clear from work carried out over many years, that landfills produce a highly polluting leachate which can have a significant effect on groundwater. However, it is also apparent that many geological environments can be very effective in neutralising and cleansing leachates by geochemical processes. Carbonate aquifers in particular are likely to be much more effective than many sandstone aquifers. This is because of their ability to buffer acidity to a neutral pH, precipitating heavy metals and providing an ideal environment for the development of bacterial populations, which attenuate the total organic carbon load (Mather 1989).

Land contaminated by industrial development gives rise to a range of potential sources of groundwater pollution. These include the disposal of wastes on site, leachate from made ground, leakage from sewers and storage tanks and accidental spillages. Pollution arises either from point sources or, in heavily industrialised areas, from a multitude of point sources such that the whole of an aquifer is subject to contamination by diffuse pollution whose exact source is impossible to identify.

Probably the most common type of incident is spillage of hydrocarbons. As these have a lower density than water they are concentrated at the water table and the capillary fringe and can migrate as a separate liquid phase or dissolved in groundwater. Even more of a problem than these light non-aqueous phase liquids (LNAPLs) are their denser cousins, DNAPLs. These dense non-aqueous phase liquids will move through the unsaturated zone, but will continue to move vertically through the saturated zone displacing water rather than soil gases. The DNAPLs will move to the base of the permeable unit and will then move downwards following the topography of the basal boundary. This flow direction may be totally different to that of the groundwater. The movement of DNAPLs is difficult to predict in porous environments, but the problems will be compounded in hard rock systems. Trying to predict the depth to which the dense fluid will move will be exceedingly difficult, never mind the direction in which fluids will move once they reach that depth.

The light hydrocarbons are more readily degradable and less persistent than the dense solvents. Their presence stimulates the growth of bacteria and in situ biodegradation appears to be limited by the rate of oxygen supply. The solvents are chemically stable and resistant to biodegradation in subsurface environments. Freeze & Cherry (1979) suggested that at sites where DNAPLs were a problem, the local groundwater has terminal cancer and that a cure in the form of returning the aquifer to drinking water standards was not achievable at any cost.

Another form of industrial pollution is that which arises from the mining industry. Groundwater pollution can arise from the extraction process itself, the processing of the commodity or ore, the disposal of spoil and as a result of the abandonment of the mine. Many areas of igneous and metamorphic rocks have metalliferous mineralisation associated with them and are, or have been, subject to intense mining activity. During operation mines are actively dewatered allowing the access of oxygen. This allows the oxidation of sulphides and other minerals containing metals. Once mining ceases and the workings are allowed to flood, the oxidised phases pass into solution as ferrous sulphate and sulphuric acid. These acid solutions also contain high levels of trace metals and metalloids such as cadmium, mercury, lead and copper.

Once the mine dewatering pumps are turned off, the water table will rise or 'rebound' close to its original natural level. The acid mine waters may be discharged through man-made adits or drainage networks or through natural fracture systems resulting in serious water pollution. For example, at Dalquaharran in Ayrshire, Scotland, pumping ceased in 1977 and surface discharge began in 1979. Groundwater discharge rose to 150 l/s with an initial iron concentration of 1200 mg/l, pH 4, aluminium concentration of 100 mg/l and sulphate 6000 mg/l. This water entered the River Girvan from the entrance of the Dalquaharran no. 1 mine entrance and caused irreparable damage to fish stocks and water users downstream. This and a number of other incidents in Scotland are reviewed by Robins (1990).

More recently, in the UK, serious pollution has arisen following the abandonment of the Wheal Jane Mine in Cornwall. The Mine was the last of a number of tin mines in the area and latterly was dewatering an extensive network of underground workings. Groundwater rebound resulted in acid groundwater, containing high levels of metals, polluting a number of surface water courses. An expensive remediation programme is currently in place to minimise the effects of the acid discharges.

Investigation of such mine water breakouts prior to mine closure is frustrated by the difficulty in predicting the position of discharge points. This can be complicated by changes which may take place in the groundwater flow regime as a result of mining subsidence (Mather et al. 1969).

11.8.3 *Acid rain*

Over the last twenty years there has been widespread concern over the acidification of surface waters in Europe and North America. The pioneering work was carried out in Scandinavia in the 1970's and there has since been much debate over whether acid rain is the major cause or, alternatively, whether the conifer development of moorland is to blame. The scientific debate continues, but there is now strong evidence to implicate acidic oxide emissions as a major causal factor (Neal et al. 1991). As well as surface waters, shallow groundwaters are also at risk, particularly those with acid sensitive lithologies. The presence or absence of calcium carbonate is the single most important factor governing acid susceptibility and hard-rock terrains where the geology consists of non-carbonate igneous and metamorphic rocks are likely to be particularly susceptible. For example, shallow groundwaters in granite rocks in Scotland are generally found to be acidic and groundwaters from metasediments frequently give rise to low alkalinity groundwaters (Edmunds & Kinniburgh 1986).

There are few long term records in existence of water quality in hard rock terrains to demonstrate the effects of acid deposition on shallow systems. Perhaps the best data set is that reported by Hrkal (1992) from the metamorphic rocks and granites of the Bohemian Massif. Bohemia – the western part of the Czech Republic – is one of the most polluted regions of the world. Extensive use of poor quality fossil fuels has led to a reduction in the pH of precipitation and a high rate of atmospheric deposition of nitrogen and sulphur species. Comparison of two groundwater chemistry data sets from the years 1955-1969 and 1980-1990 respectively, shows a five fold increase in the average concentration of NO_3^- and a halving of the average alkalinity. The most acidified groundwaters were found at the highest altitudes, but the most rapid rate of change in the groundwater chemistry was found lower down, where NO_3^- concentrations increased by a factor of ten over the study period.

Changes in Norway and Sweden appear to be less advanced than those in Bohemia and a study in Värmland, in the crystalline bedrock of southern Sweden, suggests that no regional acidification of importance is in progress (Lång & Swedberg 1989). However, results suggest that the groundwaters which are most sensitive to acidification are those where discharge from wells in small bedrock aquifers induces the rapid recharge of acidic groundwater from overlying Quaternary deposits.

11.9 CONCLUSIONS

Each aquifer has a characteristic chemistry which is determined by interaction between meteoric water, soil and rock. As a result of intensive land use in recharge areas as well as point-source and diffuse pollution, this chemistry is often masked. Except in deep environments, it is in fact difficult at the present day to find any aquifer where the influence of man is still undetectable (Edmunds 1986). The chemistry of a selection of groundwaters of various origins are presented in Table 11.6 and compared with rainfall and sea water. The data demonstrate the trend in increasing dissolved solids from precipitation; through shallow groundwaters and potable aquifer waters to deep basinal brines which have dissolved solids contents in excess of seawater.

Table 11.6. Representative groundwater analyses compared with precipitation and sea water. Species above the line are quoted in mg/l, below the line in μg/l. n.d. not determined.

Elements	1	2	3	4	5	6	7	8
Na	3.24	9.3	25	14.8	7.14	14.2	19,700	10,800
K	0.15	1.4	3.5	3.5	0.69	2.0	300	392
Ca	0.21	13.7	145	85	97.9	61.5	4,800	411
Mg	0.41	4.6	5.5	8.4	5.5	21.9	1,600	1,290
Cl^-	6.08	13.0	41	17.7	15.6	26.5	41,840	19,400
HCO_3^-	0	30	253	301	258	232	88	142
SO_4^{2-}	1.85	13.0	94	20.7	34.5	40	<1.0	1,356
NO_3^-	0.80	11	51	*	3.2	3.16	n.d.	n.d.
Si	*	3.0	n.d	8.1	2.2	5.2	n.d.	n.d.
Ba	1.02	7	n.d	51	232	88	425,000	21
Sr	2.63	45	47	2000	160	131	230,000	8,100
B	3.71	120	n.d	56	13	16	n.d.	n.d.
Fe	11.9	<15	n.d	98	1.1	34	570	3
Mn	1.97	7	n.d	3	n.d.	<1	2,900	<1
Zn	8.93	9	n.d	10	69	4	45	5
Br^-	21.0	n.d.	n.d	67	68	80	245,000	67,300

*Wide spread of results. Origin of analyses: 1. Precipitation at Plynlimon, Wales, mean for period May 1983 – May 1984 (Neal et al. 1986). 2. Median values of 37 samples of shallow groundwaters from West Devon (Edmunds et al. 1989). 3. Mean of three analyses of groundwater from alluvial gravels in the Thames Valley (Edmunds 1986). 4. Median values for 22 samples of Chalk groundwater from Berkshire, England (Edmunds et al. 1989) 5. Median values for nine samples of Carboniferous Limestone groundwaters from Derbyshire, England (Edmunds et al. 1989) 6. Median values for sixteen samples of Permo-Triassic sandstone groundwaters from northwest England (Edmunds et al. 1989) 7. Brine from Coal measures, northeast England (Edmunds 1975). 8. Typical sea water.

Although groundwaters may become contaminated, certain geochemical characteristics are still likely to be determined by reaction with the hot rock. Thus the
Ca^{2+}/Mg^{2+} ratio, and concentrations of Sr^{2+} and Ba^{2+} will be geochemically controlled. Other species such as Cl^- and F^-, may be dominated by atmospheric inputs
and still others, including SO_4^{2-}, NO_3^-, K^+ and H_3BO_3 may be influenced more by
anthropogenic activity. The hydrochemistry will also change across the aquifer due
to sequential changes resulting from oxidation/reduction reactions, ion exchange etc.
Seasonal fluctuations may occur and longer term changes resulting from drought or
overabstraction. Geochemical changes may occur suddenly as a result of a rapid discharge of contaminant or gradually as a result of the incipient leakage of a diffuse
pollutant.

It can be seen from this chapter that the composition of any groundwater arises
through a complex interaction of factors. The ultimate composition is not always directly predictable as the water evolves along intergranular and fracture pathways
taking part in both inorganic and organic reactions. These reactions can lead to the
introduction of new species by dissolution or their removal by precipitation or ion
exchange. An understanding of these reactions enables the geochemistry of a groundwater to be unravelled in terms of mixing and water/rock or water/soil interactions.

11.10 FUTURE RESEARCH

Over the past two decades major advances have been made in our understanding of
the hydrogeochemistry of groundwater systems. In shallow environments this has
developed in response to the need to assess the impact of agricultural pollution in
lowland areas and the effects of acidic precipitation in upland terrains. In deep environments, research into the development of geothermal energy and the disposal of
radioactive waste have led to the development of extensive hydrochemical databases.
Despite these advances, the means of measuring, quantifying and modelling the flow
of water from the atmosphere, through the vegetation, and then through the soil, unsaturated and saturated zones to its discharge point is still in its infancy. Even with a
simple species such as chloride there is difficulty in obtaining an input/output budget
for even a small permeable catchment never mind a major groundwater basin. As a
further example, the origin of chloride-rich waters in granites is still the subject of
much discussion and several origins have been suggested. These include migration of
sedimentary formation brines, marine transgressions, residual hydrothermal fluids,
breakdown of fluid inclusions, radiolytic decomposition of water and water/rock interaction (Edmunds & Savage 1991).

In the future, it will be necessary to develop predictive as well as interpretive
models in order to describe the evolution of groundwaters and their response to anthropogenic inputs. This has been done already to predict the rise in nitrate concentrations resulting from the application of nitrate fertilisers to arable land, but should
be applied to many other species.

One could argue that hydrogeologists have established a significant database concerning the range of composition of groundwaters in various rock types and geological environments. What is now required is a better understanding of basic processes
including:

- The role of micro-organisms;
- Transport of species in colloidal form;
- In situ oxidation/reduction conditions;
- Characterisation of ion-exchange sites;
- The importance of the soil as a buffer zone for trace metals.

There remains much work to be done in this relatively new field in order to quantify the relationship between rock, soil and groundwater compositions.

REFERENCES

Back, W. & Hanshaw, B.B. 1970. Comparison of the chemical hydrogeology of the carbonate peninsulas of Florida and Yucatan. *Journal of Hydrology* 10: 330-68.

Bricker, O.P. 1993. The geochemistry of groundwaters in fractured rocks: a geologic perspective. Paper presented at 24th Congress, *International Association of Hydrogeologists, Ås, Norway* June 1993.

Bricker, O.P. & Jones, B.F. 1995. Main Factors Affecting the Composition of Natural Waters. In: Salbu, B. & Steinnes, E. (eds), *Trace Elements in Natural Waters*, CRC Press, Boca Raton, 1-20.

Domenico, P.A. & Schwartz, F.W. 1990. *Physical and Chemical Hydrogeology.* J. Wiley & Sons Inc., International Edition.

Edmunds, W.H. 1975. Geochemistry of brines in the Coal Measures of northeast England. *Trans. Instn. Min. Metall.* (Sect. B: Appl. earth sci) 84: B39-52.

Edmunds, W.M. 1983. Hydrogeochemical investigations. In: Lloyd, J.W. (ed.), *Case studies in groundwater resource evaluation.* Oxford University Press Chapter 6: 87-112.

Edmunds, W.M. 1986. Groundwater Chemistry. In: T.W. Brandon (ed.), *Groundwater: occurrence, development and protection*, Chapter 3, Inst. Water Eng. Sci., Water Practice Manual 5: 49-107.

Edmunds, W.M., Cook, J.M., Darling, W.G., Kinniburgh, D.G., Miles, D.L., Bath, A.H., Morgan-Jones, M. & Andrews, J.N. 1987. Baseline geochemical conditions in the Chalk aquifer, Berkshire, UK: A basis for groundwater quality management. *Applied Geochemistry* 2: 251-274.

Edmunds, W.M., Cook, J.M., Kinniburgh, D.G., Miles, D.L. & Trafford, J.M. 1989. Trace-element occurrence in British groundwaters. *British Geological Survey Research Report* SD/89/3.

Edmunds, W.M. & Kinniburgh, D.G. 1986. The susceptibility of UK groundwaters to acidic deposition. *Journal of the Geological Society of London* 143: 707-20.

Edmunds, W.M. & Morgan-Jones, M. 1976. Geochemistry of groundwaters in British Triassic Sandstones. The Wolverhampton – East Shropshire area. *Quarterly Journal of Engineering Geology* 9: 73-101.

Edmunds, W.M. & Savage, D. 1991. Geochemical characteristics of groundwater in granites and related crystalline rocks. In: R.A. Downing & W.B. Wilkinson (eds), *Applied Groundwater Hydrology.* Clarendon Press, Oxford, 266-282.

Fetter, C.W. 1994. *Applied Hydrogeology 3rd Ed.* Macmillan, New York.

Foster, S.S.D. 1993. The Chalk aquifer – its vulnerability to pollution. In: Downing, R.A., Price, M. & Jones, G.P. (eds), *The Hydrogeology of the Chalk of North-West Europe.* Oxford Science Publications, 91-112.

Foster, S.S.D., Chilton, P.J. & Stuart, M.E. 1991. Mechanisms of groundwater pollution by pesticides. *Journal of the Institution of Water Environmental Management* 5: 186-93.

Foster, S.S.D. & Crease, R.I. 1974. Nitrate pollution of Chalk groundwater in East Yorkshire – a hydrogeological appraisal. *Journal of the Institution of Water Engineers* 28: 178-94.

Freeze, R.A. & Cherry, J.A. 1979. *Groundwater.* Prentice Hall, New York.

Hrkal, Z. 1992. Acidification of groundwater in the Bohemian Massif. Nor. geol. unders. *Bull.* 422: 97-102.

Johansen, O.J. & Carlson, D.A. 1976. Characterisation of sanitary landfill leachates. *Water Research* 10: 1129-34.

Kinniburgh, D.G. & Edmunds, W.M. 1984. The susceptibility of UK groundwaters to acid deposition. Report to the Department of the Environment by the British Geological Survey, 211 pp.

Knox, K. 1990. A review of co-disposal. *Proc.1990 Harwell Waste Management Symposium – The 1980's A decade of progress?* 54-76.

Lång, L.O. & Swedberg, S. 1989. Occurrence of acidic groundwater in Precambrian crystalline bedrock aquifers, Southwestern Sweden. In: Swedberg, S. (ed.), *Groundwater acidification in Southwestern Sweden; long-term changes in groundwater chemistry.* Geologiska Inst., Chalmers Tekniska Högskola/Göteborgs Universitet, Publ. A67, Göteborg, Sweden.

Marriotti, A., Landreau, A. & Simon, B. 1988. Isotope biogeochemistry and natural denitrification process in groundwater; application to the Chalk aquifer of northern France. *Geochimica et Cosmochimica Acta* 52: 1869-78.

Mather, J.D. 1989. The attenuation of the organic component of landfill leachate in the unsaturated zone: a review. *Quarterly Journal of Engineering Geology* 22: 241-6.

Mather, J.D. 1992. The pollution of groundwater by diffuse and point source contaminants. *Teaching Earth Sciences* 17(1): 3-12.

Mather, J.D., Gray, D.A. & Jenkins, D.G. 1969. The use of tracers to investigate the relationship between mining subsidence and groundwater occurrence at Aberfan, South Wales. *Journal of Hydrology* 9: 136-54.

Neal, C., Kinniburgh, D.G. & Whitehead, P.G. 1991. Shallow groundwater systems. In: R.A. Downing & W.B. Wilkinson (eds), *Applied Groundwater Hydrology – a British Perspective.* Clarendon Press, Oxford, 77-95.

Neal, C., Smith, C.J., Walls, J. & Dunn, C.S. 1986. Major, minor and trace element mobility in the acidic upland forested catchment of the upper River Severn, Mid Wales, *Journal of the Geological Society, London* 143, 635-48.

Price, M., Downing, R.A. & Edmunds, W.M. 1993. The Chalk as an aquifer. In: Downing, R.A., Price, M. & Jones, G.P. (eds), *The Hydrogeology of the Chalk of North-West Europe.* Oxford Science Publications 33-58.

Raiswell, R.W., Brimblecombe, P, Dent, D.L. & Liss, P.S. 1980. *Environmental Chemistry – The Earth-Air-Water Factory.* Edward Arnold, London.

Robins, N.S. 1990. *Hydrogeology of Scotland.* HMSO, London, 90pp.

Thurman, E.M. 1985. *Organic geochemistry of natural waters.* Martinus Nijhoff/Dr. W. Junk, Dordrecht, 497 pp.

UK Nirex Ltd. 1993. *The geology and hydrogeology of the Sellafield area.* Vol. 3. The Hydrogeology. Nirex Report No. 524.

CHAPTER 12

Towards coupling hydrological, soil and weathering processes within a modelling perspective

COLIN NEAL & ALICE J. ROBSON
Institute of Hydrology, Maclean Building, Crowmarsh Gifford, Wallingford, Oxon, UK

NILS CHRISTOPHERSEN
Department of Informatics, University of Oslo, Blindern, Oslo, Norway

12.1 INTRODUCTION

12.1.1 *General*

Computer simulation models increasingly have been relied upon to describe a variety of environmental impacts over the past few decades. These have ranged from limnological models relating nutrient input to eutrophication of freshwaters, to catchment models describing the impact of acid deposition on soil and surface water chemistry. Regional acidification models, such as the RAINS model (Hordijk 1991), moved to the continental scale and related not just deposition, but energy-use scenarios to acid deposition effects. A further increase in scale is occurring with the extension of general circulation models, which operate at the global level, to assess the effects of greenhouse gases on climate.

In this chapter, the use of models to depict changes at the catchment scale is examined in the context of geochemical processes, weathering and recharge in catchments. The theme of how to couple hydrological, soil and weathering processes within a modelling perspective, to provide both a scientific and an applied goal, is examined. Examples are provided of successes and failures over the past decade from stream and catchment acidification research. To do so, some of the main findings of our own and our colleagues' endeavours in Europe are drawn upon to provide a coherent theme for the presentation and then to link this to associated work. The aim is to show how strong the link must be between observation and modelling and the need for scientific objectivity. For the presentation, the field-based studies considered are centred at Plynlimon in mid-Wales, as the results are characteristic of many acidic and acid-sensitive sites subjected to acidic deposition. Also, the Plynlimon catchments are specifically referred to as they relate to both acidic deposition and conifer afforestation/deforestation – in Scandinavia there has been much debate over what the relative effects have been, while in Britain, both issues are of definite importance. Reference will also be given to the detailed studies at the Birkenes catchment in southern Norway. For the final part of the work, relating to modelling and to what is and what is not achievable, the views expressed are those given by Christophersen et al. (1993).

12.1.2 *Acidic deposition*

From the late 1960's, concerns were raised over acidic deposition of industrially emitted acidic oxides (SO_x in particular) adversely affecting stream and lake ecology in Scandinavia. The environmentally acid sensitive areas are those that possess acidic soils. Indeed, the main problem of acidification rests with the lack of weathering reactions of primary rock minerals, to release base cations and promote bicarbonate production from biogenic CO_2 within the soils, to neutralise the acidic inputs. The concerns generated heated political debate across Europe, which led to considerable scientific research to characterise the nature of the acidification process and to identify potential remedial treatments. Even from the onset, there was much debate over the relative importance of acidic deposition and changing land use. Nowadays, the importance of atmospheric pollution is generally recognised on a continental scale, with, for example, acidification problems being identified across Europe and North America. For Britain, the relative importance of acidic deposition and land use change remains a major issue, and both aspects will be covered here as the UK has a particular problem associated with both aspects. For example, forest development has been the most important land use change in the British uplands this century. A third of the UK is comprised of upland soils that are often acidic and acid sensitive, and pollutant deposition levels are as high as those in southern Scandinavia, where acidic deposition has clearly lead to environmental problems.

12.1.3 *Acidification in the British uplands*

Considerable evidence has been accumulated showing that acidification of the UK upland environment has occurred. The first phase of soil acidification probably started several to ten thousand years ago with the development and subsequent loss of birch, hazel, alder and oak forests (Dimbleby 1952; Taylor 1974; Pennington 1984). The deforestation occurred during Neolithic and Bronze age times due to a deteriorating micro-climate about 2700 years ago and local deforestation by man from 5000 years ago up to present day. As a consequence of this, thin acidic moorland soils developed and these are characteristic of much of the British uplands today. In waterlogged areas, where reducing conditions ensured limited breakdown of organic matter, very acidic conditions occurred and peat deposits accumulated. Set against this long term change in the upland environment, further acidification has occurred during the past hundred years and this has an adverse effect on upland ecology.

Major concern has been raised over the deterioration of stream water quality in the British uplands, as associated with this second and ecologically potent phase of acidification. Acidic oxide deposition and conifer planting have been implicated as the major factors in this decline (Stoner & Gee 1985; UKAWRG 1988; Whitehead et al. 1988, Jenkins et al. 1990). Both may lead to the generation of more acidic and aluminium bearing stream water and conditions unhealthy for stream biota. However, it is widely believed that, for the UK, as elsewhere, the driving force is the former: conifers enhance the capture of acid pollutants, thereby increasing their impacts; acidification has occurred even for the most extensive tracts of the uplands which had no tree cover. Nonetheless, the relative importance and the degree of interaction of the two factors still needs to be clarified (Rosenqvist 1978, 1990; Miller 1985; Hornung et al. 1989; Jenkins et al. 1990; Nisbet 1990; Krug 1991): for example, under pristine conditions there is evidence both for and against the deleterious effects to stream water of conifer development. In a modelling study of afforestation, Jenkins et al. (1990) comment that 'afforestation in the absence of acidic deposition, has had a lesser effect on surface water acidification even though the nutrient demands of forest growth have caused significant soil acidification'. While acidification of the soils occur by 'natural processes', it is the added constraint of changing atmospheric inputs and/or land disturbance that can lead to major changes in the stream. Much of these upland plantations are reaching the harvesting and replanting phases and hence such programmes will accelerate over the next decade (Adamson & Hornung 1990). These practices could themselves lead to further deterioration in water quality.

12.2 A CASE STUDY FROM WALES

In the spring of 1983, a research programme was initiated to study the hydrogeochemistry of an upland UK environment. The aims were to characterise the effects of forestry management on stream water quality and build on a detailed long term hydrological study at Plynlimon in mid-Wales. From the hydrochemical project's onset, the underlying approach was to study a wide range of chemical constituents. The reasons for this were twofold. First, it was essential to assess hydrological flow pathways at the catchment level. To attain this end, it was decided to use chemical fingerprinting techniques for components with very different chemical properties. The idea, analogous to that used for the development of the periodic table of the elements, was to see if one could find associations of behaviour for the chemical constituents that provide an insight into the underlying hydrogeochemical functioning of the environment. Second, environmental issues concerning the influence of man on the upland environment were being addressed: the monitoring of a wide range of chemical elements was essential – without using this approach, how could one assess those environmental impacts if one was unsure which components were changing under man's influence and which components were detrimental to the environment? Two main sub-catchments of the headwaters of the River Severn in the Hafren forest were selected for study. They were chosen since (1) a deforestation programme was planned for one of the two sub-catchments, (2) detailed hydrological data for these catchments is available since they comprise an integral part of the Institute of Hydrology's water balance studies (Newson 1976; Kirkby et al. 1991).

The overall project objective was, and remains, to continue and develop further studies on the movement through forested catchments of key elements that directly affect water quality or are relevant to the understanding of processes affecting water quality. Within this context, the following questions were being addressed:

1. What are the variations in water quality with hydrology and land use change?
2. How does deforestation affect stream water chemistry?

3. Can hydrochemical models be used to forecast how natural processes are affected by different land management practices?

12.2.1 *Description of the study area*

The work relates to two streams, the Afon Hafren and the Afon Hore, which form the main head water drainage of the River Severn in mid-Wales (Fig. 12.1).

Bedrock geology consists of lower Palaeozoic mudstones, greywackes, sandstones and grits. This is covered by a thin soil, typically 70 cm thick, with organic-rich 'L' and 'O' horizons (3 to 10 cm). The lower soils consist of a leached 'E' horizon (10 to 20 cm) and a fine textured podzolic 'B' horizon (about 40 cm thick) merging into a stone 'C' horizon. The predominant soil is a stagnopodzol, but peat, brown earth and stagno-gley soils also occur within the two sub-catchments. At the top of each sub-catchment, acid grassland is dominant. On the lower parts of the two sub-catchments, plantation forestry (predominantly Sitka spruce, *Picea sitchensis*) was introduced in various phases between 1937 and 1964, onto acid moorland.

Apart from minor 'thinning' of the forest stand, harvesting did not begin until late spring 1985 when clear felling of the lower half of the Afon Hore commenced. This

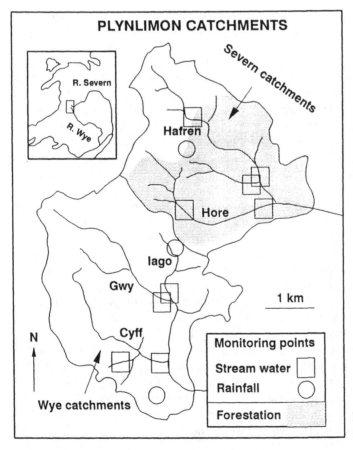

Figure 12.1. The Plynlimon catchments.

felling took 3 years to complete. In the process, extensive brash and tree stumps were left to decompose in situ. Soon after harvesting, replanting of the slopes with juvenile Sitka spruce (< 1 m high) proceeded.

Rainfall averages about 2500 mm/yr with evapotranspirational losses typically amounting to 500 and 650 mm/yr for the Afon Hafren and Afon Hore respectively. Being next to each other, of similar altitude range (360-470 m), and similar size (about 340 ha), the streamflow responses to storm events are similar and in phase for the two sub-catchments. The hydrograph response to storms is both rapid and 'flashy'; flows vary between 0.01 and 4.5 m^3/s (Newson 1976; Kirkby et al. 1991).

12.2.2 *Hydrochemistry*

12.2.2.1 *Atmospheric inputs*
Plynlimon rainfall is derived from a variety of sources and therefore it has a chemical composition that is highly variable (Table 12.1). Being within 60 km of the Irish sea and the North Atlantic Ocean, westerly winds provide rainfall enriched with sea-

Table 12.1. Flow weighted means and ranges for rainfall, cloud water and Afon Hafren and Afon Hore (pre-felling) stream water. All concentrations are given in mg/l except for pH, which is dimensionless, and alkalinity, which is given in µEq/l units.

	Rainfall	Cloud water	Hafren	Hore
Na	2.6	123	3.9	4.3
	(0-24)	(3-419)	(2.3-5.7)	(3.0-5.1)
K	0.13	4.6	0.2	0.2
	(0-0.9)	(0.2-16)	(0.1-1.0)	(0.1-0.6)
Ca	0.2	5.1	0.8	1.3
	(0-2)	(1-15)	(0.3-2.0)	(0.6-4.2)
Mg	0.3	15	0.8	0.9
	(0-3)	(0-51)	(0.3-1.5)	(0.6-1.5)
SO$_4$	1.6	38	4.4	5.3
	(0-9)	(8-106)	(2-12)	(3-11)
Si	0.34	0.03	1.3	1.5
	(0-20)	(0-0.3)	(0.5-4.6)	(1-5.6)
DOC	0.4	1.6	1.8	1.5
	(0-3)	(0-12)	(0-4.4)	(0-5.6)
NO$_3$	0.8	11	1.5	1.6
	(0-75)	(2-69)	(0-6)	(0.4-3.3)
NH$_4$	0.3	3.3	0.02	0.04
	(0-5)	(1-24)	(0-1)	(0-0.8)
PO$_4$	0.04	0.03	0.05	0.06
	(0-2)	(0-0.02)	(0-1)	(0-0.02)
F	0.03	0.06	0.05	0.05
	(0-0.23)	(0-0.41)	(0-0.2)	(0-0.2)
Cl	5.1	220	7.3	8.24
	(0-44)	(2-740)	(4-12)	(5-19)
pH	4.79	4.28	4.46	4.84
	(3.4-7.7)	(3.5-6.0)	(4.1-6.8)	(4.3-7.4)
Alk	−6.7	−48.1	−22	−0.8
	(−437-201)	(−398-31)	(−63-69)	(−47-197)

spray. In contrast, winds from other directions bring air that has passed over agricultural land and industrial areas: this provides rainfall low in sea salt but enriched with lithogenic (e.g. silica) and pollutant (heavy metals, sulphate, nitrate and ammonia) components. For sodium, magnesium, strontium, bromide and chloride, these components of rainfall come from marine sources; they are transferred from the sea to the atmosphere as sea-spray. Although the rainfall concentrations of these components are very variable, the ratios of their concentrations remain close to those of sea-water. For the non-sea salt components, rainfall concentrations are highly variable, and have low correlations due to their diverse sources: only dissolved organic carbon (DOC), ammonia, aluminium and iron are inter-correlated: a composite pattern is observed for components that have both maritime and industrial sources (e.g. calcium, sulphate and potassium): concentrations are at their lowest at intermediate sea salt levels (Fig. 12.2).

Mist and dry deposition provides an important source of chemicals to the catchment. For example, about 30% of the sea salts and 60% of the sulphate comes from sources other than wet deposition. In mist, concentrations of sea salt and acidic oxide components are high but they are low for the heavy metals.

The total pollutant loadings entering the Plynlimon catchments cannot be accu-

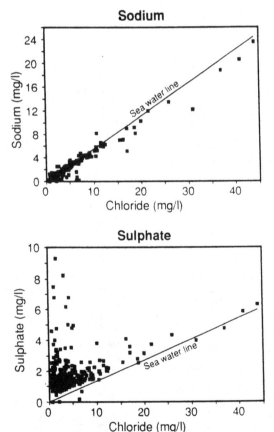

Figure 12.2. Sodium and sulphate concentration relationships with chloride, in rainfall.

Table 12.2. Wet deposited non-marine sulphate, nitrate and ammonium in Plynlimon rainfall.

	Plynlimon	Mid Wales	UK range
Concentrations (µEq/l)			
Sulphate	18.9	20-40	<20 to >100
Nitrate	12.9	10-20	<10 to >40
Ammonium	16.7	20-30	<10 to >50
Fluxes (gm^{-2} y^{-1})			
Sulphate	0.76	0.6-0.9	<0.3 to >1.2
Nitrate	0.45	0.3-0.4	<0.2 to >0.5
Ammonium	0.58	0.3-0.6	<0.2 to >0.6

rately gauged owing to uncertainties in the amount of dry and mist deposition. In terms of the pollutant inputs of sulphate, nitrate and ammonium as wet deposition, they are relatively low for the UK in terms of concentrations, although the high annual rainfall volume means that the fluxes entering the catchment are relatively high (Table 12.2). The trace metal concentrations in rainfall at Plynlimon are low compared with many areas of the UK due to the lack of heavy industry in the region. For nitrogen, sulphur and ammonium, concentrations are moderate for the UK and are associated with long range transport of industrial and farming emissions.

12.2.2.2 *Streams draining undisturbed forest*
The Plynlimon catchments experience rainfall with very variable chemistry while the streams exhibit a 'flashy' flow response to rainfall inputs. Therefore, one might expect large variations in stream water concentrations that mirror the rainfall signal. This is not usually the case even for components such as chloride and ^{18}O, which are chemically unreactive (Neal & Rosier 1990). This shows that the catchments have the ability to smooth the rainfall's variable chemical imprint to a very considerable degree. In other words, rainfall does not usually pass directly through the catchment to provide the major volume of water in the stream during the hydrograph response. During the major storm events, the rainfall signal sometimes influences the stream response and affects the response for the next few events.

In spite of the weak relationship between rainfall quality and stream chemistry, large fluctuations in many chemical species occur in the stream because of hydrological variations (Table 12.3). Baseflow waters have higher calcium and silicon concentrations and alkalinities than the corresponding stormflow waters. Stormflow waters, in contrast, have higher aluminium, yttrium, manganese, cobalt, nickel, beryllium and hydrogen ion concentrations than their baseflow counterparts. The major concentration changes occur at relatively low flows: at intermediate to high flows, concentrations remain either constant or decline as flow increases. This difference reflects the large chemical gradients between soil and bedrock within the catchment and the different flow paths followed by water under drier and wetter conditions (see Chapter 3). The soil zones, being organic and aluminium oxide/hydroxide rich, produce acidic, aluminium and transition metal rich soil water. The bedrock, consisting of weatherable and acid-soluble inorganic components such as calcite and layer lattice silicates (chlorite and illite), has the capacity to neutralise these acid waters and

Table 12.3. Major ion chemistry for baseflow and stormflow waters of the Afon Hafren and the upper and lower stretches of the Afon Hore for the period September 1984 to October 1990. High flow and low flow correspond with values greater than 0.1 mm/15 min and less than 0.01 mm/15 min. All units are mgl^{-1} except for alkalinity (μEql^{-1}) and pH.

Element	Afon Hafren		Upper Afon Hore		Lower Afon Hore	
	Base flow	Storm flow	Base flow	Storm flow	Base flow	Storm flow
Na	3.9	3.9	3.8	4.2	4.2	4.1
K	0.1	0.2	0.1	0.1	0.2	0.3
Ca	1.0	0.8	3.1	0.7	3.1	1.0
Mg	0.8	0.8	0.9	0.4	1.2	0.8
Total Al	0.1	0.4	0.1	0.5	0.1	0.5
DOC	0.8	2.2	1.0	2.0	1.0	1.9
Cl	6.8	7.2	7.0	8.7	7.5	7.9
SO$_4^{2-}$	3.5	4.7	3.1	4.1	4.8	5.1
NO$_3^-$	0.8	1.7	0.8	1.2	1.2	2.5
Alkalinity	24.2	−28.3	145.7	−27.7	126.6	−16.0
pH	6.3	4.6	7.1	4.6	7.1	4.8

precipitate the easily hydrolysable transition metals and aluminium. Such reacted water degasses carbon dioxide on its passage to the stream and produces the characteristic low acidity base flow water rich in calcium and silica, but depleted in aluminium and transition metal. During storm events, when the catchment soils have wetted up and the groundwater table is high, more water entering the stream is derived from the soil zone. With intense and long duration storms, more of the stormflow water is derived from the upper and most acidic soil layers and this change accounts for the declining concentrations of beryllium and aluminium at very high flow values.

Under baseflow conditions, water drains mainly from near the stream bank and from the groundwater area. Variations in the bedrock composition lead to differences in the stream water chemistries of the Afon Hafren and Afon Hore (Reynolds et al. 1986). The presence of calcite and sulphides of lead and zinc in the Hore catchment leads to high calcium concentrations and measurable lead and zinc concentrations in the stream (a factor of four or more difference between the two streams). Other differences in bedrock composition lead to contrasting behaviours of strontium and magnesium for the two catchments. Concentrations of lead and zinc in the Afon Hore are lowest under baseflow conditions. Thus, either the parent sulphides in the bedrock are not oxidized to soluble forms or there is an unidentified ion exchange or solubility control in the groundwater. Differences in the bedrock are reflected in the correlation structure of the chemical data.

12.2.2.3 *Streams draining disturbed forest*
The effects of felling were dominated by the hydrologically induced variability of the system and compounded because the streams are not simultaneously sampled. To resolve this, time series plots were constructed for smoothed data by using a 'running mean' covering periods of 4 months. The major changes occurring are as follows (Fig. 12.3).

The stream response to major sea salt rainfall events has changed following fell-

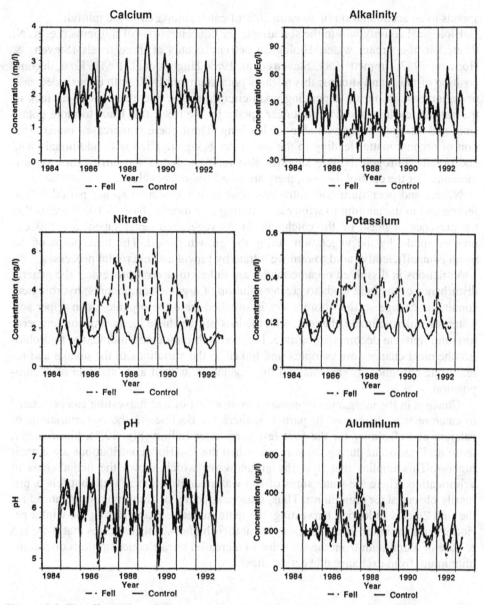

Figure 12.3. The effects of tree felling on nitrate, potassium, aluminium, calcium, alkalinity and pH (data from Neal et al. 1992b,c). Felling began in late spring 1985 and was completed in the summer of 1989.

ing. The concentrations of sea salts in the stream are reduced, relative to the control response, and over time the difference between the responses widens. Since the decline in flow is relatively small following harvesting, evapotranspirational variations cannot explain this difference. The results fit well with the notion that trees capture sea salt components as mist and dry deposition. This enhanced deposition corresponds to an accumulation of an extra 20% of catch compared to the rainfall.

Biological activity within the soil affects the concentrations of nutrients (i.e. K, N, P, etc.) in the stream water. Biological activity occurs at three levels (Stevens & Hornung 1987; Emmett 1989; Stevens et al. 1989; Hughes et al. 1990). First, decomposition of brash and stumps: this releases potassium and DOC with either a decrease or an increase in nitrate depending upon circumstances. Second, there is a break in the nutrient cycle as there is no longer uptake by the trees: this leads to more potassium and nitrogen being available for leaching. Third, there is increased mineralization of organic matter leading to the soil water being supplied with additional DOC and organic nitrogen. Nitrate supplies also increase due to the nitrification of ammonium. For the control data set, there are small seasonal cycles.

Nitrate and potassium concentrations peak in the winter to spring period before felling and in the autumn to winter after felling (a 3 month phase shift). It seems that the presence of brash on the catchment limits vegetation development and this depresses uptake by the vegetation during the growth period. The breakdown of the brash is insufficiently rapid to remove nitrate by fungal and microbial processes.

Variations in dissolved organic matter are linked to several processes: the general disturbance of the land leads to greater solution of organic components; hydrological conditions change so that more water is supplied from the organic-laden upper soil water; the micro-climate changes, with the loss of tree shade increasing soil temperature and thus the decomposition rate. For boron, bromide and iodine, the hydrobiogeochemical changes are complex and linked to the variations in the uptake and release rates by micro-organisms and the vegetation as well as to the rate of decomposition.

Changes in the inorganic components as an effect of tree harvesting can be related to catchment acidification. In particular there are declines in the concentrations of calcium and alkalinity for the first few years after felling (Fig. 12.3). Surprisingly, these declines occur during the summer, when the baseflow contributions are at their highest. This implies that it is the groundwater which shows the major signs of acidification. There are some signs of soil water acidification as well, but this is primarily observed for aluminium. Thus, seasonal cycling of aluminium is enhanced for the first 2 years following harvesting: the main increases occur during the winter period when aluminium concentrations are at their highest. This indicates that there is a release of aluminium from the soils due to increased nitrate concentrations displacing aluminium from exchangeable cation sites.

12.3 SOURCES CONTRIBUTING TO STREAM FLOW

12.3.1 *Rationale*

The variations in stream water chemistry are linked, at a gross scale, to varying contributions from soil and ground water inputs. To fully interpret the hydrochemical processes operating, there is a need to examine in more detail the chemical hydrograph in order to determine the dynamics of stream water supply. Unfortunately, water mixing relationships are difficult to assess for the streams. Although a neutral bicarbonate-bearing groundwater component is inferred from the stream data, such an end-member has not been sampled until very recently with the introduction of boreholes within the catchment. Most chemical components that exhibit response to flow are chemically reactive and hence conservative mixing formulations cannot be used. Nonetheless, basic inferences can be made that provide some insight into the nature of the hydrological and hydrochemical processes operating. This is considered here by examining the chemical signals in the light of both weekly samples and continuous measurements of pH and conductivity.

12.3.2 *Hydrograph separation*

Hydrograph separation encompasses a range of techniques aimed at elucidating the hydrograph storm response by subdividing the storm waters into two or more components (Neal et al. 1990b,d; Robson & Neal 1991). Early methods of hydrograph separation were either graphical (Sherman 1933), or mathematically based black-box methods (Hewlett & Hilbert 1967), both relying entirely on rainfall-runoff data. The results of such exercises are usually a *slow* component, assumed to tie in with subsurface flows, and a *quick* component, linked to the rain waters. Unfortunately, there is no single unique separation, so that *quick* and *slow* components are method, not process, dependent. As Hewlett and Hilbert (1967) stated, such methods are *one of the most desperate analysis techniques in use in hydrology.* Any linkage with physical processes must therefore be considered highly dubious (Beven 1991).

Chemical separation of stream hydrographs was carried out by Pinder & Jones (1969) using total dissolved solids, and was followed, in the late 1970's, by the use of isotopes (Sklash & Farvolden 1979). Here, the separation was explicitly made between *old* or pre-event water and the *new* rainwater. This lead to the unexpected discovery of the dominance of pre-event water in the storm hydrograph response – a feature of the Plynlimon catchments as well. Many other isotope studies have made similar findings (Rohde 1981; Hooper & Shoemaker 1986; Pearce et al. 1986; Sklash et al. 1986; Sklash 1990) indicating the dominance of the pre-event water. However, isotopic separation between rain and pre-event water fails to address many of the variations that are important in terms of stream water quality. Along with many non-isotopic hydrograph separations (Pilgrim et al. 1979; Hooper & Shoemaker 1986; Wells et al. 1991), the studies neglect the upper soils and any interaction between rain water and soil water; they fail to address within-catchment chemical gradients (Kennedy et al. 1986). Yet, the soil water component has been shown, using three-component isotope analysis, to be of a different composition to baseflow waters (DeWalle et al. 1988).

Chemical end-member mixing techniques provide a basis allowing a subdivision of pre-event waters into distinctive components. End-member mixing techniques were initially used in the estuarine sciences (see Liss 1976, for summary). Here, the aim was to explain observed chemical relationships for estuarine waters, by comparing selected chemical determinands against a conservative measure, the salinity. The highly saline sea waters and the fresh river waters provided two very distinct water types or end-members, the relative proportions of which were estimated from the resultant salinity. A straight line relationship between salinity and a determinand was taken as an indication of conservative mixing whereas a non-linear relationship was thought to imply non-conservative mixing (Boyle et al. 1974; Liss 1976).

The starting point for assessing stream water mixing relationships is to examine concentration changes for the most variable, but chemically conservative, tracer, the acid neutralisation capacity (ANC).

ANC is a measure of the acidity of a solution and is closely related to pH. Like pH, the ANC shows substantial variations between high and low flows that reflect chemical gradients within the catchment. In contrast to pH, ANC behaves conservatively on mixing; it is unaffected by CO_2-degassing/$Al(OH)_3$-precipitation-solution changes (Reuss & Johnson 1986), and by precipitation/solution of H-organic and Al-organic ('humic') substances. The primary restriction on the use of ANC is that, during mixing, ion exchange and weathering, reactions involving H^+ and/or inorganic aluminium with other ions must not be important. ANC is defined here (according to Reuss et al. 1986) as:

$$ANC = \Sigma \text{ strong base cations} - \Sigma \text{ strong acid anions} \qquad (12.1)$$

Thus, for the Plynlimon waters, ANC is approximately given by:

$$ANC \approx [Na^+] + [K^+] + 2[Ca^{2+}] + 2[Mg^{2+}] + [NH_4^+]$$
$$- [Cl^-] - [NO_3^-] - 2[SO_4^{2-}] - [F^-] \qquad (12.2)$$

or, alternatively, using the charge balance constraint, as:

$$ANC = \Sigma \text{ weak acid anions} - \Sigma \text{ weak base cations} \qquad (12.3)$$

The baseflow is assumed to be representative of the groundwater chemistry. Groundwater has not been directly sampled and average baseflow ANC values have been used to depict the end-member value. Within the soil zones, mean ANC differs for the grassland and forested parts of the catchment and areally averaged values have been used for the calculations. Under such circumstances, the percent hillslope (or soil water) ($\%_{hs}$) in a water sample of measured ANC (ANC_m) is estimated from the simple mixing equation:

$$\%_{hs} = 100(ANC_m - ANC_g)/(ANC_{hs} - ANC_g) \qquad (12.4)$$

where subscripts g and hs denote groundwater and hillslope water values respectively.

Results from this exercise show that as flow increases the proportional contribution of soil water from the hillslope increases and then levels off at high flows (Fig. 12.4). Even under high flow conditions the groundwater component can form a significant contribution to the total flow.

The use of ANC only allows separation of the hydrograph into two end-members. Studying the behaviour of other chemical species provides a means of resolving further end-members even though mixing may no longer be conservative. Different species show increased contributions in particular flow regimes, indicating when chemically distinct soil horizons begin to flow.

Manganese, cobalt and nickel behave fairly conservatively except during drought conditions when they may be adsorbed onto the stream bed. The conservative-mixing nature of these components is most clearly shown when the concentrations are plotted against the ANC since a straight line relationship is observed.

Aluminium concentration changes take place over a wide range of flows. When

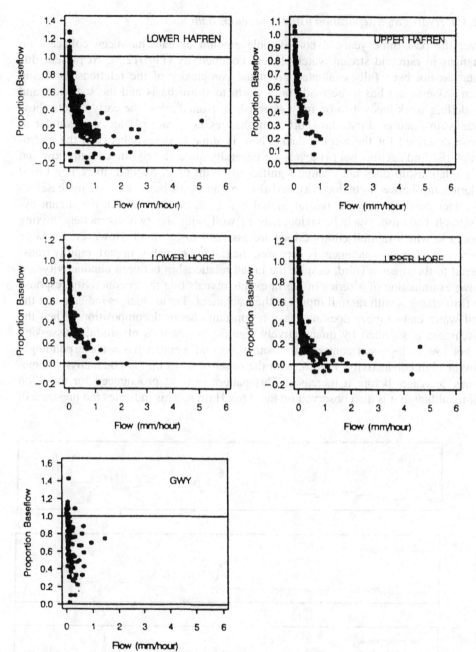

Figure 12.4. The ANC hydrograph split for the Plynlimon catchments (data from Robson 1993).

concentrations are plotted against ANC, a curvilinear graph results. This behaviour is explained by non-conservative mixing involving aluminium solution/precipitation either in the stream or in the near stream bank areas and this is discussed later in this paper.

12.3.3 *Hydrograph separation using continuous data*

Over the past three years, a considerable amount of data has been collected on changes in rain and stream water pH and conductivity (Fig. 12.5). At present this data has not been fully evaluated given the complexity of the relationships being seen. Assessment has to be made on a storm to storm basis and the statistical and modelling work has yet to be fully established. Nonetheless, the early results show four main features. First, the short term changes in stream pH are very similar to those described for the weekly data. There is some hysteresis between the hydrograph rise and decline, but pH levels are generally quite closely linked with flow and this relationship does not change significantly with time. Second, there is a broad relationship between conductivity and flow on an event basis: as flow increases so too does conductivity. The rainfall signal is not usually seen within the stream hydrograph response. Such behaviour fits in well with the two component mixing model in which rainfall contributions are assumed to be small. However, for large storms or when the catchment is very wet, the rainfall signal is, in part, rapidly transferred to the stream. Third, despite the broad relationship between conductivity and flow, examination of a series of storm events reveals that the conductivity response to flow changes with rainfall inputs to the catchment. These changes indicate that the soil water end-member does not have a constant chemical composition. When the catchment is supplied by quantitatively significant amounts of rainfall possessing either low or high salt contents, the soil water end-member has a correspondingly low or high conductivity. Fourth, with the onset of a storm, the conductivity sometimes decreases before it increases. This pattern is most pronounced for the Afon Hore, although it is also observed on the Afon Hafren. This indicates the presence of

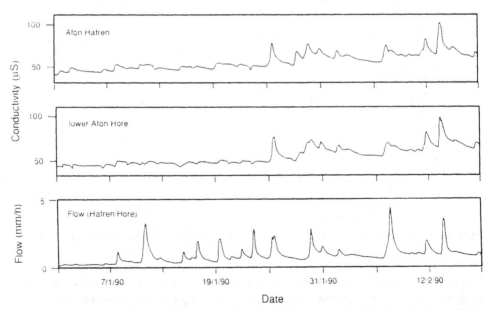

Figure 12.5. Short term conductivity changes in Afon Hafren and Afon Hore stream water (data Robson 1993).

a second groundwater end-member for the Afon Hore and a differential behaviour between the Afon Hafren and the Afon Hore. Given the rapidity and the short lived nature of the response, the results indicate the input of a water from near the stream bank. Such differences could not be spotted with the weekly data: they are minor and occur over a very short time period.

12.3.4 *Conclusions concerning water mixing relationships*

A broad explanation of the chemical variations within the Plynlimon stream waters can be formulated using weekly spot sample data, but the system is of sufficient complexity that more studies are required to describe mechanistically the detailed changes occurring. The results indicate that the water supplied to the stream comprises a mixture of at least two types of ground water and two types of soil water as well as rain water.

12.4 CHEMICAL MIXING, SPECIATION AND SOLUBILITY CONTROLS FOR ALUMINIUM

12.4.1 *Rationale*

An important criteria for a healthy stream ecology is that inorganic forms of aluminium remain low (Stoner & Gee 1985; Egglishaw et al. 1986; UKAWRG 1988). The environmentally harmful forms of aluminium within the streams are associated with trivalent aluminium, hydroxy-aluminium complexes and freshly formed polymeric hydrolysis products (Muniz & Leiverstad 1980). During the acidification process, concentrations of these environmentally harmful forms increase in the streams (UKAWRG 1988). Reliable estimates of the levels of the different aluminium complexes within solution are needed since some forms present in upland waters, such as the organic, fluoride and silicate complexes, reduce the harmful effect (e.g. Chappell & Birchall 1988; Birchall et al. 1989).

  In this section, the subject of aluminium solubility and speciation controls is addressed. This is provided to indicate the extent of environmentally harmful and inert levels of aluminium complexes in Plynlimon stream waters. To achieve this objective, it has been necessary to simplify the analysis as the speciation is influenced by a complex series of interactive factors such as hydrology, variable source area and temperature. This has been achieved by considering an averaged situation where the stream water is considered as a mixture of soil and ground water end-members of fixed composition. Subsequently, the speciation of Plynlimon stream waters are examined to characterize the overall behaviour.

12.4.2 *Modelling chemical speciation*

The two component mixing model used here is an extension to that described in Section 12.3. During the mixing of soil and ground water, microcrystalline gibbsite $(Al(OH)_3)$ can precipitate in the river if the mixed waters become oversaturated with respect to this phase. No allowance is made for dissolution of microcrystalline gibbsite under the conditions where the mixed waters are undersaturated with respect to this phase. Within the calculation, account is taken of aluminium complexation with fluoride, sulphate, hydroxide, organics and silicate. Following mixing and on transit to the stream and within the stream itself, water chemistry is modified by degassing of carbon dioxide and the extent of aluminium complexation varies in response to pH change. Partial pressures of carbon dioxide were set to constant values: 3 and 30 times the atmospheric value for the stream and soil zones, respectively.

The detailed speciation of aluminium is evaluated using the latest version of the programme ALCHEMI (Schecher & Driscoll 1987). This involves using (a) output data, on inorganic aluminium, pH and bicarbonate, from the mixing model described above, (b) mean concentrations for the other components needed in the ALCHEMI computations (as determined from the hydrochemical data set) and (c) an average value for organically bound aluminium in the stream.

12.4.3 *Mixing model results*

There is a strong relationship between hydrogen ion and total dissolved aluminium concentrations for Afon Hafren stream water. The relationship is non-linear (Fig. 12.6). At low to moderate hydrogen ion concentrations, total aluminium concentrations increase sharply with increasing hydrogen ion concentration. At higher hydrogen ion concentrations, total aluminium concentrations increase only moderately with increasing hydrogen ion concentration. This pattern is well characterised by the mixing model. Four effects produce the curve shown. First, microcrystalline gibbsite is precipitating at low hydrogen ion concentrations – this produces a cubic relationship between trivalent aluminium and hydrogen ion concentrations. Second, trivalent aluminium is supplemented by the presence of aluminium hydroxide and fluoride complexes at low hydrogen ion concentrations. Third, on entering the stream, the mixed waters degas carbon dioxide and this causes an increase in pH but no change in the total aluminium concentration. The pH change is most noticeable at the lowest hydrogen ion concentrations when bicarbonate buffering of acidity is most important. Fourth, at moderate to high hydrogen ion concentrations, both hydrogen ions and total aluminium are unreactive and a simple linear mixing pattern results – carbon dioxide degassing does not significantly change the pH while aluminium hydroxide precipitation does not come into play.

The analysis shows the following:

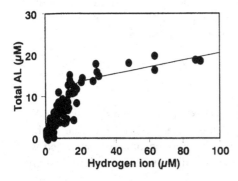

Figure 12.6. Concentration relationships between total aluminium and hydrogen ions for the Afon Hafren: Observed (solid circles) and modelled (curve) values (data from Neal 1996).

1. The relative concentrations of trivalent aluminium and aluminium complexes with hydroxide, fluoride, dissolved organic matter and silicate are all significant for the water end-members and their mixtures: they comprise over 15% of the total aluminium in solution in part of the pH range (Fig. 12.7): Aluminium sulphate and aluminium hydroxy fluoride complexes make up less than 1% of the total aluminium concentration.

2. The highest concentrations of trivalent aluminium occur at the lowest pH considered; concentrations decline to insignificant levels at pH's greater than 5.

3. At intermediate pH's a mixture of complexes occur in similar proportions.

4. In the higher pH range aluminium – hydroxy complexes becomes more prevalent and aluminate starts to dominate.

5. The proportions of the various aluminium complexes vary markedly as a function of temperature, although the same basic patterns are maintained. This variation is due to the pronounced temperature dependence of the equilibrium constants for aluminium hydrolysis and is most important for the aluminate ion $(Al(OH)_4)$. Consequently, the highest deviations occur at the highest pH's.

12.4.3.1 *Field results*

Analysis of the stream water data shows similar features of aluminium complexation as described in the previous section. The results show:

1. Trivalent aluminium predominates at low pH.

2. Although most fluoride is bound to aluminium at low pH, there is insufficient total fluorine present to complex large amounts of aluminium present at low pH.

3. Aluminium hydroxide complexation increases in importance at intermediate to high pH.

4. Trivalent aluminium and aluminate, while important at intermediate pH values, do not dominate the system in this range.

5. While organic aluminium concentrations vary with pH, the variation is small – at low pH the organic acids are essentially undissociated while at high pH trivalent aluminium is extremely low.

The field results exhibit considerable scatter for the various aluminium species. This scatter represents the variability in the stream water chemistry and temperature. Temperature-related scatter occurs for the higher pH waters: The greatest temperature dependent equilibrium constants are for those species most prevalent at higher pH; baseflow occurs during both the summer and winter periods and therefore temperature is at it's most variable (2-20°C).

12.4.3.2 *Aluminium regulation within Plynlimon soils*

One of the features of the acidification research has been the assumption that the soils undergo simple cation exchange reactions and that the transfer of acidity and aluminium to the stream are linked to the movement of 'mobile anions' such as sulphate, nitrate and chloride. However, hydrogen ions and trivalent aluminium concentrations conform neither to the theoretical cubic relationship nor the associated temperature dependent co-relationship as discussed in more detail later in this chapter (see also Neal 1988a, b; Neal et al. 1990b; Mulder et al. 1989; McMahon & Neal 1990). These findings are reinforced by hydrochemical features associated with the catchment scale perturbations. The simple ion exchange equations commonly em-

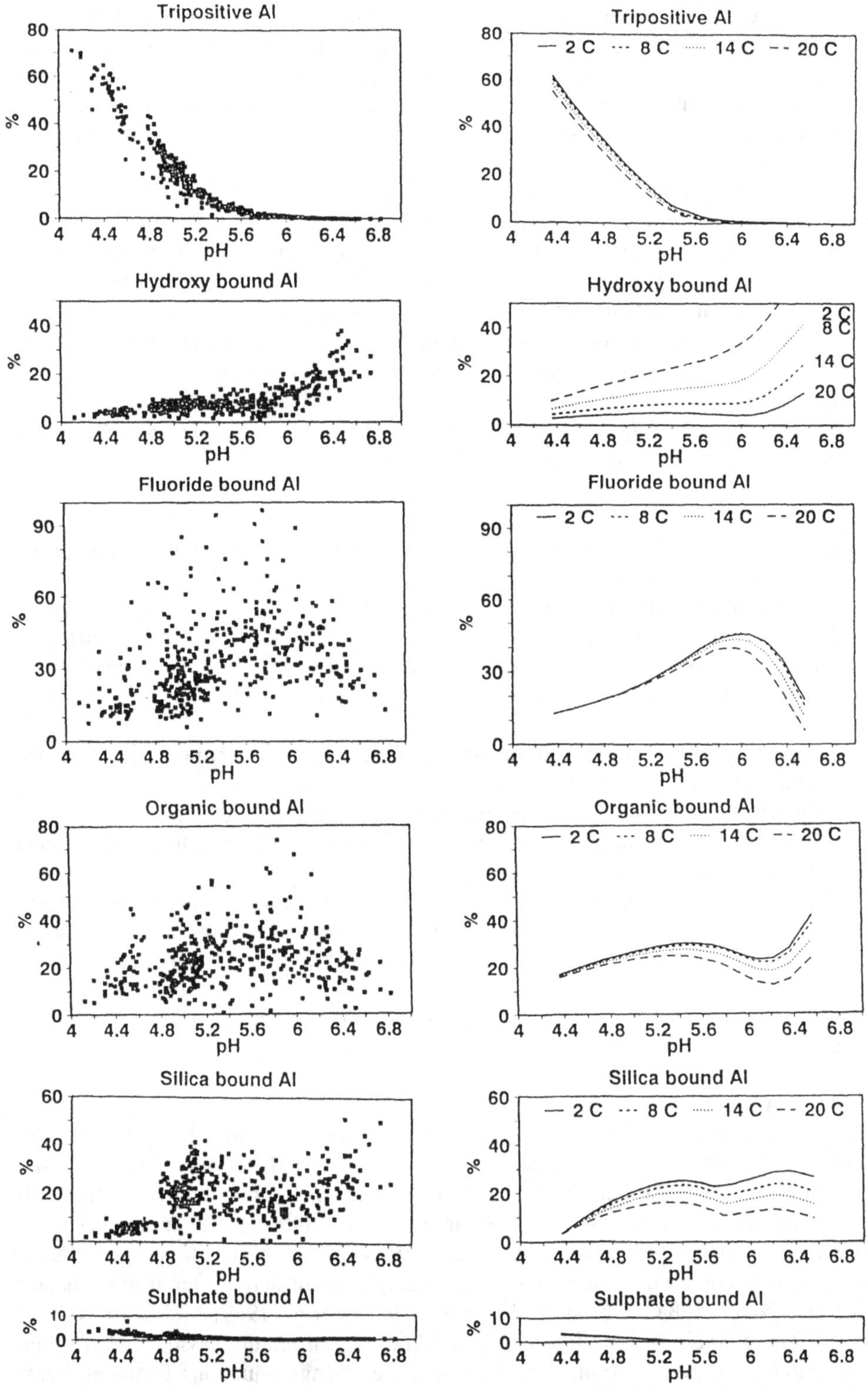

Figure 12.7. Aluminium speciation in Afon Hafren stream water: Field data (left) and chemical mixing model results at 2, 8, 14 and 20°C (right) (data from Neal 1996).

ployed, indicate that all the base cations increase as the anion concentration increases (Reynolds et al. 1988; Neal et al. 1989); this is not the case with calcium and sodium as their concentrations decrease in the stream under high flow conditions following felling. Furthermore, the sea salt events which occasionally characterize the stream chemistry have typical acidity and aluminium levels in contradiction to the simple cation exchange formulation where abnormally high values would be predicted. Consequently, one of the most important features of the Plynlimon study has been the recognition that, despite over a decade of intensive research world-wide, the mechanisms determining hydrogen ion and aluminium mobility remain obscure.

12.5 MODELLING STUDIES

12.5.1 *Modelling background*

Modelling work has been undertaken for the Plynlimon catchments to describe both the short and long term variations in stream water quality. These modelling studies constituted part of the 'Surface Water Acidification Programme' and the European Economic Communities 'Encore' project. A brief summary of the findings so far are presented and the future direction of the work is considered.

12.5.2 *Short term modelling studies*

12.5.2.1 *The Birkenes model*
As one outcome of the extensive acid deposition modelling effort, the Birkenes model was formulated (Christophersen et al. 1982). The model was conceptually simple (Fig. 12.8). It comprised a two reservoir hydrological model (plus a top snow melt layer) operating on a daily time-step, upon which have been imposed what was at the time deemed the important chemical processes controlling catchment acidification. Inputs to the model are precipitation, mean daily soil and air temperature and sulphate deposition rates. The model represented stream flow as a mixture of soil water from the top reservoir and groundwater areas by the lower reservoir. The importance of the model was that it forever dispelled the myth that acid rain passed straight through catchments to give acid surges in streams, thus destroying fish and other aquatic life. It provided the main sounding board for recognising that within catchment processes (cation exchange and mineral weathering) coupled with hydrological flow pathways are critical in determining runoff chemistry.

Each of these reservoirs obeyed simple equations that were representative of classic hydrological rainfall-runoff modelling and chemical reactions were depicted using equations that conformed with the classic soil science literature. Within the model, the processes depicted for the soil zone (the upper reservoir) were:

1. Water – precipitation, evapotranspiration, infiltration to the lower reservoir and discharge to the stream.

2. Sulphate – wet plus dry deposition, adsorption/desorption and mineralization.

3. Calcium and magnesium – cation exchange.

4. Hydrogen ion – cation exchange, equilibrium with $Al(OH)_3$ (gibbsite) and bicarbonate.

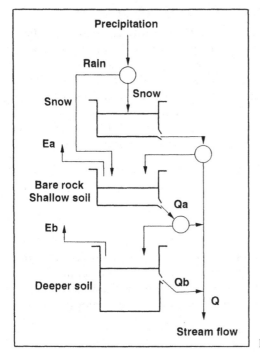

Figure 12.8. The Birkenes model.

5. Trivalent aluminium – equilibrium with $Al(OH)_3$ (gibbsite).

6. Inorganic carbon – equilibrium between H_2CO_3 and HCO_3^- in the soil solution and seasonally varying CO_2 in the soil atmosphere.

Within the groundwater areas (the lower reservoir), the reactions were deemed to be:

1. Water – infiltration, evapotranspiration, piston flow and discharge to the stream.

2. Sulphate – adsorption/desorption and reduction.

3. Calcium and magnesium – release by weathering.

4. Hydrogen ion – consumption by weathering and equilibrium with $Al(OH)_3$ (gibbsite) and bicarbonate.

5. Trivalent aluminium – equilibrium with $Al(OH)_3$ (gibbsite).

6. Inorganic carbon – equilibrium between H_2CO_3 and HCO_3^- in the soil solution and the soil atmosphere.

12.5.2.2 *Birkenes model application*

Initially it was intended that the Birkenes model would be applied to the full suite of major element chemistries. At the time of application, Birkenes type modelling was mainly confined to flow and ions typically associated with the acidification process (hydrogen, base cations, inorganic aluminium and sulphate). However, here, the first step taken was to examine chloride. This component can be considered to be a conservative tracer and the large variations in chloride in rainfall provides a good chemical tracer to test the model.

The model was first calibrated for hydrology using data for the period April 1983 to May 1985. The results of this exercise showed an extremely good fit (Fig. 12.9b). Indeed early simulations showed a poor fit for two months during the calibration period and further examination showed the original hydrological data to be incorrect at the anomalous time. However, when the model was run for chloride using weekly bulked rainfall data with weekly grab sampling of stream water, serious discrepancies occurred (Fig. 12.9a). Thus, the model predicted increases in chloride concentrations during baseflow periods and large fluctuations in concentrations due to variations in atmospheric inputs and flow, which were not observed. In order to ex-

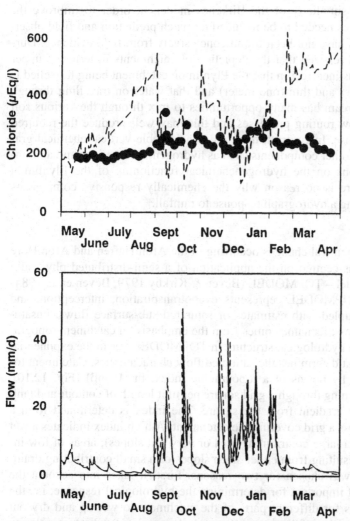

Figure 12.9. Chloride (top) and flow (bottom) simulations of the Afon Hafren using the Birkenes model (data from Neal et al. 1988). a) The dashed lines represent model fit and the circles depict the measured values. b) The dashed and solid lines represent modelled and actual values respectively.

plain these differences within the current model structure, modifications are needed. On a qualitative level, these could include:

1. Increasing the amount of water in each reservoir or adding a partially mobile chloride store in the upper soil reservoir: this would increase the dampening response to the rainfall signal.

2. Decreasing evapotranspiration from the lower reservoir: This would reduce the modelled increase of stream chloride concentration during the summer months.

3. Increasing the dry deposition of sea salts during the summer months or introducing bypass mechanisms whereby rainwater passes directly through the upper soils, without chemical mixing, before entering the groundwater zone. Again, this will negate the modelled stream chloride concentration increase during the summer.

Despite the elegant simplicity of the Birkenes model, in order to improve the model fit, extra processes needed to be included to match prediction and field observation. These processes have not been found conclusively from field evidence. During the analysis, it was realized that the depiction of catchments in terms of upper and lower reservoirs was incorrect in that the Plynlimon catchment being modelled is both long (one kilometer and thin (one meter) and that water on travelling through the catchment to the stream has many opportunities to mix through the various soil layers by stochastic flow routing processes and this may well produce the required damped chloride response. In contrast, there are considerable vertical chemical gradients that can regulate other components such as hydrogen, base cations and aluminium (see earlier sections on the hydrogeochemical functioning of the Plynlimon catchments). Thus, there is no reason why the chemically responsive components should be damped during a hydrograph response to rainfall.

12.5.2.3 *TOPMODEL*

Modelling studies of chemical changes occurring in the Afon Hafren and Afon Hore during storm events has centred on the application of a semi-distributed physically based hydrological model – TOPMODEL (Beven & Kirkby 1979; Beven et al. 1984) by Robson (1993). TOPMODEL represents evapotranspiration, interception, and transpiration losses coupled with estimates of saturated sub-surface flows, unsaturated flow and quick flow: its name comes from the emphasis on catchment topography. The essence of the hydrological structure in TOPMODEL lies in the exponential transmissivity curve that determines the saturated flow characteristics. Catchment topography is examined by means of a topographic index, $\ln (A/\tan\beta)$ (Fig. 12.10), where A is the area draining through a grid square per unit length of contour and $\tan\beta$ is the average outflow gradient from the square. The index is determined from a digital terrain map across a grid covering the catchment: A high index indicates a wet part of the catchment (a large contributing area or very flat slopes); areas of low index are usually drier resulting from either steep slopes or a small contributing drainage area. This model was chosen for two reasons. First, it takes into account the catchment feature most important for determining the hydrological response, i.e. the topography, and allows for different parts of the catchment to wet up and dry out over time with changes in rainfall amount. Second, it is an intermediate complexity model lying between that of the Birkenes case described above and much more elaborate fully distributed models.

The work has focused on trying to link the hydrological and the chemical re-

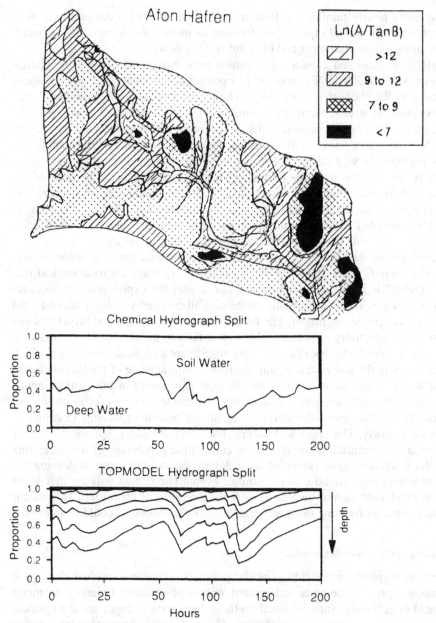

Figure 12.10. TOPMODEL application: The topographic index map and hydrograph split comparisons between model and ANC evaluations (data from Robson 1993).

sponse of the stream on an episodic basis. TOPMODEL was adapted so that the contributions to the stream from different soil horizons could be estimated. To do so, TOPMODEL was used to obtain contributions from different soil layers within the catchment by estimating the depth to the water table in the banks and the flow contributions were integrated along the length of the stream. TOPMODEL was then calibrated using hourly rainfall and flow records. TOPMODEL estimates two components of streamflow – (1) quickflow, thought to move by macropore-flow, overland-flow or piston displacement and (2) a subsurface flow.

The pH/ANC ratio and conductivity signals have been modelled for the Hafren stream. For ANC, TOPMODEL was used to predict components of flow, and these were related to the chemistry of the soils. The model structure was found to be capable of explaining stream chemistry assuming mixing relationships to be appropriate (Fig. 12.10). The results suggested that at least four end-members are required to explain the lower and upper Hafren stream chemistry. In general, the parameters were not particularly well defined, and cross validation did not produce very good results. It is not possible to judge exactly why such problems arise, because of the problems with pH measurements. Although ANC is modelled successfully here, the results are not considered to provide evidence for mixing hypotheses – the ANC is very highly correlated with flow and this may be the real reason why ANC predictions have been possible (as in the Birkenes case presented above).

The conductivity data have been modelled using a layered structure within a one-dimensional simplification of TOPMODEL, TOP1D. It retains the hydrological features of TOPMODEL, but it is semi-distributed: it uses the exponential transmissivity curve as its base. This structure allows the rainfall conductivity signal to be traced as it moves through the catchment. The model is able to reproduce the broad features of the stream conductivity response – although it fails to match all of the short-term dynamics. The model implies that there are significant chemical stores with which rainwater mixes in the soil zones. Again, the results mimic those of the Birkenes case provided above. These stores account for the high dampening of the rainfall inputs. The conductivity results suggest that the ranges of residence times for rainwater are very wide. The oldest stores of water, seen mainly at baseflows, change in composition only very slowly. The upper soil waters change more rapidly, but are still highly damped relative to rainfall. Estimates of the cumulative percentage of the tracer flux leaving the catchment give values of only 2/3 of the rainfall input as leaving the catchment within three months of the rainfall event. The results indicate that water within the catchment cannot be treated as having a uniform chemistry. This means that isotope separations using rainwater and baseflow components could be in error.

12.5.3 *Long term modelling studies*

Long term atmospheric deposition of acidic pollutants can cause gradual changes in the chemical stores of the soils and ground waters of sensitive upland catchments (Whitehead et al. 1988). Superimposed on these long term changes are the episodic acidic pulses occurring during storm events. These episodic chemical pulses, during which hydrogen and aluminium concentrations may increase by orders of magnitude, are critical in determining whether streams can sustain healthy fish and stream invertebrates (UKAWRG 1988). Fish kills may occur not only from long term chemical changes (chronic effects), but also as a result of short term variations in the stream chemistry during storm events (acute effects: Haines 1986; Turnpenny et al. 1987; UKAWRG 1988). To understand the effects of acidification, the short term episodic response and the slowly varying long term trends must be assessed alongside one another.

A long term modelling approach is provided to consider both the average changes and the hydrologically induced short term changes. To do so the MAGIC model (Model of Acidification of Groundwaters In Catchments) is employed here.

12.5.3.1 *Model of Acidification of Groundwaters In Catchments (MAGIC)*

MAGIC is a model designed to examine the long term changes in stream and soil waters occurring in response to acid inputs to the system. MAGIC assumes that atmospheric deposition, mineral weathering and cation exchange processes in the soil zone are responsible for the observed stream water chemistry in a catchment. Full descriptions of the conceptual basis of the model and the equations on which it is based can be found in Cosby et al. (1985a,b). The model comprises one or two reservoirs. In the simplest form, a one box representation, the catchment is considered as a single soil/groundwater unit in which cation exchange and weathering zones are combined. In the two reservoir case, the structure can either be like that for the Birkenes model or as two adjacent sub-catchments. Unlike the Birkenes model, the hydrological sub-model operates over monthly or yearly time steps and does not incorporate individual storm events. For the MAGIC model there is also an above-ground compartment to allow for element uptake by and release from the biomass. A sequence of atmospheric deposition and mineral weathering is assumed in the model. Current deposition levels of base cations, sulphate, nitrate and chloride are needed, along with some estimate of how these levels have varied historically. Historical deposition variations may be scaled to emissions records or may be taken from other modelling studies of atmospheric transport into a region. Weathering estimates for base cations are extremely difficult to obtain. Nonetheless, it is the weathering process that controls the long term response and recovery of catchments to acidic deposition, and some estimate of weathering rate is required. Several chemical, biological and hydrological processes control stream water chemistry. These are often interactive and not easily identifiable from field observation and as a consequence of this not all the factors are determinable from field measurements: recent developments with kinetic formulations of weathering may well allow direct estimates of weathering, given information on the mineral composition of the soil/groundwater-matrix, water residence time and moisture content (Sverdrup 1990). A summary of the main processes represented in the model is as follows:

1. Strong acid anion concentrations are calculated for the soils and stream waters. Sulphate adsorption is assumed to follow the Langmuir isotherm:

$$E_s = E_{max} (SO_4^{2-}) / [C + (SO_4^{2-})] \qquad (12.5)$$

where E_{max} is the maximum adsorption capacity of the soils (μEq/kg) and C is half the saturation concentration (μEq/m^3). This allows for the lags that can occur between atmospheric deposition and the resultant changes in stream water sulphate concentrations. Chloride, nitrate and fluoride are assumed not to have an adsorbed phase. Loss of ammonia and nitrate to the vegetation is included within the model by estimating a percentage uptake.

2. Cation exchange processes involving aluminium, sodium, calcium, magnesium and potassium are assumed to be operative in the soils. The general form for the exchange reactions between the ions M^{m+} and N^{n+} and their adsorbed states MX_m and NX_n, is:

$$nM^{m+} + mNX_n = mN^{n+} + nMX_m \qquad (12.6)$$

The total cation exchange capacity is then defined as:

$$\text{CEC} = \Sigma \text{ Exchangeable cations (mEq/m}^3) \qquad (12.7)$$

Equilibrium expressions for the cation reactions are approximated by:

$$S_{MN} = \{N^{n+}\}^m \, E/\{M^{m+}\}^n \, E \qquad (12.8)$$

where S_{MN} is the selectivity coefficient between M and N, $\{ \ \}$ refers to the chemical activity of an ion in solution, and E is the exchangeable fraction for the appropriate ion on the soil complex, e.g.:

$$E_N = X/\text{CEC} \qquad (12.9)$$

with X equalling the amount of adsorbed cation N (μEq/m^3); the cation exchange aspect is covered in more detail later in this chapter.

3. Within the soil, aluminium concentrations are assumed to be in equilibrium with the solid phase $Al(OH)_3$:

$$3\,H^+ + Al(OH)_3 = Al^{3+} + 3\,H_2O \qquad (12.10)$$

with equilibria constant:

$$K_{Al(OH)_3} = \{Al^{3+}\}/\{H^+\}^3 \qquad (12.11)$$

Sulphate and fluoride complexes with aluminium in solution are also included.

4. Allowance is made for dissolution of carbonic acids caused by the elevated pCO_2 levels in the soils, and for the effects of degassing as water moves from the soils to the stream.

5. It is assumed that there are constant long term net inputs of base cations from mineral weathering. These are difficult to measure and must therefore be estimated.

6. Organics are modelled within MAGIC using a triprotic representation, by allowing for the species H_3Org^0, H_2Org^-, $HOrg^{2-}$ and Org^{3-}.

For the calculation, thermodynamic constants used are those given in the ALCHEMI programme.

12.5.3.2 *Long term application of the MAGIC model*

The MAGIC model can be applied to describe average changes in stream water chemistry in relation to atmospheric deposition and conifer afforestation.

Early attempts to model the forestry system were confined to a one box application of the MAGIC model: The upper soil and groundwater areas were lumped together. This application, for the Dargall Lane catchment in southwest Scotland (Neal et al. 1986), modelled the effects of changing atmospheric pollution levels of catchments of moorland, afforested moorland and deforested plantation forest. For the simulations only the effects of changing atmospheric scavenging of pollution by the change in vegetation and the associated changes in evapotranspiration were examined: Planting trees enhances the scavenging of pollutants from the atmosphere and increases the degree of evapotranspiration. The study showed that (Figs 12.11 and 12.12):

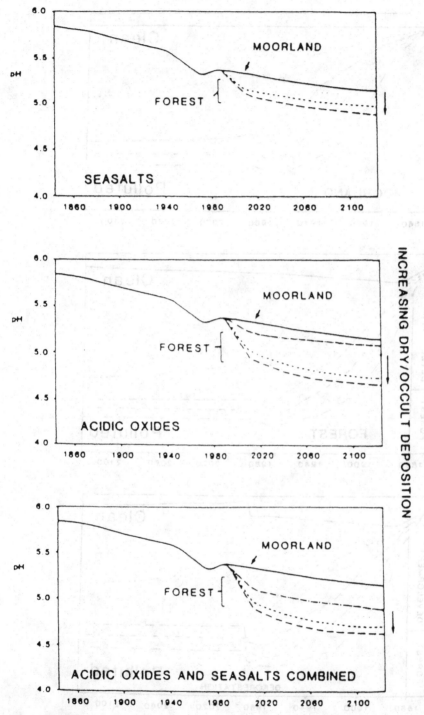

Figure 12.11. MAGIC simulation of the pH response from the Dargall Lane catchment, comparing moorland and afforested moorland responses to changing long term atmospheric deposition patterns appropriate for Scotland (data from Neal et al. 1986).

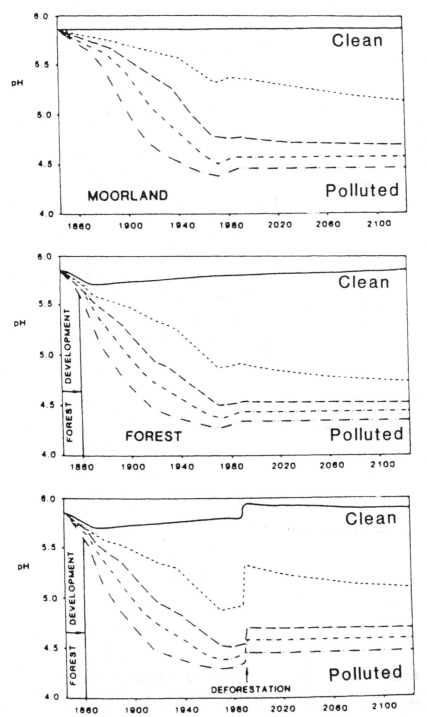

Figure 12.12. MAGIC simulation of the pH response from the Dargall Lane catchment, comparing moorland, afforested moorland and deforestation responses to changing long term atmospheric deposition patterns covering no too high pollution loadings (data from Neal et al. 1986).

1. Acidic oxide deposition results in stream water acidification for both moorland and forested areas: The degree of the acidification increased with increasing acidic oxide deposition.

2. The introduction of trees enhanced the acidification effect because of the scavenging of atmospheric pollutants.

3. Harvesting of trees in an environment polluted by atmospheric deposition results in a reduction in the stream acidity and an increase in stream pH (provided that biological effects are small – i.e. nitric acid is not generated).

4. Afforestation does not cause acidification unless there are pollutant inputs to the catchments (unless base cation uptake from the growing second generation forest depletes the exchangeable cation store within the soil).

12.5.3.3 *Event application of the MAGIC model*

The modelling approach is applied to the Gwy catchment (the southernmost of the two Plynlimon catchments), a typical example of a British acid-moorland upland system. Here a more detailed account of the modelling work is presented as it contains elements of the important water tracing, hydrograph splitting and hydrogeochemical framework described above. The site is selected because of its acid moorland vegetation and soils; this allows investigation of the long term effects of acidic precipitation without the added complexity of examining the effects of afforestation, felling and reafforestation. The results presented here update earlier work presented in Robson et al. (1990) and Neal et al. (1992a) in the light of further data (Robson 1993). The analysis is enhanced by the inclusion of more detail on the organic species in solution. Here, the MAGIC model is used to estimate these gradual changes in the end-members caused by acidic rainfall inputs.

The aluminium-hydrogen relationship used in MAGIC does not allow directly for exchangeable hydrogen [HX], even though this may be a significant component of the exchange fraction. However, in practice [AlX$_3$] is used to represent both the exchangeable hydrogen and the exchangeable aluminium – since [AlX$_3$] is taken to be the difference between the CEC and the exchangeable base cations. This simplification is reasonable because the relative proportions of exchangeable hydrogen and aluminium will remain fairly constant; the $\{Al^{3+}\}/\{H^+\}^3$ ratio in solution is defined to be constant.

A two layer version of MAGIC (Jenkins & Cosby 1989) is used in the application, enabling the simulation of two chemically distinct waters. The layers are chosen to correspond with the flow components identified from the mixing considerations outlined earlier; the top layer represents the lumped O, E and B and C soil horizons, whilst the bottom layer represents the deeper till layers and other water stores producing baseflow. The flow proportions included in the model are those suggested by the mixing approach. Concentrations of chloride and sulphate in the rainfall were adjusted for occult and dry deposition of sea-salts and anthropogenic sulphur compounds. Chloride is assumed to be chemically unreactive within the catchment, whereas sulphate is currently assumed to be in near-equilibrium, so that the total sulphate input is equal to the stream output. Differences between the rainfall concentrations and the stream outflow concentrations (allowing for evaporation effects) are assumed to result from dry and occult (mist, etc.) deposition, and these components are incorporated within the adjusted rainfall. Different reaction mechanisms dominate in each of the two layers in the model. The water from the upper soils is known to be acidic and organic and aluminium rich, indicating that ion exchange mechanisms and organic acid deprotonation reactions are the most important influence on soil water chemistry. Water from the deeper sources is rich in base cations, but low in exchangeable cations, as it comes from a high weathering zone. In this application, it is therefore assumed that ion-exchange occurs at a significant level only in the top layer. On the other hand, weathering is assumed to take place predominantly in the deeper layers. The thickness of the upper soil (0.9 m) corresponds to the average combined thicknesses of the O, E and B and C horizons and of the peats. An average depth of 1 m is assumed for the lower layer. The bulk density and cation exchange of the upper soils is calculated from field data for the Gwy soils. The bottom layer is represented as being denser and of lower porosity. The partial pressure of carbon dioxide is assumed to be 30 times atmospheric pressure for both layers. For the stream, the pCO_2 is set to 2.5 times atmospheric pressure, in line with average observed values for the Plynlimon streams (Neal & Hill 1994).

Very little is known about the sulphate characteristics of the Gwy soils. The selected constants describing the isotherms were chosen to be typical of values used in earlier MAGIC applications to the uplands (Jenkins & Cosby 1989; Whitehead et al. 1988). The same sulphate absorption isotherm was used in the two layers. The model is calibrated to the present day chemistry of the stream water, the soil water and the deeper groundwater by adjusting weathering rates, uptake rates and the initial soil base saturation. Given the limitations of the assumptions required within the mixing approach, and the analytical error in the chemical measurements, a perfect match between the modelled and the observed chemical species in the stream is unlikely. The ANC gives a good indication of the overall composition of a water sample and, whereas some deviations in individual determinands are acceptable, it is important that the modelled ANC is accurate.

12.5.3.4 *Long term modelling and forecasting*

The model was run for the period 1844 to 1984. The selectivity coefficients and initial base saturation were calibrated so as to match present day base saturation characteristics of the upper soils and peats. Weathering/uptake rates were then adjusted for the two layers in order to match stream and end-member chemistries as closely as possible. Nitrate and ammonia were modelled by calibrating catchment uptake to match the difference between the inputs and the outputs. Weathering-inputs/biological-uptake-rates were used to match observed base cation concentrations. Optimisation was performed manually, by adjusting parameters so as to match the present day stream chemistry as closely as possible.

The model has also been used to estimate future changes in stream, soil and deep water components. This was done by continuing the already calibrated run for 1844 to 1984, using two future sulphate deposition reduction scenarios; a linear reduction in non-marine sulphate to either 30% or 60% of present day levels, starting in 1984 and being completed by the year 2000, and held at a constant level thereafter. The time trends of reconstructed soil, ground and stream water chemistry from 140 years ago, that is pre-acidification, up to 2124, are shown in Figure 12.13. The ANC shows a decline from 1844 to the 1980's and recovery thereafter. Recovery under 60% reduction is more substantial than with 30% reduction, although the recovery takes

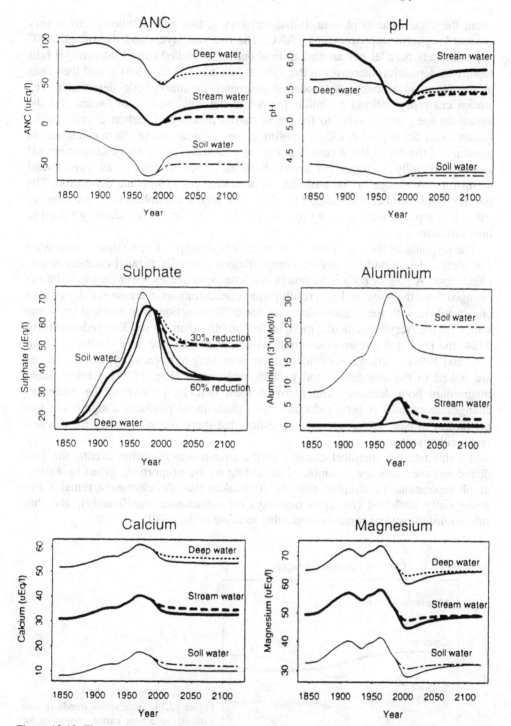

Figure 12.13. Time series trends in stream and soil chemistry for the two box MAGIC application for the Gwy catchment (data from Robson 1993). Solid lines represent 60% reduction and dashed lines indicate 30% reduction of sulphur deposition from 1984 to 2000.

about the same time to plateau. Initial recovery is fast and is followed by a very gradual long-term improvement in ANC. The two soil layers and the stream ANC approximately parallel one another throughout the modelled record. Stream pH falls rapidly as deposition increases in the 1950's, levels out in the 1980's, and then rises in the future in response to modelled decreases in atmospheric deposition. The stream chemistry follows a similar pattern to the soils and groundwater, but the variations are greater owing to the lower partial pressure of carbon dioxide in the stream. As a consequence, CO_2 degassing occurs as water moves from catchment to stream, and the pH of the stream is not always bounded by the two end-member pH values, especially at higher pH levels. Sulphate concentrations in all three model compartments change in line with the assumed changes in sulphate deposition. The baseflow end-member takes longer to adjust to changes in sulphate concentration because it is replenished from the upper box and is therefore only indirectly linked to the rainwater.

The response of the base cation concentrations to sulphate reduction is somewhat complex. Calcium and magnesium concentrations generally parallel changes in sulphate input. A 30% or 60% reduction in sulphate input causes fewer cations to be exchanged from the soils, and as a result cation concentrations decrease rapidly in both end-members with the greatest decrease for 60% reduction. For magnesium, some long term recovery is predicted, more so for the 60% than for the 30% reduction. By 2124, the predicted stream concentrations of magnesium are very similar for both 60% and 30% reductions. For calcium there is no long term recovery. These changes are linked to the changes in soil exchangeable bases (Fig. 12.14). Exchangeable magnesium both decreases and recovers most quickly, relative to the other exchangeable cations. A 60% reduction in sulphate input produces a significant improvement in base saturation for magnesium, but there is less recovery for calcium under either scenario.

To simulate the chemical changes in the stream waters during events, the predicted end-members are recombined according to the proportions given by hydrograph separation. To simplify matters, it is taken that the catchment remains hydrologically unaltered (i.e. flow pathways do not change significantly), and that future rainfall patterns remain comparable to those of the 1980's.

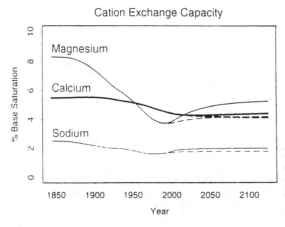

Figure 12.14. Time series trends in soil exchangeable base cation chemistry for the two box MAGIC application for the Gwy catchment (data from Robson 1993).

Variations in flow are matched by altered proportions of the end-members in the streams, and thus by altered stream chemistry. The chemical mixing reactions for the end-members are complex because non-conservative processes come into play for determinands such as hydrogen and aluminium as shown earlier in this chapter (Section 12.2.6), and by Christophersen & Neal (1989). The ALCHEMI chemical speciation programme has been used, again, to assess the chemical changes occurring when the end-member waters are mixed. The end-member compositions were then mixed in varying proportions (0-100% in 1% steps) and the resulting equilibrium chemical composition was found using ALCHEMI: Conditions were set so that dissolved carbon dioxide levels corresponded with those observed in the stream (approximately 2.5 times the atmospheric value). Within the calculation, allowance was made for aluminium hydroxide to precipitate from stream waters when they become oversaturated with respect to natural gibbsite following CO_2 degassing.

The results obtained from the mixing analysis allow the reconstruction/prediction of time series of estimated concentrations. These are produced using a time series of proportions of mix, estimated from current day continuous stream chemistry. The end-members are combined using these proportions according to the ALCHEMI results. In general, the basic patterns of chemical storm response remain essentially the same, but the peak and trough values vary. For ANC, the time series for the different years are parallel to one another. The most non-linear response of concentration variation with flow is the total aluminium charge which is low at baseflows in all years, but at high flows, is very different in each of the years considered. Figure 12.15 shows the model assessment of acid events with predictions for future scenarios involving different atmospheric pollutant loadings.

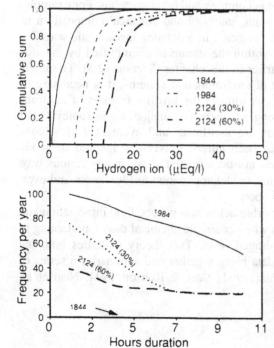

Figure 12.15. MAGIC model assessment of the frequency of acid events for the Gwy catchment, hydrogen ion duration (above), frequency incidence diagram (below) (data from Neal et al. 1992a).

The changes predicted by MAGIC suggest that the catchment end-member chemistry is likely to improve relative to present conditions, provided sulphate reduction is enforced. A 60% reduction is likely to result in a much greater improvement than 30% reduction, in the sense that the stream water chemistry will be more ecologically favourable especially at high flows. For example, the model predicts that whereas the difference in cation concentrations at high and low flows will remain the same, relative to each other, the predicted recovery in the upper soil water chemistry, will lead to a decrease in the peak concentrations of hydrogen and aluminium seen during high flows. Baseflow concentrations of both hydrogen and aluminium are currently low, and this situation will continue.

The approach of combining MAGIC and mixing techniques has been repeated in a similar form for Panola, USA (Hooper & Christophersen 1992). Here, the model was calibrated to soil water chemistry alone and was able to explain stream water chemistry on this basis. Three end-members were used, one of which was assumed constant over time. A similar feature of both applications was the need to incorporate a base cation source into the upper soil layer.

12.6 TOWARDS INTEGRATED MODELS

So far, the model development has been made in a step wise fashion based on the regular patterns of water quality response seen in the stream in response to rainfall events. However, in the process, the wide variability in water quality seen on or within the catchment has not been considered. This additional consideration is now made here, as it has many ramifications to how environmental systems can be modelled. In reality, the natural environment is complex and heterogeneous. For example, rain falling onto a region varies in amount, intensity and chemical composition in time and space in a way that cannot be predicted in a detailed mechanistic way. Indeed, this complexity can be observed within the stream as exemplified by the aluminium speciation studies described earlier in this chapter (Section 12.4). To understand the complex and variable nature of environmental systems, it is necessary to average data in a way that simplifies and mimics the broader features of environmental response. Unfortunately, in doing so, there is a major inconsistency (Neal 1992): Most of the chemical reactions are non-linear and average($F(x)$) seldom equals F(average(x)). This problem has been almost universally ignored by modellers who use lumped models and it has not been treated above in any serious way above, except for the case of the Birkenes model application to Plynlimon, and even there, the subject was only just touched upon.

In this section, chemical variability within acidic soil waters, field information and a modelling exercise are used to assess whether simple chemical theory pertaining to cation exchange may apply at the averaged level. This theory underlies both the Birkenes and MAGIC models. These data bring together and summarise a series of modelling and field based studies (Neal 1992; Neal & Robson 1994; Neal et al. 1994; Neal et al. 1994).

12.6.1 *Developing a cation exchange model accounting for soil heterogeneity*

Soils exhibit large spatial variation, in physical make up, and in their chemical and hydrological properties (Mulder et al. 1991; Neal 1992; Avila et al. 1995; Taugbøl & Neal 1994). Hydrochemical models of catchments standardly 'lump' such variability so that the model structure remains manageable (Christophersen et al. 1982; Cosby et al. 1985a,b). This means that the soils/soil-waters are assumed to be of constant and homogeneous composition at a given time or over a given time span. For catchment-based hydrochemical models, it is essential to establish how the soil variability influences the stream chemistry. This is required because water draining from the soil provides a chemically and hydrologically important component of storm runoff. To tackle this issue, the effects of heterogeneity within the catchment need to be examined. Synthetically generated data are used to examine how heterogeneity affects the properties of averaged quantities. This is exemplified by the detailed and extensive hydrochemical work at the Birkenes catchment in southern Norway (Christophersen et al. 1982, 1984, 1990a,b). This Norway spruce catchment on acidic and acid-sensitive soils has experienced acidic deposition typical of many parts of southern Scandinavia. The variability of the cation exchange fraction in localized areas of the soil (Mulder et al. 1991) is incorporated into the analysis to provide a field-based measure of the actual heterogeneity. The Birkenes soil data consists of measurements of the organic horizons of a matrix of podzols, peat and shallow organic soils collected from a 200 by 200 m grid. These organic horizons are common to many acidified areas: the results are therefore likely to be representative of many acidic and acid sensitive regions of environmental concern. Only the principal cations which make up the soil water and soil exchangeable fractions are considered here; i.e. Na^+, H^+, Mg^{2+}, Ca^{2+} and Al^{3+}.

For a soil region in which simple cation exchange equilibrium processes operate, then for any cation pair:

$$m\text{Ads}(N) + nM^{m+} = n\text{Ads}(M) + mN^{n+} \tag{12.12}$$

where M and N are two cations with respective charge m and n. Ads() refers to an adsorbed phase.

The cation concentration relationships are described by the equation:

$$K_{M,N} = \frac{\text{Ads}^n(M) \cdot \{N^{n+}\}^m}{\text{Ads}^m(N) \cdot \{M^{m+}\}^n} \tag{12.13}$$

where $K_{M,N}$ is an equilibrium constant, { } refers to ion concentrations (in equivalents units), and Ads( ) is dimensionless.

To distinguish between the variations in solution and in the soil cation exchange stores, Equation (12.13) is recast by introducing two terms, the ion activity ratio (IAR) and the adsorbed cation ratio (ACR). These terms are defined by:

$$\text{IAR}_{M,N} = \frac{\{M^{m+}\}^n}{\{N^{n+}\}^m} \tag{12.14}$$

and

$$\text{ACR}_{M,N} = \text{Ads}^n(M)/\text{Ads}^m(N) \tag{12.15}$$

and

$$K_{M,N} = \frac{\text{ACR}_{M,N}}{\text{IAR}_{M,N}}$$ (12.16)

To examine how the variations in micropore water chemistry (areas in equilibrium with the soil cation exchange sites) relate to the average chemistry of the soil water being transported to the stream (here termed average or macropore water), there is a need to (1) assess how, within the catchment, variability in ACR translates to the corresponding variation in IAR, and (2) examine how cation concentrations behave within the averaged soil water. Here, only one case is considered: K is assumed to be a constant, average, value throughout the soil, and anionic concentrations are constant within the soil micropores. Thus, at each localized micropore area of the soil, IAR varies only in response to the localized variability in ACR (Equation 12.16). Average K is determined from the average soil water chemistry determined in the field and from average exchangeable cation data. Inevitably there will be a bias owing to the heterogeneous nature of the situation being represented. None the less, the bias is unimportant for the theme being developed in this presentation. The analysis is described in detail by Neal & Robson (1994a,b), but, in brief:

1. Average K is estimated for Na–H, Na–(Ca+Mg) and Na–Al exchange.

2. Localized ACR values are approximated using the exchangeable cation data from the individual sites and the corresponding local IAR values are found using Equation (12.16).

3. The average catchment anion concentration is varied (as for example during the course of a storm event): All points in the catchment are assumed to have the same anion concentration at any one time and each contributes equally to the stream: ACR and IAR values at each point in the catchment are taken as constant. Cation concentrations are estimated for selected levels of total anion concentrations ($\{A^{\Sigma-}\}$) in the soil micropores using the IAR values. These concentrations are determined using Equations (12.13) to (12.16) and the charge balance equation.

12.6.1.1 *Results*

Variation in field ACR values introduces a large spatial variation in local soil-micropore water chemistry even when the anion concentration is fixed and uniform across the whole catchment. The results show large scatter at each anion concentration and many of the patterns, if not random, are near random (Figs 12.16 and 12.17). For the monovalent cations, different features are seen at low and high $\{A^{\Sigma-}\}$. When $\{A^{\Sigma-}\}$ is low, there is a negative and linear relationship between sodium and hydrogen: precisely the opposite of the theoretical relationship for the stream and entirely a consequence of these cations making up the main proportion of the positive charge at low $\{A^{\Sigma-}\}$. In contrast, when $\{A^{\Sigma-}\}$ is very high, there will be no statistically significant relationship between sodium and hydrogen ion. This is because at high ionic strength, divalent and trivalent cations dominate the positive charge. As $\{A^{\Sigma-}\}$ increases, the minimum, average and maximum cation concentrations also increase. The relationships between averaged cation and anion concentrations modelled compare reasonably well with what would be expected from a homogeneous cation exchange system, which is representative of average conditions. In

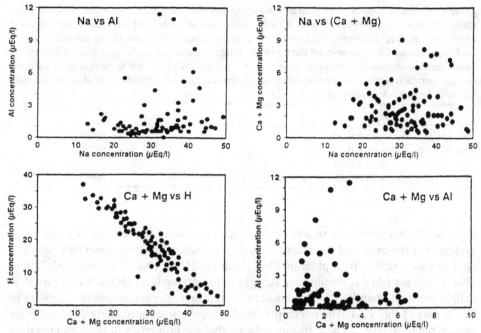

Figure 12.16. Modelled cation concentration relationships for micropore waters with anion concentrations set to 50 µEq/l (data from Neal & Robson 1994b).

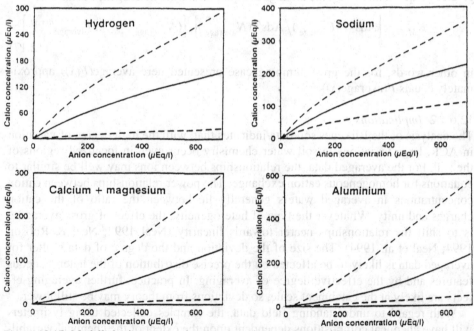

Figure 12.17. Modelled micropore cation concentration variations with changing anion concentration: Minimum, average and maximum changes (data from Neal & Robson 1994b).

Table 12.4. Comparison of theoretical and modelled power relationships for the Birkenes soil water study. Average cation concentrations at each anion concentration level are logged and regressed against one another. The regressions assess the appropriateness of the underlying theoretical cation exchange reactions to the averaged data. For the lumping theory to hold, p_{calc}, the gradient of the linear regression, should correspond to p_{theory}, the theoretical power for homogeneous cation exchange. For each regression, the left cation in column one is the independent variable and the right cation is the dependent variable.

	p_{calc}	p_{theory}	r^2
Na-H	1.03 ± 0.00	1	0.999
Na^2-Ca	2.01 ± 0.01	2	0.999
Na^3-Al	2.58 ± 0.01	3	0.999
Ca^3-Al^2	1.28 ± 0.01	1.5	0.999

other words, the curvature of the average lines is near to that for the homogeneous system. For the averaged (or macropore) data to match the homogeneous cation exchange case exactly, then plots of $\{Na^+\}$ against $\{H^+\}$, $\{Na^+\}^2$ against $\{M^{2+}\}$ and $\{Na^+\}^3$ against $\{Al^{3+}\}$, should give straight lines passing through the origin (as ACR is held constant, then IAR is also constant). To a first approximation, this appears to be the case (Table 12.4: The intercepts are statistically insignificant by different from zero at the 95% level). These results indicate that the properties of the averaged data are such that they can be represented by homogeneous cation exchange – provided the equilibrium constants for cation exchange do not vary considerably within the soil (Neal & Robson 1994; Neal et al. 1994). Thus, for the homogeneous reaction given by Equation (12.13), the analogous equation for the heterogeneous case is:

$$K_{M,N} = \left[\mathrm{Ads}^n(M_{average}) \middle/ \mathrm{Ads}^m(N_{average}) \right] \cdot \left[(N_{average}^{n+})^m \middle/ M_{average}^{m+})^n \right]$$
(12.17)

in other words, for the gross lumping case presented here average$(F(x))$ approximately equals F(average(x)).

12.6.1.2 *Implications*
The analysis of the Birkenes soil data indicates that, with the high degree of variation in ACR, large variations in soil water chemistry occur even in localized regions of the soil. For the averaged data, the relationships between ions may well be similar to equations for homogeneous cation exchange. The power relationships between cation concentrations in averaged waters generally lie between the ratio of the cation charges and unity. Whatever the type of heterogeneity, the effect of gross averaging is to shift the relationships nearer towards linearity (Neal 1992; Neal & Robson 1994; Neal et al. 1994). The size of the deviation and the degree of data scatter for averaged data is likely to be affected by the precise distribution of the heterogeneous features and by the effective degree of averaging. In practice, further averaging effects are likely at the catchment scale, so deviations from theory may be still lower.

With regard to understanding field data, the samples collected using lysimeters will have chemical compositions dependent upon the extent of the chemical variability within the soil micropores and the volumetric contribution of each water type.

Lysimeters may thus collect water from micropores which all have very similar water types, or may collect a mixture of radically different micropore waters. Thus, the lysimeter sample could have properties similar to micropore waters, or properties similar to averaged water types, or something in between. This type of unpredictable behaviour is what is actually observed in the field (Neal 1992; Avila et al. 1995; Taugbøl & Neal 1994). Unfortunately, such patterns are seldom published owing to the reluctance of many authors and reviewers alike to publish 'messy' data (Taugbøl & Neal 1994). With such field data, the relationships between cations in the lysimeters can vary from straight lines, to curves, to random patterns. By acknowledging the heterogeneity of the system, it becomes obvious that such data may still be consistent with a micropore cation exchange system – soil lysimeter data cannot be used as a test for cation exchange theory.

The hydrochemical models presently in use, address either the short term episodic variability of stream water chemistry or the long term year-to-year trends, concerned primarily with the effects of changing cation exchange stores in the soil and in the prediction of average stream water chemistry. The results of this study support the use of averaging techniques for the short term models, providing the underlying cation exchange formulation is correct and the heterogeneity considered here is not excessively large. For the long term there is more difficultly. Localized differences in hydrological flux, chemical input and weathering rates will influence the changes through time of the adsorbed cation stores. Quantifying such changes is nearly impossible even though there may be systematic elements to these changes. If such changes result in ACR distributions that are strongly skewed, there could be marked deviations away from idealized behaviour. For example, Hooper & Christophersen (1992) showed clearly that a small surface horizon was critical in the evolution of stream water quality in response to acidic deposition.

12.6.1.3 *Field testing*
The theory described above is based on the assumption, used almost universally in modelling of acidic catchments, that simple cation exchange reactions provide an appropriate description of the reactions occurring in soil micropores. However, is this assumption valid? And what, really, are the underlying mechanisms? There is little evidence for suggesting that the theoretical power relationships between cations in the soil solution hold, and that soil and stream waters are in equilibrium with solid aluminium hydroxide, not just at Plynlimon as mentioned above, but for many catchments (Neal et al. 1990a). Also, there is an unproven assumption that the soil system can be described simply by inorganic reactions: the soil cation exchanger is dominated by organic exchange sites (which are not of fixed charge) and disassociation of organic acids in the soil solution affects the soil water acidity (Krug & Frink 1983; Hendershot et al. 1991).

To test the validity of the cation exchange equilibria as applied on a catchment scale, standard techniques (e.g. collecting waters from soil lysimeters) are inappropriate. Rather, measurements need to be made at a more appropriate and better focused level. This requires catchment based experiments involving field sites where drainage is primarily from the soil zone. Consider such a case for a small drainage area in mid-Wales.

Since 1988, weekly sampling of a first order ephemeral stream draining a spruce forested hill-slope was undertaken at Plynlimon (Neal et al. 1994). This stream, south-2-Hore, drains thin stagnopodzol acid moorland soils (< 1 m) overlying slates and shales. The area, afforested in the 1930's with Sitka spruce (*Picea Sitchensis*) was clearfelled during three months in the autumn of 1989.

The South-2-Hore streamlet run-off is characterised by acidic and aluminium-bearing waters similar to those described above for Birkenes and, except for base-flow, the main Plynlimon forest streams (Neal et al. 1994). For Na^+, Ca^{2+}, Mg^{2+}, H^+, total-aluminium and trivalent aluminium there is a variable inter-correlation (Table 12.5). The major cations (Na^+, Ca^{2+} and Mg^{2+}) are highly correlated and these ions are also highly correlated with the total anion concentrations ($r^2 > 0.7$, $N = 205$). These patterns are essentially the same for the periods prior to felling, during felling and post felling. There are long term patterns of change. For example, during the winter of 1989 concentrations of chloride and cations were at their maximum. The changes observed do not relate directly to varying yearly inputs of the sea salts as, for example, there is no relationship between chloride concentrations in rainfall and chloride concentrations in stream water ($r^2 = 0.01$, $N = 195$). Rather, they reflect hydrological changes and possibly a reduction in the scavenging of sea salts with tree harvesting. In contrast, there is a poor correlation between the major cations with H^+ and total-aluminium (Table 12.5). It seems that H^+ variations are related to the deforestation phase, which seems to lead to increased acidification. The relationship between total aluminium and the other cations is unclear: it may be related to deforestation and nitrate production, but data scatter is high and the relationship is weak.

The results, even for the major cations, do not conform with cation exchange theory when examined in detail: the power relationships observed between cations of

Table 12.5. Linear regression data for logged concentration data for the South-2-Hore (UK) and Storgama (Norway) streams. The gradients correspond to the power relationships between ions and the bracketed terms are twice the standard error: $N = 205$.

	r^2	Gradient (regression)	Gradient (theory)
South-2-Hore			
$Na^+ - Ca^{2+}$	0.42	0.73 ± 0.12	1/2
$Na^+ - Mg^{2+}$	0.82	0.78 ± 0.05	1/2
$Na^+ - Al^{3+}$	0.12	0.54 ± 0.20	1/3
$Na^+ - H^+$	0.08	0.24 ± 0.11	1/1
$Ca^{2+} - Mg^{2+}$	0.64	0.61 ± 0.06	1/1
$Ca^{2+} - Al^{3+}$	0.00	0.12 ± 0.20	2/3
$Ca^{2+} - H^+$	0.01	0.10 ± 0.11	2/1
$H^+ - Al^{3+}$	0.27	0.94 ± 0.21	1/3
Storgama			
$Na^+ - Ca^{2+}$	0.67	0.78 ± 0.04	1/2
$Na^+ - Mg^{2+}$	0.73	0.86 ± 0.04	1/2
$Na^+ - Al^{3+}$	0.20	0.89 ± 0.14	1/3
$Na^+ - H^+$	0.03	0.20 ± 0.04	1/1
$Ca^{2+} - Mg^{2+}$	0.75	0.91 ± 0.02	1/1
$Ca^{2+} - Al^{3+}$	0.28	1.10 ± 0.14	2/3
$Ca^{2+} - H^+$	0.01	0.12 ± 0.08	2/1
$H^+ - Al^{3+}$	0.46	1.33 ± 0.11	1/3

different charge are well away from those given by theory for homogeneous cation exchange (Table 12.5). In the case of H^+ and Al^{3+} no simple relationship occurs between each other or the major cations and the data is highly scattered and, as with the major cations, the power relationships differ substantially from those given by theory (Table 12.5). While there is a broad relationship between the cation concentrations and the sum of the anions, there is no proper compatibility with the cation exchange formulations used in most acidification models. The Plynlimon results are hardly exceptional because nearly identical results are found for an acidified moorland site in southern Norway (Storgama, Table 12.5) and for a Holm oak catchment in northeastern Spain (Avila et al. 1995). For the major cations, the power relationships describing their co-variations lie between the theoretical values and unity at both sites. This is compatible with the theory described above for spatially and temporally heterogeneous cation exchange. However, the decoupling of H^+ and Al^{3+} from the major cations, implies that these components are not determined primarily by simple cation exchange reactions in the soil. Indeed, previous work at Plynlimon indicates that aluminium release to drainage water can be associated with kinetic 'weathering' type reactions involving proton consumption (Reynolds & Hughes 1986; Reynolds et al. 1989). These findings together with the evidence from long term trends of a decoupling of the base cations from hydrogen ion and aluminium concentrations (Dillon et al. 1987; Driscoll et al. 1989; Christophersen et al. 1990a), cast doubt upon the value of the commonly used cation exchange equations. Thus the mobile anion theory, which is almost universally evoked in acidification models, seems irrelevant.

12.6.2 *Summary of findings*

The above studies illustrate the difficulties of dealing with heterogeneous chemical systems. For the soils case, heterogeneity introduces problems both in the collection of appropriate data to determine the underlying chemical processes and in the production of lumped models. Despite this, the use of synthetic data has explained why field data may not match the underlying theory and helped to formulate the experimentation to test both the theory and the applicability of the lumped models. The second case provides a very disturbing example of incorrect conclusions resulting from averaging procedures and of how the acquisition of detailed experimental information (in this case thermodynamic details of the inorganic carbon system) is of no use for the modelling purpose.

12.7 DISCUSSIONS

In terms of the objectives of this chapter on modelling, the main message is that, while systems may be modelled in relatively simple ways by approaches and concepts that seem eminently sensible at the time, wrong conclusions may well result.

The problem for both hydrologic and chemical modelling is our inability to measure the key properties at the same scale that is needed to model the catchment. For example, the chemical models include non-linear equations, such as chemical equilibria, that have been derived for homogeneous environments, not for lumped averages over heterogeneous conditions (as discussed above and elsewhere, e.g. Neal

1992; Rastetter et al. 1992; Neal & Robson 1994; Neal et al. 1994) but needs to be more fully explored in quantitative terms. For hydrology, a description of streamflow generation, starting from first principles, is not available because the simplifications generally used for describing flow through porous media cannot be applied to the shallow soils in upland catchments of most interest in acidification studies. For example, these soils are anisotropic and contain macropores as important conduits. Such complications are solvable in principle if the properties of the porous media could be measured. But, the lack of definitive in situ measurements of soil properties, such as unsaturated conductivity, at the appropriate scale has so far made the problems insurmountable. There are long-standing and ongoing fundamental questions in hydrology as to how water is moving in a catchment (Dooge 1986; Beven 1989; see Chapter 3). For the chemical aspects, there is no adequate thermodynamic theory that partitions the chemical transfers between the mobile and immobile soil water environments.

The types of averaged properties identified here have wide and profound ramifications. This work demonstrates that averaging procedures may distort the picture of possible environmental impacts; simple synthetic data exercises can highlight the nature and direction of potential distortion. For example, terms such as the classical mass action equations which apply in a homogeneous situation, no longer apply in a heterogeneous world: only the mass and charge balance terms remain strictly applicable. The work on testing the cation exchange models at a representative level highlights a missing or incorrectly posed process fundamental to acidification research. In some ways, the methods used here resemble those associated with a very active and developing area of scientific research called complexity theory. Here one is concerned with how properties at the local scale determine global characteristics. Recent research has highlighted many areas of the environmental sciences where simple patterns occur at the macroscopic scale, but where immense complexity is observed at the microscopic level even though the determining mechanisms may well be described by simple rules (Cohen & Stewart 1994). However, the simple emergent rules identified, such as the straight line relationships shown in this paper, must be determined from field observations: theory simply explains the broad reasons why a particular phenomenon occurs. Thus, given a mismatch between field observation and modelling work, the model must be rejected or changed or challenged. This is the opposite path to the way the acidification research has developed: A key issue for environmental impact assessment must be to decide when a model can or cannot be used with confidence.

There is no easy route to a next generation of catchment models based more on field evidence. More data, per se, does not solve the problem. Although future advances in instrumentation may provide measurements at a larger scale (such as the effective conductivity of a hillslope), there is much that can be done today on the modelling side to bridge the difference in scale between models and observation. Here, two approaches are suggested that should be useful in this respect and this is followed by considering modelling of environmental systems in a larger setting.

In the first approach, an ad hoc technique is presented to make catchment modelling more measurement based. This approach, however, is specific to catchment research. The second approach comprises synthetic data analyses that can be used to test how the variables and concepts in a model can be related to measurable quantities and to explore how measurements, taken at one scale, may be related to the physics and chemistry on other scales. Although discussed in the context of catchment acidification, this approach applies generally to models of environmental systems.

The ad hoc approach correlates observed temporal variations in streamwater chemistry with the chemistry of observed soil water solutions. Soil waters often have a distinct chemical fingerprint because they arise from horizons that have chemically different solid phases due to natural soil formation. At some catchments, streamwater can be described as a dynamic mixture of these soil solutions, which are called end-members (Christophersen et al. 1990b; Hooper et al. 1990; Neal et al. 1990b,d).

By viewing soil solutions relative to streamwater, a measure for their variation can be established. Suppose the end-member chemical variations in time and space are small relative to the temporal variations in streamwater, and that the end-members are well separated in a chemical sense. Then, an end-member and the soil environment from which it arises, can be considered homogeneous enough from the point of view of explaining the origins of streamwater. The end-member units may then form the basis for a hydrological model as the relative contributions from the end-members to the stream can be computed and used to infer the temporal changes in hydrological routing through the catchment (Hooper & Christophersen 1992; Neal et al. 1992a). The hydrological structure is constrained by measured soil solution chemical properties, instead of imposing a priori a 'conceptual' lumped hydrological model on the catchment. Information on the hydrology is therefore obtained implicitly by using the chemical signals. An analogy from hydrology is the use of stochastic analysis for river basin planning (Loucks et al. 1981). In both cases, a process-based hydrological model was unnecessary to solve the problem at hand.

In the end-member approach, nothing is stated about how the end-member chemistry arises or how it will evolve through time. However, because these soil environments are defined through measurements, their actual physical locations are known and they can be examined to determine the within-end-member variability, and hence the appropriateness of applying an equilibrium chemical model. The end-member approach has only been applied to catchments with areas on the order of 1 km^2. Its applicability to larger catchments is not known, although it is an active area of research. Nonetheless, this approach provides an example of a measurement-based method that gives information on the internal catchment hydrochemistry at the appropriate scale.

The second approach uses synthetic data to establish more scientifically based models. In hydrology, for example, the approach could involve generation of an 'artificial' catchment as defined by the model under investigation for a given set of parameters. From the model, one could then generate 'observations' comprising, for example, runoff and water table elevations. Binley and Beven (1989) provide an example of this approach for a headwater hillslope. The necessary, but not sufficient, condition from the scientific viewpoint, is that the model is well enough posed so that all the parameters to be calibrated can be determined unambiguously from the 'observations' (properly error corrupted) given another set of initial calibration parameter values. The synthetic data exercise determines the type and density of measurements required. The practicality of implementing the required field program can be assessed, or, conversely, the utility of the information gained from a certain level of effort in the field can be assessed. After passing this test, the hydrological model is applied to the field. One is then in a much better position to perform a more stringent test of the relevance of the model structure and its parameters to the real environment.

A synthetic data exercise for the hydrological part of the Birkenes model along these lines was carried out by Hooper et al. (1988). Here, the model could not pass the first test since some of the parameters, even in this simple model, turned out to be ambiguous given only the runoff and chemically conservative tracer signals. Synthetic data have been used in hydrology elsewhere (Ibbitt & O'Donnell 1971; Gupta & Sorooshian 1985; Wheater et al. 1986). Our proposal is simply that environmental models routinely be subjected to such tests before they are applied and that efforts be directed towards development of the necessary software tools including parameter identification methods.

On the chemical side, synthetic data have proved useful in assessing the methods for determining the presence or absence of chemical equilibria in natural waters as shown above (Neal 1988 a,b; Christophersen & Neal 1989). In addition, such procedures are beginning to be used to assess the reliability of a lumped representation by investigating the properties of measurements representing averages over a non-homogenous environment. One example is the chemical properties of soil water collected in lysimeters (Neal 1992; Taugbøl & Neal 1994; Neal & Robson 1994; Neal et al. 1994). These devices (typically on the decimeter scale) are made of a porous medium that allows sampling of the soil solution by applying suction. If the soil is not homogenous on the decimeter scale, the solution sampled will be drawn from a heterogeneous environment. By assuming it is a mixture of water from homogenous micro-environments, each obeying equilibrium chemistry, one can investigate the properties of the average using the computer. In such situations, the results so far do provide some basis for the lumping procedure used to date whereby average soil properties are related to average soil water chemistry through the same type of equilibrium equations used in the homogenous case (Neal 1992; Taugbøl & Neal 1994; Neal & Robson 1994; Neal et al. 1994). The results of Kirchner (1992) and Kirchner et al. (1992), using real data, point in a similar direction. These authors have tested the runoff chemistry from several catchments, and state that chemical equilibrium relationships for homogenous systems could still be applicable at the catchment scale. The use of large scale field studies perturbing catchments, will also be useful in assessing these questions. At present, however, more work is needed before one can say from a process based viewpoint how lumped chemical properties should be related at the catchment scale.

The difficulties discussed here are not limited to hydrochemistry. Peters (1990) gives an account of related problems in limnology. In that area, long term field studies and modelling efforts aimed at quantifying the processes behind lake eutrophication have been carried out. The development of process based models is problematic for many of the same reasons as discussed here. However, from the management viewpoint, one has in limnology reasonably reliable methods for predicting lake response to nutrient (i.e. phosphate) inputs. These techniques, developed by Vollenweider and others, are empirically based (cf. OECD 1982). Given enough observations, empirical models will often be sufficient from a management viewpoint – after all, the methods used in managing the complex systems encountered in agriculture and forestry are empirical in nature.

Within acidification research, a related empirical approach was put forward by Henriksen (1980). However, in the acidification case, the important processes take place on larger spatial and temporal scales than in limnology. If one has to wait some years to see the effects of a change in phosphate inputs to a lake, the response time of a stream, regarding a change in sulphate deposition, is rather measured in decades. This implies that considerably more time and effort are needed to arrive at useful empirical relationships. Considering global climatic change, these concerns are again magnified – see Hauhs (1990) for a discussion of empirical versus process based approaches to the problem of ecosystem management.

Given this backdrop, one might well ask to what degree process based models will be able to quantify the behaviour of lakes, catchments and other environmental systems at the scale of interest. There could be fundamental limitations that cannot be overcome, given the measurements one can perform – i.e. can the environment be modelled? Probing this and related questions is vital to the future of environmental modelling.

12.8 RECOMMENDATIONS

From this presentation and associated work (Christophersen et al. 1993), the following recommendations are advanced:

a) When fitting a model to observations, the limited information content of the calibration signals must be recognized. For hydrology, this implies that modelling advances can only be made through the incorporation of chemical signals, groundwater elevations or other 'internal' state variables in addition to runoff. Thus fitting a model to the output signal of a complex environmental system (e.g. runoff measurements for catchments) is a necessary, but weak, condition and does not give scientific credibility to the mechanisms contained within the model. Such a fit only represents a starting point in model development;

b) The internal structure of current hydrochemical models has not been tested at the catchment scale. Recognition of this may serve to focus research efforts to advance the understanding of how ecosystems function. Extensions of current models without addressing their internal structure (for example, adding nitrate cycling to existing acidification models) should not be construed as scientific advances;

c) To progress within catchment science, studies of intensively and extensively monitored sites are needed where the internal structure of the catchment is examined;

d) To progress within environmental modelling, fuller use of synthetic data must be made both for tying models to observations and for exploring the observations themselves. The computer simulations will offer important learning tools in their own right;

e) Improved modelling of environmental systems is not a technological problem in that more data will necessarily give us better models. Rather there could be fundamental limitations to the modelling of these systems that are critical to unravel;

f) The fundamental equations depicting the acidification of catchments needs substantial re-examination;

g) And finally, environmental impacts must be related to biological response. And yet none of the present models address such an issue directly. Indeed, as with many biological processes, environmental impacts can only be addressed in a descriptive sense. Biological systems change in response to environmental pressures often in an obscure way. For example, at Plynlimon, deforestation has led to a deterioration in water quality characteristic of acidification, and yet acid sensitive may-flies have returned to the stream. Perhaps the greatest challenge within the environmental modelling studies is to produce an environmental impact model of direct relevance to the biology.

12.9 CONCLUSIONS

Over the past decade, we have tried to observe within catchments the dominant hydrochemical processes operative and we have attempted to relate the within catchment observations to the chemical variations within the stream. However, in trying to do so, as more and more data accumulated, we finished up with more questions than we started with. The problems encountered were associated with the complex and highly heterogeneous nature of catchments and the ease of producing catchment scale hydrochemical model fits with almost inevitably model overparameterisation. At the catchment level, there seemed to be no problem in that relatively simple patterns of stream behaviour occurred and this behaviour could be well represented by simple conceptual and mathematical model constructs which seemed eminently sound and convincing. However within the catchment, either random or straight line relationships occurred between ions that did not conform with the theory which so well fitted at the catchment level.

Within catchments, a complex interplay between chemical, biological and hydrological processes is encountered. This, coupled to the highly heterogeneous nature of catchments, means that reliable, quantitative, predictive mechanistic models of catchment systems are either very difficult or, most probably, impossible to produce. Because of this, there is a limit to how researchers can describe the hydrochemical functioning of catchments. However, at present the position of this limit cannot be clearly defined. Certainly, (1) the true value of hydrochemical models needs to be established so that more representative and reliable environmental impact models can be produced, and (2) no catchment model can counteract the need for high quality field observation.

12.10 FUTURE RESEARCH DIRECTIONS

It is essential for the development of catchment research and environmental impact studies at the catchment to regional scale to fully recognise that a highly complex and heterogeneous state exists within catchments and that fundamental research on such heterogeneity is required. Researchers must not be reluctant to look for and present results which do not conform to present concepts of how catchments function. Within this light, it is urged that more appropriate field experimentation is made to characterise the processes operative at a representative scale and to test for model assumptions and predictions. At the same time, it is essential that more appropriate models be developed to take into account spatial and temporal variability and to depict model uncertainty in some way. Within this context, the models should allow for biology, which has a fundamental role in determining chemical transfers. This requires new field experimentation and a fundamental reappraisal of how one describes and models such living systems.

Presently, hydrochemical models have acted as focuses for environmental research, and less and less funding has been given to field and long term monitoring. This is a very important mistake, as ultimately mechanistic environmental water quality models describing such complex heterogeneous environments cannot be relied upon with any certainty. It is urged that long term field programs continue to produce information which (1) can be used for testing environmental models, and (2) allows the development of an expert systems approach to environmental research – an expert in this context being a person who has observed environmental change already.

REFERENCES

Adamson, J.K. & Hornung, M. 1990. The effect of clearfelling a Sitka spruce (*Picea sitchensis*) plantation on solute concentrations in drainage waters. *Journal of Hydrology* 116: 287-298.

Avila, A., Bonilla, D., Rodà, F., Pinõl, J. & Neal, C. 1995. Soilwater chemistry in a Holm oak (*Quercus ilex*) forest: inferences on biogeochemical processes for a montane-mediterranean area. *Journal of Hydrology* 166: 15-35.

Beven, K.J. & Kirkby, M.J. 1979. A physically based variable contributing area model of basin hydrology. *Hydrology Science Bulletin* 24(1): 43-69

Beven, K.J. 1989. Changing ideas in hydrology: the case of physically based models. *Journal of Hydrology* 105: 157-172.

Beven, K.J. 1991. Hydrograph separation? *BHS Third National Symposium, Southampton University, September 1991.* Institute of Hydrology, 3.1-3.8.

Beven, K.J., Kirkby, M.J., Schoffield, N. & Tagg, A. 1984. Testing a physically based flood forecasting model TOPMODEL for three UK catchments. *Journal of Hydrology* 69: 119-143.

Binley, A. & Beven, K.J. 1989. Modelling heterogenous Darcian headwaters, in British Hydrological Society, *2nd National Symposium, University of Sheffield, 4-6 Sept.* Published by Institute of Hydrology, Wallingford, Oxon, UK, pp 1.17-1.22.

Birchall, J.D., Exley, C., Chappell, J.S. & Phillips, M.J. 1989. Acute toxicity of aluminium to fish eliminated in silicon-rich waters. *Nature* 338: 146-148.

Boyle, E., Collier, R., Dengler, A.T., Edmond, J.M., Ng, A.C. & Stallard, R.F. 1974. On the chemical mass-balance in estuaries. *Geochimica et Cosmochimica Acta* 28: 1719-1728.

Chappell, J.S. & Birchall, J.D. 1988. Aspects of the interaction of silicic acid with aluminium in dilute solution and its biological significance. *Inorganic Chimica Acta* 153: 1-4.

Christophersen, N. & Neal, C. 1989. A rational approach to the assessment of aluminium solubility controls in freshwaters. *Science Total Environment* 84: 91-100.

Christophersen, N., Neal, C. & Hooper, R.P. 1993. Modelling the hydrochemistry of catchments: a challenge for the scientific method. *Journal of Hydrology* 152: 1-12.

Christophersen, N., Neal, C., Hooper, R.P., Vogt, R.D. & Andersen, S. 1990b. Modelling streamwater chemistry as a mixture of soilwater end-members – A step towards second generation acidification models. *Journal of Hydrology* 116: 307-320.

Christophersen, N., Robson, A.J., Neal, C., Whitehead, P.G., Vigerust, B. & Henriksen, A. 1990a. Evidence for long-term deterioration of streamwater chemistry and soil acidification at the Birkenes catchment, southern Norway. *Journal of Hydrology* 116: 63-76.

Christophersen, N., Rustad, S. & Seip, H.M. 1984. Modelling streamwater chemistry with snow-melt. Philosophical Transactions of the Royal Society, London, series B, 305: 427-439.

Christophersen, N., Seip, H.M. & Wright, R.F. 1982. A model for streamwater chemistry at Birkenes, Norway. *Water Resourcese Research* 18: 977-996.

Cohen, J. & Stewart, I. 1994. *The Collapse of Chaos: Discovering simplicity in a complex world.* Viking, London, UK, 495 pp.

Cosby, B.J., Wright, R.F., Hornberger, G.M. & Galloway, J.N. 1985a. Modelling the effects of acidic deposition: assessment of a lumped-parameter model of soil water and stream water chemistry. *Water Resources Research* 21: 51-63.

Cosby, B.J., Wright, R.F., Hornberger, G.M. & Galloway, J.N. 1985b. Modelling the effects of acid deposition: estimation of long-term water quality responses in a small forested catchment. *Water Resources Research* 21: 1591-1601.

DeWalle, D.R., Swistock, B.R. & Sharpe, W.E. 1988. Three component tracer model for stormflow on a small appalachian forested catchment. *Journal of Hydrology* 104: 301-310.

Dillon, P.J., Read, R.A. & de Grosbois, E. 1987. The rate of acidification of aquatic ecosystems in Ontario, Canada. *Nature* 329: 45-48.

Dimbleby, G.W. 1952. Soil regeneration on the north-east Yorkshire moors. *Journal of Ecology* 40: 331-341.

Dooge, J.C.I. 1986. Looking for hydrologic laws. *Water Resources Research* 22: 46S-58S.

Driscoll, C.T., Likens, G.E., Hedin, L.O., Eaton, J.S. & Bormann, F.H. 1989. Changes in the chemistry of surface waters. *Environmental Science and Technology* 23: 137-143.

Egglishaw, H.J., Gardener, R. & Foster, J. 1986. Salmon catch decline and forestry in Scotland. *Scottish Geological Magazine* 102: 57-61.

Emmett, B. 1989. The effects of harvesting intensity on soil nitrogen transformations in Sitka spruce (*Picea sitchensis* (Bong.) Carr.) plantation at Beddgelert Forest (N. Wales). PhD thesis, University of Exeter, UK.

Gupta, V.J. & Sorooshian, S. 1985. The automatic calibration of conceptual catchment models using derivative-based optimization algorithms. *Water Resources Research* 21: 473-485.

Haines, T.A. 1986. Fish population trends in response to surface water acidification. In: *Acid Deposition, Long-term trends.* National Academic Press, Washington, DC, 300-334.

Hauhs, M. 1990. Ecosystem modelling: Science or technology?. *Journal of Hydrology* 116: 25-34.

Hendershot, W.H., Warfvinge, P., Courchesne, F. & Sverdrup, H.U. 1991. The mobile anion concept – time for a reappraisal? *Journal of Environmental Quality* 20(3): 505-509.

Henriksen, A. 1980. Acidification of freshwaters – a large scale titration, In: Drabløs, D. & Tollan, A. (eds), *Ecological Impact of Acid Deposition.* Norwegian Institute for Water Research, 68-74.

Hewlett, J.D. & Hilbert, A.R. 1967. Factors affecting the response of small watersheds to precipitation in humid areas. In: Sopper, W.E. & Lull, H.W. (eds), *Forest Hydrology.* Pergamon, Oxford, UK, 275-290.

Hooper, R.P. & Christophersen, N. 1992. Predicting episodic stream acidification in the Southeastern United States: Combining a long-term acidification model and the end-member mixing concept. *Water Resources Research* 28: 1983-1990.

Hooper, R.P. & Shoemaker, C.A. 1986. A comparison of chemical and isotopic hydrograph separation. *Water Resources Research* 22(10): 1444-1454.

Hooper, R.P., Christophersen, N. & Peters, N.E. 1990. Modelling streamwater chemistry as a mixture of soilwater end-members – An application to the Panola Mountain catchment, Georgia, USA. *Journal of Hydrology* 116: 321-343.

Hooper, R.P., Stone, A., Christophersen, N., de Grosbois, E. & Seip, H.M. 1988. Assessing the Birkenes model of stream acidification using a multisignal calibration methodology. *Water Resources Research* 24: 1308-1316.

Hordijk, L. 1991. Use of the RAINS model in acid rain negotiations in Europe. *Environmental Science and Technology* 25: 596-603.

Hornung, M., Adamson, J.K., Reynolds, B. & Stevens, P.A. 1989. Impacts of forest management practices in plantation forests. *Air Pollution Research Report* 13: Effects of land use in catchments on the acidity and ecology of natural surface waters, 91-106. Commission of the European Communities, CEC-DG XII/E-1, 200 Rue de la Loi, B-1049, Brussels, Belgium.

Hughes, S., Reynolds, B. & Roberts, J.D. 1990. The influence of land management on concentrations of DOC and its effects on the mobilisation of aluminium and iron in podzol soils in mid-Wales. *Soil Use and Management* 6: 137-144.

Ibbit, R.P. & O'Donnell, T. 1971, Fitting methods for conceptual catchment models. *J. Hydraulic Division American Society of Civil Engineering* 97: 1331-1342.

Jenkins, A. & Cosby, B.J. 1989. Modelling surface water acidification using one and two soil layers and simple flow routing. In: Kamari, J., Brakke, D.F., Jenkins, A., Norton, S.A. & Wright, R.F. (eds), *Regional Acidification Models*. Springer-Verlag, Berlin, Heidelberg, Chapter 19.

Jenkins, A., Cosby, B.J., Ferrier, R.C., Walker, T.A.B. & Miller, J.D. 1990. Modelling stream acidification in afforested catchments: an assessment of the relative effects of acid deposition and afforestation. *Journal of Hydrology* 120: 163-181.

Kennedy, V.C., Kendall, C., Zellweger, G.W., Wyerman, T.A. & Avanzino, R.J. 1986. Determination of the components of stormflow using water chemistry and environmental isotopes, Matolle river basin, California. *Journal of Hydrology* 84: 107-140.

Kirchner, J.W. 1992. Heterogenous geochemistry of catchment acidification. *Geochimica et Cosmochimica Acta* 56: 2311-2328.

Kirchner, J.W., Dillon, P. & LaZerte, B.D. 1992. Catchment geochemical buffering predictions tested with acid episodes. *Nature* 358: 478-482.

Kirkby, C., Newson, M.D. & Gilman, K. 1991. Plynlimon research: the first two decades. Institute of Hydrology Report 109: 1-187.

Krug, E.C. & Frink, C.R. 1983. Acid rain on acid soil: a new perspective. *Science* 122: 520-525.

Krug, E.C. 1991. Review of acid deposition-catchment interaction and comments on future research needs. *Journal of Hydrology* 128: 1-27.

Liss, P.S. 1976. Conservative and non-conservative behaviour of dissolved constituents during estuarine mixing. In: Burton, J.D. & Liss, P.S. (eds), *Estuarine Chemistry*. Academic Press, London, 93-130.

Loucks, D.P., Stedinger, J.R. & Haith, D.A. 1981. *Water Resource Systems Planning and Analysis*. Prentice-Hall, Inc., Englewood Cliffs, NJ, 559 pp.

McMahon, R. & Neal, C. 1990. Aluminium dis-equilibrium solubility controls in Scottish acidic catchments. *Hydrological Sciences Bulletin* 35(1): 21-28.

Miller, H.G. 1985. The possible role of forests in stream acidification. *Soil use and management* 1: 28-35.

Mulder, J., Pijpers M. & Christophersen, N. 1991. Water flow paths and the spatial distribution of soils and exchangeable cations in an acid rain-impacted and a pristine catchment in Norway. *Water Resources Research* 27: 2919-2928.

Mulder, J., van Breemen, N. & Eijck, H.C. 1989. Depletion of soil aluminium by acid deposition and implications for acid deposition. *Nature* 337: 247-249.

Muniz, I.P. & Leiverstad, H. 1980. Acidification effects on freshwater fish. In: Drablos, D. & Tollan, A. (eds), *Ecological Impacts of Acid Precipitation*. Oslo-As:SNSF, 84-92.

Neal, C. & Robson, A.J. 1994. Integrating soil water chemistry variations at the catchment level within a cation exchange model. *Science of the Total Environment* 144: 93-102.

Neal, C. 1988a. Aluminium solubility in acid waters. *Earth and Planetary Science Letters* 86: 105-112.

Neal, C. 1988b. Aluminium solubility relationships in acid waters; a practical example of the need for a radical reappraisal. *Journal of Hydrology* 104: 141-159.

Neal, C. 1992. Describing anthropogenic impacts on streamwater quality: the problem of integrating soil water chemistry variability. *Science of the Total Environment* 115: 207-218.

Neal, C. 1996. Towards lumped integrated models of complex heterogenous systems. Science of the Total Environment, in press.

Neal, C. & Hill, S. 1994. Dissolved inorganic and organic carbon in moorland and forest streams: Plynlimon, Mid-Wales. *Journal of Hydrology* 153: 231-243.

Neal, C. & Rosier, P. 1990. Chemical studies of chloride and stable oxygen isotopes in two conifer afforested and moorland sites in the Britsih uplands. *Journal of Hydrology* 115: 269-283.

Neal, C., Christophersen, N., Neale, R., Smith, C.J., Whitehead, P.G. & Reynolds, B. 1988. Chloride in precipitation and streamwater for the upland catchment of River Severn, Mid-Wales; Some consequences for hydrochemical models. *Hydrological Processes* 2: 155-165.

Neal, C., Fisher, R., Smith, C.J., Hill, S., Neal, M., Conway, T., Ryland, G.P. & Jeffrey, H.A. 1992b. The effects of tree harvesting on stream water quality at an acidic and acid sensitive spruce forested area: Plynlimon, mid-Wales. *Journal of Hydrology* 135: 305-319.

Neal, C., Mulder, J., Christophersen, N., Neal, M., Waters, D., Ferrier, R.C., Harriman, R. & McMahon, R. 1990a. Limitations to the understanding of ion exchange and solubility controls for acidic Welsh, Scottish and Norwegian sites. *Journal of Hydrology* 116: 11-23.

Neal, C., Neal, M., Ryland, G.P., Jeffery, H.A., Harrow, M., Hill, S. & Smith, C.J. 1994. Chemical variations in near surface drainage water for an acidic spruce forested UK upland area subjected to timber harvesting: inferences on cation exchange processes in the soil. Science of the Total Environment, in press.

Neal, C., Neal, M., Waters, D., Mulder, J., Christophersen, N., Ferrier, R.C., McMahon, R & Harriman, R. 1990c. Limitations to the understanding of ion-exchange and solubility controls for acidic Welsh, Scottish and Norwegian sites. *Journal of Hydrology* 116: 11-24.

Neal, C., Reynolds, B., Smith, C.J., Hill, S., Neal, M., Conway, T., Ryland, G.P., Jeffrey, H.A., Robson, A.J. & Fisher, R. 1992c. The impact of conifer harvesting on stream water pH, alkalinity and aluminium concentrations for the British Uplands: an example for an acidic and acid sensitive catchment in mid-Wales. *Science of the Total Environment* 126: 75-87.

Neal, C., Reynolds, B., Stevens, P. & Hornung, M. 1989. Hydrogeochemical controls for inorganic aluminium in acidic stream and soil waters at two upland catchments in Wales. *Journal of Hydrology* 106: 155-175.

Neal, C., Robson, A.J. & Smith, C.J. 1990b. Acid Neutralization Capacity variations for Hafren Forest streams: inferrences for hydrological processes. *Journal of Hydrology* 121: 85-101.

Neal, C., Robson, A.J., Reynolds, B. & Jenkins, A. 1992a. Prediction of future short-term stream chemistry – a modelling approach. *Journal of Hydrology* 130: 87-103.

Neal, C., Smith, C.J., Walls, J., Billingham, P., Hill, S. & Neal, M. 1990d. Hydrogeochemical variations in Hafren forest stream waters, mid-Wales. *Journal of Hydrology* 116: 185-200.

Neal, C., Whitehead, P.G., Neale, R. & Cosby, B.J. 1986. Modelling the effects of acidic deposition and conifer afforestation on stream acidity in the British Uplands. *Journal of Hydrology* 86: 15-26.

Newson, M.D. 1976. The physiography, deposits and vegetation of the Plynlimon catchments. Institute of Hydrology Report, 30: 1-59.

Nisbet, T.R. 1990. Forests and surface water acidification. *Forestry Commission Bulletin* 86: 1-7.

OECD 1982. Eutrophication of waters. *Monitoring, assessment and control.* OECD, Paris, 154 pp.

Pearce, A.J., Stewart, M.K. & Sklash, M.G. 1986. Storm runoff in humid headwater catchments 1. Where does the water come from? *Water Resources Research* 22(8): 1263-1272.

Pennington, W. 1984. Long term natural acidification of upland sites in Cumbria: evidence from post-glacial lake sediments. Report of the Freshwater Biology Association, 52: 28-46.

Peters, R.H. 1990. Pathologies in limnology. In: de Bernardi, R., Giussani, G. & Barbanti, L. (eds), Scientific Perspectives in theoretical and applied limnology. *Mem. Ist. Ital. Idrobiology*, 47: 181-217.

Pilgrim, D.H., Huff, D.D. & Steele, T.D. 1979. The use of specific conductance and contact times relations for separating flow components in storm runoff. *Water Resources Research* 15(2): 329-339.

Pinder, G.F. & Jones, J.F. 1969. Determination of the groundwater component of peak discharge from the chemistry of total runoff. *Water Resources Research* 5: 438-445.

Rastetter, E.B., King, A.W., Cosby, B.J., Hornberger, G.M., O'Neill R.V. & Hobbie, J.E. 1992. Aggregating fine-scale ecological knowledge to model coarser-scale attributes of ecosystems. *Ecological Applications* 2: 55-70.

Reuss, J.O. & Johnson, D.W. 1986. Acid Deposition and the acidification of soils and waters. *Ecological Studies* 59. Springer Verlag. 1-119.

Reuss, J.O., Christophersen, N. & Seip, H.M. 1986. A critique of models for freshwater and soil acidification. *Water, Air and Soil Pollution* 30: 909-931.

Reynolds, B. & Hughes, S. 1986. An ephemeral forest drainage ditch as a source of aluminium to surface waters. *Science of the Total Environment* 80: 185-193.

Reynolds, B., Hornung, M. & Hughes, S. 1989. Chemistry of streams draining grassland and forest catchments at Plynlimon, mid-Wales. *Hydrological Sciences Bulletin* 34(6): 667-686.

Reynolds, B., Neal, C., Hornung, M. & Stevens, P.A. 1986. Baseflow buffering of stream water acidity in five mid-Wales catchments. *Journal of Hydrology* 87: 167-185.

Reynolds, B., Neal, C., Hornung, M. & Stevens, P.A. 1988. Impact of afforestation on the soil solution chemistry of stagnopodzols in mid-Wales. *Water, Air and Soil Pollution* 38: 55-70.

Robson, A.J. & Neal, C. 1991. Chemical signals in an upland catchment in mid-Wales – some implications for water movement. In: *BHS Conference Proceedings*, Southhampton University. Institute of Hydrology. 3.17-3.24.

Robson, A.J. 1993. The use of continuous measurement in understanding and modelling the hydrochemistry of the uplands. PhD dissertation, University of Lancaster, Lancaster, UK, 278 pp.

Robson, A.J., Neal, C. & Smith, C.J. 1990. Hydrograph separation using chemical techniques: an application to catchments in mid Wales. *Journal of Hydrology* 116: 345-365.

Rodhe, A. 1981. Spring flood. Meltwater or groundwater? *Nordic Hydrology* 43: 45-65.

Rosenqvist, I.T. 1978. Alternative sources for acidification of river water in Norway. *Science of the Total Environment* 10: 39-49.

Rosenqvist, I.T. 1990. From rain to lake; pathways and chemical change. *Journal of Hydrology* 114: 3-10.

Schecher, W.D. & Driscoll, C.T. 1987. An evaluation of uncertainty associated with aluminium equilibrium calculations. *Water Resources Research* 23: 525-534.

Sherman, L.K. 1933. Stream flow from rainfall by the unit-hydrograph method. *Engineering News Necord* 108: 501-505.

Sklash, M.G. & Farvolden, R.N. 1979. The role of groundwater in storm runoff. *Journal of Hydrology* 43, 45-65

Sklash, M.G. 1990. Environmental isotope studies of storm and snowmelt runoff generation. In: M.G. Anderson & T.P. Burt (eds), *Process studies in Hillslope Hydrology*. Wiley, Chichester, 401-435.

Sklash, M.G., Stewart, M.K. & Pearce, A.J. 1986. Storm runoff generation in humid headwater catchments. 2 A case study of hillslope and low order stream response. *Water Resources Research* 22: 1273-1282.

Stevens, P.A. & Hornung, M. 1987. Nitrate leaching from a felled Sitka spruce plantation in Beddgelert Forest, north Wales. *Soil Use and Management* 4: 3-8.

Stevens, P.A., Hornung, M. & Hughes, S. 1989. Solute concentrations, fluxes and major nutrient cycles in a mature Sitka spruce plantation in Beddgelert Forest, North Wales. *Forest Ecology and Management* 27: 1-20.

Stoner, J.H. & Gee, A.S. 1985. Effects of forestry on water quality and fish in Welsh rivers and lakes. *Journal of the Institute of Water Engineering Sciences* 39: 27-45.

Sverdrup, H.U. 1990. *The kinetics of base cation release due to chemical weathering.* Chartwell Bratt Ltd., Bromley, Kent, UK, 246 pp.

Taugbøl, G. & Neal, C. 1994. Soil and stream water chemistry variations on acidic soils. Application of a cation exchange and mixing model at the catchment level. *Science of the Total Environment* 149: 83-96.

Taylor, J.A. 1974. Organic soils in Wales. In: Adamson, W.A. (ed.), *Soils in Wales*. Welsh soils discussion group. Report 15: 30-43.

Turnpenny, A.W.H., Sadler, K., Aston, R.J., Milner, A.G.P. & Lyman, S. 1987. The fish population of some streams in Wales and Northern England in relation to acidity and associated factors. *Journal of Fish Biology* 31: 415-434.

UKAWRG 1988. United Kingdom Acid Waters Review Group, second report. Acidity in United Kingdom fresh waters. Her Majesty's Stationary Office, London, 1-61.

Wells, C., Cornett, R.J. & Lazerte, B.D. 1991. Hydrograph separation: A comparision of geo-chemical and isotopic tracers. *Journal of Hydrology* 122: 253-274.

Wheater, H., Bishop, K.H. & Beck, M.B. 1986. The identification of conceptual hydrologic models for surface water acidification. *Hydrological Processes* 1: 89-108.

Whitehead, P.G., Bird, S., Hornung, M., Cosby, B.J., Neal, C. & Paricos, P. 1988. Stream acidification trends in the Welsh Uplands: a modelling study of the Llyn Brianne catchments. *Journal of Hydrology* 101: 191-212.

SUGGESTED READING

Bronowski, J. 1973. The Ascent of Man. British Broadcasting Corporation, 35 Marylebone High Street, London, W1M 4AA, UK.

Cohen, J. & Stewart, I. 1994. The Collapse of Chaos: Discovering simplicity in a complex world. Viking, London, UK.

CHAPTER 13

Chemical changes attending water cycling through a catchment – An overview

PATRICE DE CARITAT & OLA M. SAETHER
Geochemistry and Hydrogeology Section, Geological Survey of Norway, Trondheim, Norway

Just like every drop of rain,
Many drops can someday make a river,
And many rivers go down to the sea,
And the sea rolls on forever, forever   Chris De Burgh, 1979

13.1 INTRODUCTION

In the previous chapters, we have delved, to greater or lesser depth, into the main categories of processes that affect water quality in a catchment; these processes can be (geo)chemical in nature (e.g. weathering), physical (groundwater recharge) or physico-(bio)chemical (soil formation). This compartmentalisation of catchment studies is, to a large extent, both unavoidable and useful: we can not understand the mechanisms driving a so complex and inter-related entity as a catchment without first focusing on how its numerous parts function. But such an approach also drives us inexorably away from a more encompassing generalised overview of the total catchment, as a unit of the Earth's surface environment. Casting the various gems of scientific erudition gleaned from the different sub-disciplines of catchment studies into a single, generalised master scheme of integrated catchment study would be nearly impossible, and certainly well beyond the scope of this work. We think, however, that it would be appropriate for this closing chapter to attempt to present, within the limits of practicality, a brief, unifying overview of the catchment processes governing water chemistry. We propose to do this by considering the fate of a unit volume of water as it makes its way through a catchment (located in a cool, humid region such as northern Europe), based mostly on the previous chapters of this textbook and on some of the works listed at the end of this chapter. To grasp the context in which chemical changes occurring to this volume of water take place, let us first consider the interacting geochemical and hydrological cycles.

13.2 THE GEOCHEMICAL CYCLE

The elements in the periodic table are, at the outset, present in major, minor or trace quantities in the magma of the earth's interior. Magma ultimately is the source material of oceanic crust (at mid-oceanic ridges) and continental crust (in subduction zones). Rocks within the oceanic and continental crust are divided into igneous, metamorphic and sedimentary types, depending on whether they formed by cooling of a molten phase (T>1100°C), by (re)crystallisation of partially melted rock (200-1100°C), or by deposition of particulate or dissolved matter generated during chemical and physical weathering (<200°C) at the earth's surface. The conveyor belt analogy in plate tectonics explains how the two major types of crust are formed and recycled by the interaction of moving lithospheric plates. This analogy is one of the major contributions to the understanding of how the various elements are cycled within the earth's crust (the endogene environment) and at the earth's surface (the exogene environment). The latter environment is of particular interest here, especially with regard to how it interacts with the water cycle.

The compartments and reservoirs of the (exogene) geochemical cycle are the atmosphere, hydrosphere (oceans, surface and ground waters, and glaciers), pedosphere/biosphere (including soil, vegetation and organic matter), and geosphere (crust) (Fig. 13.1).

The residence time of a given element in each reservoir or compartment is given by the ratio of its abundance there to the rate at which it circulates through the compartment. This concept implies the idea of flux of an element through the reservoir, underscoring that the concentration level of an element determined in any sample always should be seen in a dynamic perspective, at various time scales.

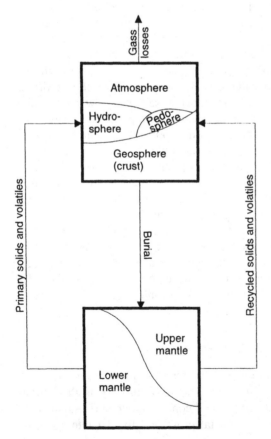

Figure 13.1. Outline of the geochemical cycle (modified after Mason & Moore 1982). The upper box represents the exogene environment, the lower one the endogene environment. Compartment boundaries (thin lines) are open for mass flow in both directions.

13.3 THE WATER CYCLE

Water is a mobile medium within any landscape, and, as such, it comes into contact with many parts of a catchment that do not themselves directly interact with one another, for instance humus and till, or the atmosphere and buried bedrock. Thus, one of the most fundamental driving mechanisms for geochemical and hydrological processes in a catchment is the global water cycle: precipitation, which falls as rain, snow, fog, etc., is partly evaporated (or evapotranspired), partly stored in living organisms and soil, partly recharged into the groundwater, and partly transferred to surface water streams by surface (or shallow subsurface) runoff. Whatever the size of the catchment, all of these paths, and others that occur in the ocean and atmosphere, are integral parts of the global water cycle (Fig. 13.2). If the purely physical or hydrological aspects of this dynamic 'loop' are relatively well understood, at least macroscopically, at the catchment level, the same can not be said of its geochemical implications. We have seen, for instance, how complicated predicting stream water chemistry through time can become when an attempt is made to take into account the heterogeneity of a catchment's soil. Of course, the water cycle does not by itself control entirely the nature, rate and extent of geochemical processes occurring in any catchment: the influence of the energy cycle (seasons), climate change and, on a much longer time scale, the geological cycle also is of importance. However, inasmuch as one can decide that the time scale of interest in catchment studies typically is the year or decade, rather than the day or aeon, these latter controlling

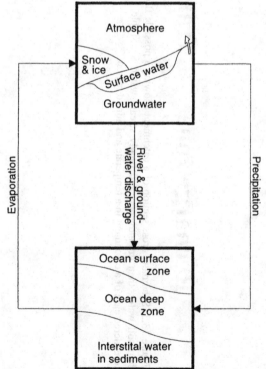

Figure 13.2. Outline of the hydrological cycle (modified after Drever 1988). The upper box represents the continental environment, the lower one the marine environment. Compartment boundaries (thin lines) are open for mass flow in both directions, except where marked by an arrow.

NOT TO SCALE

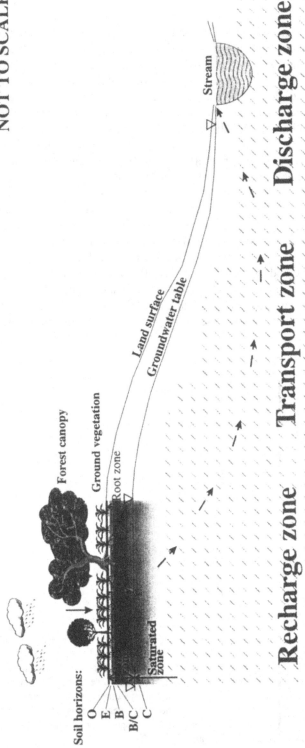

Figure 13.3. Schematic cross section of a catchment showing a possible flowpath for water as discussed in the text: rain water may become throughfall, then soil water, then groundwater, and finally stream water. See Table 13.1 for an overview of the chemical changes attending such cycling in the considered case of a catchment in a temperate and humid region.

mechanisms can be set aside (temporarily). So then, is it possible to describe, even in the most simplistic way, how the chemical composition of a drop of water changes from the time it falls on the upslope parts of a catchment to the time it leaves it?

13.4 EVERY DROP OF RAIN

Let us consider a light diurnal rain shower on a natural, vegetated landscape set in a relatively remote humid temperate region (Fig. 13.3). The rain water typically is dilute (even concentrations of chloride, of the base cations sodium, calcium, magnesium and potassium, and of sulphate, nitrate and ammonium are low), poor in particle content (little local soil dust or long-transported particles), and does not contain levels of anthropogenic acids or other pollutants exceeding background. Even so, it does not have a neutral pH: rain water in equilibrium with atmospheric carbon dioxide has a pH around 5.6 (the presence of other natural acids, such as derived from sulphur dioxide spewed out of volcanoes, may lower further the pH of rain). Many industrialised areas are in or near humid temperate regions in the world. As a result, the present background levels do reflect the ensuing atmospheric pollution, and even in our remote catchment, we may expect rain to have a pH between 4.5 and 5.0 (assuming there is no significant amount of calcite dust in the air, which would give a higher pH). If the rain water reaching the ground has been intercepted by a forest canopy first, its concentration of base cations, nitrogen, organic carbon, chloride and sulphate may have increased manyfold (these compounds may come from the plants or from dry deposition on the canopy surface, and be upconcentrated by evaporation on leaf surfaces). At the same time, the water's pH may have increased (in other cases, a pH decrease may occur). Rainfall reaching the ground through a vegetation cover is generally described as 'throughfall' water (Tabel 13.1).

Let us consider a drop of 'throughfall' water falling on a cluster of *Hylocomium splendens* (an almost ubiquitous terrestrial moss in our imaginary catchment) growing near the base of a tree close to the top of a small hill in the uplands of the catchment. While temporarily clinging to *H. splendens* by capillary forces, ion exchange occurs between the rain drop and the turgid shoots of the moss. The latter attempts to extract from the water as much of the macro-nutrients (e.g. sulphur, nitrogen, potassium) and micro-nutrients (e.g. copper) as it can, and, in doing so also takes up some undesired trace elements and even a few solid particles. Conversely, some elements may be transferred from the plant to the water. Being in summer, the air temperature is quite mild, and some evaporation takes place as soon as it stops raining. Let us rather consider the path followed by water down to the ground water. Our drop of water has now selectively lost some elements to the terrestrial moss, has undergone concentration by evaporation and dissolution of 'dry fallout' salts from the plant surface, and has undergone ion exchange. Overall, we are now dealing with a water that is clearly more concentrated and less acidic (in the case chosen here) than the original rain water before it fell on the catchment.

Table 13.1. Simplified overview of some common chemical changes occurring to water as it circulates through a pristine catchment in a humid, temperate-to-cool region with well developed podzols (see text for discussion of specific chemical pathways chosen)

Compartment	Acidity (= - Alkalinity)	Base Cations	Silica	Iron, Aluminium	Organic Carbon
Canopy	↘ leaching	↗ plants, dry deposition	= minor dry deposition	↗ or =	↗ organic material
O-horizon	↗ organic acids, CO_2	↗ weathering, ↘ plant uptake	= or ↗ if minerals present	↗	↗ micro-organisms
E-horizon	=	↗ weathering, ↘ plant uptake	↗ mineral weathering	↗ mobilisation	?
B-horizon	↘ Fe & Al hydroxides	↗ mineral weathering, =	↗ mineral weathering	↘ precipitation	↘ microbial degradation, adsorption
B/C-horizon	↘ mineral weathering	↗ mineral weathering	↗ mineral weathering	↘ precipitation	↘ microbial degradation, adsorption
C-horizon	↘ mineral weathering	↗ mineral weathering	↗ mineral weathering	=	=
Groundwater	↘ mineral weathering	↗ mineral weathering	↗ mineral weathering	Fe: ↗ reduced Al: =	↘ microbial degradation, adsorption
Stream water	= or ↘	↘ dilution	↘ dilution	↘ precipitation	↗ mixing with shallow soil water
Changes downstream	↘ mixing with older water	↗ mixing with older water	↗ mixing with older water ? de/adsorption	= ↘ precipitation	↗ mixing with shallow soil water

Changes are relative to previous compartment: ↗ increase, ↘ decrease, = similar, ? variable or unknown.

13.5 SOIL SEARCHING

Under the action of the unabating shower (and gravity), the drop of rain eventually runs down to the base of the moss and rolls off onto the top layer of the humus. It infiltrates relatively easily and rapidly through this web of intertwined fragments of dead vegetation into the denser and darker humus mass *per se*. The ambient air here is still in relatively good contact with the atmosphere, and oxygen is thus nearly as abundant as in the free air. In contact with this mass of colloidal organic compounds derived from degrading vegetal material, the drop of water begins to acquire some carbon dioxide originating from microbial respiration as well as a series of organic ions and molecules (especially hydroxyl and carboxyl compounds of low molecular weight, which are easily soluble and possess readily exchangeable protons). Exchangeable metals (esp. base cations) and free protons are also taken up. Nitrogen and sulphur also may be picked up by the water; our drop of rain water is becoming soil water, part of the pervasive soil solution. Compared to the incoming rain, its composition has changed due to ion exchange reactions in the humus layer, its concentration has increased as a result of continuing evapotranspiration, and its acidity has increased due to dissolved carbon dioxide. Because the humus layer constitutes a very large reservoir of exchangeable cations (on solid or aqueous organic compounds), the natural variation in rain water composition is dampened whilst passing through it.

Considering its continued journey downward, the drop of water percolates through the various horizons of the well-drained podzol that exists here. As it proceeds along this path, the water progressively leaves the realm dominated by organic compounds, as typified by the humus layer (O-horizon), and enters a sphere of more mineral or inorganic interactions. In the elluvial layer (E-horizon), the soil water picks up relatively large amounts of aluminium (a major contributor to the acidity of the water), iron and manganese, which precipitate in the illuvial layer (B-horizon) as oxide or oxy-hydroxide (esp. goethite and diaspore) 'skins'. If available, calcium, sodium, potassium and magnesium are also leached out of the E-horizon (though some of the base cations are taken up from the water by plants, the net effect being a balance between weathering and biomass aggradation rates), and the soil water may then contain much more of these base cations than the rain water originally did.

In the transition layer (B/C-horizon) and within the parent material (C-horizon), mineral dissolution/hydrolysis, exacerbated by the enhanced aggressivity of the soil water now laden with carbonic and organic acids, contributes many solutes, especially base cations and other metals, as well as bicarbonate ions, to the soil water. Kaolinite commonly is found as a by-product of mineral weathering in the mineral horizons (e.g. through incongruent dissolution of ubiquitous feldspars). The gasses in contact with the water drop infiltrating the soil become less and less aerobic; carbon dioxide, derived from plant decomposition and bacterial activity, can make up several percent of the gas phase (100 to 1000 times its atmospheric abundance), especially in the microenvironments near the roots. Here, sulphate can be reduced to hydrogen sulphide or pyrite, and nitrate to gaseous nitrogen or nitrogen oxide; thus, sulphur and nitrogen content in the soil water can decrease. This is one of nature's many ways to clear the environment of nitrates.

As long as roots are present, oxygen and nutrients like nitrogen compounds and phosphate are likely to be taken out of the water by the vegetation, and carbon dioxide continues to be discharged to the soil water. Transformed into carbonic acid, it will accelerate subsequent mineral weathering reactions. While percolating through the root zone, the soil water may be taken up by plants via well-developed root and rootlet systems, and subsequently may be stored in vegetation or returned to the atmosphere via (evapo)transpiration. This further increases the concentration of the soil water.

Below the root zone and above the groundwater table (i.e. within the lower part of the unsaturated zone), weathering takes place at a more sluggish pace. The soil gas is still rich in carbon dioxide here, but the organic acids and chelates from the humus layer have (nearly?) all been consumed, decomposed or have precipitated (esp. in the B-horizon). In addition, the content of base cations and other weathering products is high, hindering further massive dissolution.

13.6 THE WATER TABLE

As the unit volume of water we are tracking touches the capillary fringe and enters the underlying groundwater reservoir (i.e. the upper part of the saturated zone), it passes a major geochemical divide: it is now in limited contact with the atmosphere, and redox conditions consequently can become more anaerobic. Hydrologically too, the regime is different as groundwater is a continuous, subterranean body of water with a major lateral flow component, as opposed to the dominantly vertical direction of soil water motion. The water is now also beyond the influence of plant material (carbon dioxide input is over), and freshly recharged groundwater quickly may lose much of its organic matter (e.g. dissolved organic carbon) content through either bacterial decay, oxidation or adsorption, and the concentration organic complexes (notably iron and aluminium chelates) decreases.

Assuming we have a situation where the groundwater becomes anaerobic, oxidised (trivalent) iron is reduced, e.g. by organic carbon, to divalent iron, which is soluble in water; sulphate reduces to metal sulphides or hydrogen sulphide, which may then leave the aqueous phase. Simultaneously, many majors solutes (base cations, silica, some bicarbonate, etc.) pass into solution as a result of mineral weathering (commonly by congruent or incongruent dissolution) and pH increases as a result of this mineral dissolution. The salinity or load of dissolved solids in the groundwater thus increases progressively along the flowpath, and smectite minerals commonly precipitate in the saturated zone. Weathering slows progressively down within the aquifer, as the acidity of the groundwater is consumed, and the content of dissolved solids increases.

Groundwater, whether it flows through loose overburden material or fractured bedrock, may also pick up products of radioactive decay (e.g. radon) of parent material; in practice, this is particularly true in the case of hardrock aquifers with a high uranium or radium content. The water continues its migration down the hydraulic gradient through the aquifer; it may move relatively fast if using the macropore/macrofracture network, or much more sluggishly if it gets stuck in a micropore/microfracture. In the former case, it will undergo less chemical transformation, in the latter it will become further saline and neutralised if conditions allow (e.g. relatively reactive mineral particles). The unit volume of water we tracked through the soil and aquifer now has a much higher dissolved solids content (esp. base cations) and pH than when it fell as rain water.

13.7 DISCHARGE !

Areas along the stream bed often are natural discharge zones for groundwater, as are the occasional springs and seeps within the catchment. Groundwater movement in such zones has a vertical upward component. Because all infiltrated water must at some point resurface, and because discharge areas commonly are much smaller than recharge areas, groundwater flows vigorously out of the soil in discharge zones. Ion exchange reactions occur within the root zone of the discharge zone, attempting to reach equilibrium with the (relatively saline) groundwater. Thus, the soil will be up to three times richer in base cations in discharge than in recharge areas. The discharging groundwater can also solubilise organic substances and take in suspension small humus particles. Shallow soil water may, in periods of heavy rainfall, contribute significant amounts of dissolved organic carbon to the stream.

Both processes contribute to the brown coloration occasionally seen in stream water.

When it re-enters in contact with the atmosphere, the geochemistry of the water is affected quite profoundly. Oxidation of groundwater as it seeps into a river can result in ferrous iron being oxidised to ferric iron and precipitating as a rusty residue; manganese may follow the same fate. This results in the common brown to black coatings on boulders in river beds. In rare occurrences, precipitates of greyish aluminium hydroxides may form. Carbon dioxide will tend to degas rapidly, increasing the pH of the water (now stream water); if our catchment were rich in carbonate rocks, calcium carbonate crusts (travertine) may form at discharge points. Other gases, such as radon will also pass into the atmosphere. In periods of low flow in the stream, the groundwater and stream water have similar chemical compositions (seeing the former feeds the latter), apart for some (important) exceptions, such as carbon dioxide mentioned above. These are periods of 'baseflow' in the stream, and they occur when it has not rained for some time, typically in summer. In periods of flood, however, as caused by heavy rain or melting of the snow cover in the springtime, ground and stream waters can be chemically quite distinct. In the case of our remote catchment, the stream water is likely to be considerably more dilute during such events than the groundwater. If the snowpack contained abundant pollutants such as heavy metals or sulphate, the concentration of these elements (or compounds) may be many times higher in the stream water than in the groundwater.

13.8 ALL THE RIVERS RUN

As water flows downriver, it generally undergoes mixing with waters bearing different geochemical signatures: these can be waters from a tributary, a lake or a peatbog encountered downstream, or rain water, interflow and overland flow during and just after rainstorms, for instance. Depending on the geochemical composition of the mixing waters, changes can include dilution, concentration, precipitation, dissolution, (de)complexation, (de)flocculation, oxidation, reduction and ion exchange. The effect of water-rock interaction in the stream is very limited, compared to its effect in the soil or aquifer, because of the shorter residence time and higher ratio of water to solid surface in the stream. But organic matter, such as aquatic plants (mosses, algae, diatoms) can modify to a limited extent the stream water's chemistry by adsorption, uptake or ion exchange; nitrogen, phosphorus, potassium, silica, and certain trace elements are primarily implicated. The water in the stream also interacts with its bed and banks, where organisms living on and between boulders and stones may take up nutrients.

As a river flows from an upland area down to the lowlands, its proton and trivalent aluminium content generally decrease, whilst base cations and silica content and electrical conductivity increase. This is a consequence of the increasing time the discharging groundwater has spent within the catchment (residence time) the further downstream one considers. In the case of silica, however, concentrations in stream water may be controlled abiologically by adsorption onto, or desorption from, quartz surfaces depending on the degree of silica saturation with respect to quartz.

To close the hydrologic loop, the river runs down to the ocean, and our drop of water may eventually evaporate, become incorporated in clouds in the atmosphere, and fall as new rain on another catchment.

ACKNOWLEDGEMENTS

We are grateful to Per Aagaard, David Banks, Tim Drever and Colin Neal for their thoughtful comments about an earlier draft of this overview. Discussions with lecturers of the short course from which this book is a result, as well as with other colleagues from the Kola Ecogeochemistry project and elsewhere, have inspired us to attempt this challenging, and undoubtedly incomplete, generalisation of hydrogeochemical processes in catchments.

A NON-COMPREHENSIVE LIST OF USEFUL REFERENCES

The following list contains a selection of references that we have found handy in writing up this chapter (in addition to the preceding 12 chapters of the present book), as well as some helpful starting points, we hope, for the reader wishing to dig a little bit deeper into the topic we have, however superficially and incompletely, attempted to address here. For more detailed bibliography, please consult the reference lists of the appropriate chapters of this book.

Albu, M., Banks, D. & Nash, H. (eds), 1996. *Mineral and Thermal Groundwater Resources.* Chapman & Hall, in press.

Appelo, C.A.J. & Postma, D., 1993. *Geochemistry, Groundwater and Pollution.* A.A. Balkema, Rotterdam, 500 pp.

Berner, E.K. & Berner, R.A., 1996. *Global Environment: Water, Air and Geochemical Cycles.* Prentice-Hall, Englewood Cliffs, 376 pp.

Domenico, P.A. & Schwartz, F.W., 1990. *Physical and Chemical Hydrogeology.* John Wiley & Sons, New York, 824 pp.

Downing, R.A. & Wilkinson, W.B., 1991. *Applied Groundwater Hydrology.* Clarendon Press, Oxford, 340 pp.

Drever, J.I., 1988. *The Geochemistry of Natural Waters.* Second Edition. Prentice Hall, Englewood Cliffs, 437 pp.

Eriksson, E. & Holtan, H., 1974. Hydrokemi - Kemiska Processer i Vattnets Kretslopp Hydrochemistry Chemical Processes in the Water Cycle, in Swedish). International Hydrological Decade, Oslo. Nordic IHD Report, 7, 124 pp.

Eriksson, E., 1985. *Principles and Applications of Hydrochemistry.* Chapman & Hall, London, 187 pp.

Fleet, M.E. (ed.), 1984. *Environmental Geochemistry.* Mineralogical Association of Canada, Short Course Handbook, 10, 306 pp.

Freeze, R.A. & Cherry, J.A., 1979. *Groundwater.* Prentice-Hall, Englewood Cliffs, 604 pp.

Garrels, R.M. & Christ, C.L., 1965. *Solutions, Minerals, and Equilibria.* Freeman, Cooper, San Francisco, 450 pp.

Garrels, R.M., Mackenzie, F.T. & Hunt, C., 1975. *Chemical Cycles and the Global Environment: Assessing Human Influences.* William Kaufmann, Los Altos, 206 pp.

Grip, H. & Rodhe, A., 1994. *Vattnets Väg från Regn til Bäck* (The Water's Path from Rain to River, *in Swedish*). Third Edition. Hallgren & Fallgren Studieförlag AB, Uppsala, 155 pp.

Lasaga, A.C. & Kirkpatrick, R.J., 1983. *Kinetics of Geochemical Processes.* Second Printing. Mineralogical Society of America, Reviews in Mineralogy, 8, 398 pp.

Lerman, A. & Meybeck, M. (eds), 1988. *Physical and Chemical Weathering in Geochemical Cycles*. NATO Advanced Science Institutes Series. Kluwer Academic Publishers, Dordrecht. Series C, vol. 251, 375 pp.

Mason, B. & Moore, C.B., 1982. *Principles of Geochemistry*. Fourth Edition. John Wiley & Sons, New York, 344 pp.

Moldan, B. & Cerny, J. (eds), 1994. *Biogeochemistry of Small Catchments - A Tool for Environmental Research*. SCOPE 51, John Wiley & Sons, Chichester, 419 pp.

Morel, F.M.M. & Hering, J.G., 1993. *Principles and Applications of Aquatic Chemistry*. John Wiley & Sons, New York, 606 pp.

Nordstrom, D.K. & Munoz, J.L., 1994. *Geochemical Thermodynamics*. Second Edition. Blackwell. Scientific Publications, Cambridge, MA, 493pp.

Price, M., 1996. *Introducing Groundwater*. Second Edition. Chapman & Hall, London, 270 pp.

Sly, P.G. (ed.), 1984. *Sediments and Water Interactions*. Springer Verlag, New York.

Stumm, W. & Morgan, J.J., 1995. *Aquatic Chemistry-Chemical Equilibria and Rates in Natural Waters*. Third Edition. John Wiley & Sons, New York, 1040 pp.

Stumm, W., 1992. *Chemistry of the Solid-Water Interface*. John Wiley & Sons, New York, 428 pp.

Thurman, E.M., 1985. *Organic Geochemistry of Natural Water*. Martinus Nijhoff/Dr. W. Junk, Dordrecht, 497 pp.

Trudgill, S.T. (ed.), 1986. *Solute Processes*. John Wiley & Sons, Chichester, 509 pp.

Trudgill, S.T. (ed.), 1995. *Solute Modelling in Catchment Systems*. John Wiley & Sons, Chichester, 473 pp.

White, A.F. & Brantley, S.L. (eds), 1995. *Chemical Weathering Rates of Silicate Minerals*. Mineralogical Society of America, Reviews in Mineralogy, 31, 583 pp.

List of authors

Ole K. Borggaard, Chemistry Department, Royal Veterinary & Agricultural University, Thorvaldsensvej 40, DK-1871 Frederiksberg C, Denmark.
E-mail: okb@kvl.dk

Patrice de Caritat, Geochemistry & Hydrogeology Section, Geological Survey of Norway, P.O. Box 3006-Lade, N-7002 Trondheim, Norway.
E-mail: patrice.de.caritat@ngu.no

Nils Christophersen, Department of Informatics, University of Oslo, P.O. Box 1080, Blindern, N-0316 Oslo, Norway.
E-mail: nils@ifi.uio.no

James I. Drever, Department of Geology & Geophysics, University of Wyoming, P.O. Box 3006, Laramie, WY 82071-3006, USA.
E-mail: drever@uwyo.edu

Sylvi Haldorsen, Department of Soil & Water Sciences, Agricultural University of Norway, P.O. Box 5008, N-1432 Ås, Norway.
E-mail: sylvi.haldorsen@ijvf.nlh.no

Ånund Killingtveit, Department of Hydraulic & Environmental Engineering, Norwegian University of Science and Technology, N-7034 Trondheim, Norway.
E-mail: aanund.killingtveit@ivb.unit.no

David N. Lerner, Department of Civil & Environmental Engineering, University of Bradford, Bradford, West Yorkshire BD7 1DP, UK.
E-mail: d.n.lerner@bradford.ac.uk

Donald L. Macalady, Department of Chemistry & Geochemistry, Colorado School of Mines, Golden, CO 80401, USA.
E-mail: dmacalad@flint.mines.colorado.edu

John Mather, Department of Geology, Royal Holloway, University of London, Egham, Surrey TW20 0EX, UK.

Colin Neal, Institute of Hydrology, Maclean Building, Crowmarsh Gifford, Wallingford, Oxon OX10 8BB, UK.
E-mail: c.neal@ua.nwl.ac.uk

James F. Ranville, Department of Chemistry & Geochemistry, Colorado School of Mines, Golden, CO 80401, USA.
E-mail: jranvill@mines.colorado.edu

Gunnhild Riise, Department of Soil & Water Sciences, Agricultural University of Norway, P.O. Box 5008, N-1432 Ås, Norway.
E-mail: gunnhild.riise@ijvf.nlh.no

Alice J. Robson, Institute of Hydrology, Maclean Building, Crowmarsh Gifford, Wallingford, Oxon OX10 8BB, UK.
E-mail: a.robson@ua.nwl.ac.uk

Allan Rodhe, Department of Earth Sciences, Uppsala University, S-75309 Uppsala, Sweden.
E-mail: allan.rodhe@geo.uu.se

Knut Sand, SINTEF, Norwegian Hydrotechnical Laboratory, N-7034 Trondheim, Norway.
E-mail: knut.sand@nhl.sintef.no

Ronald S. Sletten, Quaternary Research Center, University of Washington, P.O. Box 351360, Seattle, WA 98195-1360, USA.
E-mail: sletten@u.washington.edu

Berit Swensen, Department of Soil & Water Sciences, Agricultural University of Norway, P.O. Box 5008, N-1432 Ås, Norway.
E-mail: berit.swensen@ijvf.nlh.no

Nils Roar Sælthun, Norwegian Institute for Water Research, P.O. Box 173-Kjelsås, N-0413 Oslo, Norway.
E-mail: nils.saelthun@niva.no

Ola M. Sæther, Geochemistry & Hydrogeology Section, Geological Survey of Norway, P.O. Box 3006-Lade, N-7002 Trondheim, Norway.
E-mail: ola.sather@ngu.no

Magne Ødegård, Laboratories Section, Geological Survey of Norway, P.O. Box 3006-Lade, N-7002 Trondheim, Norway.
E-mail: magne.odegard@ngu.no

Index

Printed in the United States
by Baker & Taylor Publisher Services

Printed in the United States
by Baker & Taylor Publisher Services